ADDISON-WESLEY

WESTERN CANADIAN EDITION

Applied Mathematics 12

SOURCE BOOK

Robert Alexander
Mathematics Teacher
and Consultant
Richmond Hill, Ontario

Cam Bennet
Dauphin Regional
Comprehensive
Secondary School
Dauphin, Manitoba

Carol Besteck Hope
Elmwood High School
Winnipeg, Manitoba

Garry Davis
Saskatoon Public Board
of Education
Saskatoon, Saskatchewan

Terry Kaminski
Grand Centre High School
Cold Lake, Alberta

Jim Nakamoto
Sir Winston Churchill
Secondary School
Vancouver, British Columbia

Brent Richards
Crocus Plains Regional
Secondary School
Brandon, Manitoba

David Sufrin
Ballenas Secondary School
Parksville, British Columbia

Paul Williams
Red Deer College
Red Deer, Alberta

Elizabeth Wood
Calgary Board of Education
Calgary, Alberta

Rick Wunderlich
Shuswap Junior
Secondary School
Salmon Arm, British Columbia

Leanne Zorn
Mount Boucherie
Secondary School
Kelowna, British Columbia

Addison Wesley

Toronto

PUBLISHER
Claire Burnett

MANAGING EDITOR
Enid Haley

SENIOR CONSULTING EDITOR
Lesley Haynes

PRODUCT MANAGER
Reid McAlpine

DEVELOPMENTAL EDITOR
Susan Lishman

EDITOR
Anna-Maria Garnham

EDITORIAL CONTRIBUTORS
Rosina Daillie
William Dodd
David Gargaro
Janine LeBlanc
Andrea May
Gay McKellar
Eileen Pyne-Rudzik
Alison Reiger

DESIGN/PRODUCTION/ART DIRECTION
Pronk&Associates

The publisher has taken every care to meet or exceed industry specifications for the manufacturing of textbooks. The spine and the endpapers of this sewn book have been reinforced with special fabric for extra binding strength. The cover is a premium, polymer-reinforced material designed to provide long life and withstand rugged use. Mylar gloss lamination has been applied for further durability.

ISBN 0–201–39597–5

This book contains recycled product and is acid free.

Printed and bound in Canada

Addison Wesley

2 3 4 5 GG 05 04 03 02

Pilot Teachers and Reviewers

Barry Blair
Alberni District Secondary School
Alberni, British Columbia

Steven Hardy
Harry Ainley High School
Edmonton, Alberta

Sheeva Harrysingh
J.H. Bruns Collegiate
Winnipeg, Manitoba

Claude Laflamme
Associate Professor of Mathematics
Department of Mathematics
and Statistics
The University of Calgary
Calgary, Alberta

Maureen Mills
Dr. E.P. Scarlett Senior High School
Calgary, Alberta

Ken Riise
Forestburg School
Forestburg, Alberta

Renée Sonnenberg
St. Joseph's Catholic High School
Grande Prairie, Alberta

David Sufrin
Ballenas Secondary School
Parksville, British Columbia

Colleen Tong
Strathcona Tweedsmuir School
Okotoks, Alberta

Kevin Weimer
Lester B. Pearson Senior High School
Calgary, Alberta

CONTENTS

Addison-Wesley Applied Mathematics 12 Source Book

Welcome

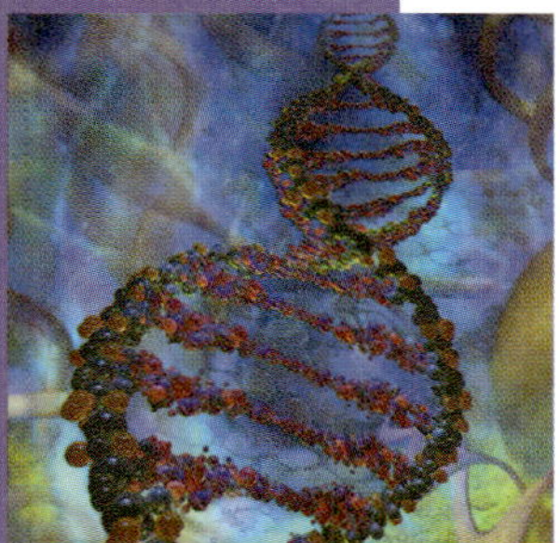

CONTENTS

Balance Due
102.00

CONTENTS

7 Vectors

Welcome to Addison-Wesley Applied Mathematics 12

Western Canadian Edition

Addison-Wesley Applied Mathematics presents mathematics as an essential tool in everyday life and in many careers. In this course, you will apply mathematics in many different contexts. You will develop new mathematical concepts and skills by examining authentic situations and investigating solutions.

To meet the requirements of the Applied Mathematics course, you need to work from both the Project Book, and the Source Book. These introductory pages illustrate how your student books work.

Source Book

The Source Book has 7 chapters and 3 reference sections. Like the Project Book, it is designed for regular use during the course, with time taken out to work on projects.

Each chapter starts with an Overview. This table tells you what mathematics you will learn.

Each chapter presents a Chapter Project, and a series of Tutorials.

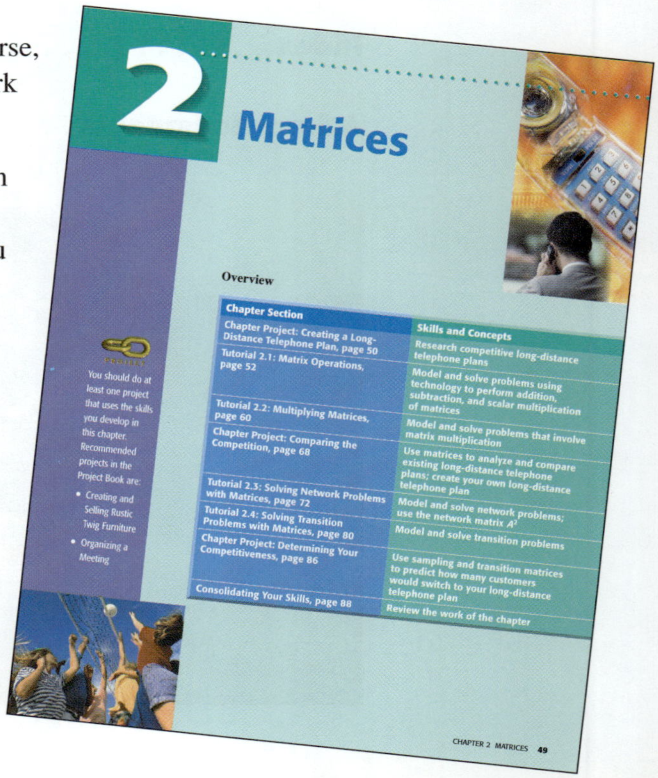
2

Matrices

Overview

Chapter Section	Skills and Concepts
Chapter Project: Creating a Long-Distance Telephone Plan, page 50	Research competitive long-distance telephone plans
Tutorial 2.1: Matrix Operations, page 52	Model and solve problems using technology to perform addition, subtraction, and scalar multiplication of matrices
Tutorial 2.2: Multiplying Matrices, page 60	Model and solve problems that involve matrix multiplication
Chapter Project: Comparing the Competition, page 68	Use matrices to analyze and compare existing long-distance telephone plans; create your own long-distance telephone plan
Tutorial 2.3: Solving Network Problems with Matrices, page 72	Model and solve network problems; use the network matrix A^2
Tutorial 2.4: Solving Transition Problems with Matrices, page 80	Model and solve transition problems
Chapter Project: Determining Your Competitiveness, page 86	Use sampling and transition matrices to predict how many customers would switch to your long-distance telephone plan
Consolidating Your Skills, page 88	Review the work of the chapter

You should do at least one project that uses the skills you develop in this chapter. Recommended projects in the Project Book are:
- Creating and Selling Rustic Twig Furniture
- Organizing a Meeting

CHAPTER 2 MATRICES 49

Chapter Project

The Chapter Projects in the Source Book are more structured than the projects in the Project Book. They will help you develop the study skills and work habits you need to complete more open-ended projects.

You will find an introduction to the Chapter Project at the start of the chapter. You revisit the Chapter Project at later points in the chapter, to complete portions of the required project work.

Each time you see the Chapter Project, you will find a Project Link. This short section tells you what mathematical concepts and skills you need for the project, and the tutorials you might complete to develop those concepts and skills.

FYI Visit refers you to our web site, where you can connect to other sites to research the Chapter Project.

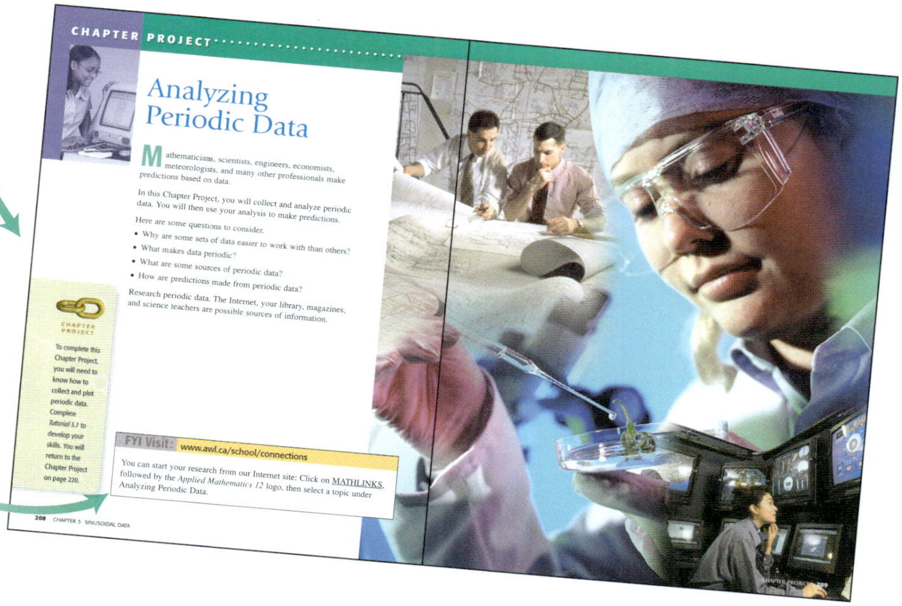

You return to the Chapter Project after completing relevant Tutorials.

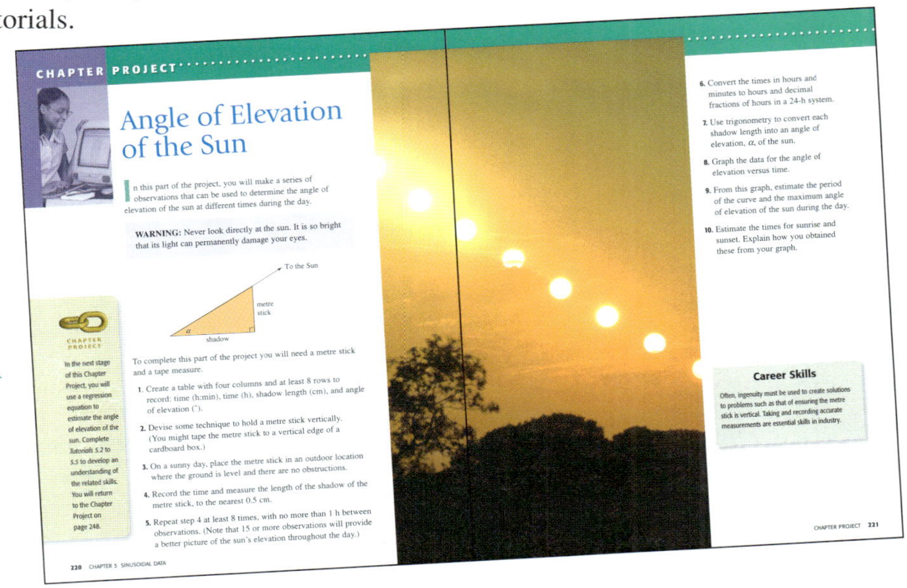

Tutorials

Each Tutorial develops a specific mathematical concept or result. Often you will reach that outcome by conducting an **Investigation**. Sometimes, you will follow **Examples** and their completed Solutions.

Here is how a typical Tutorial works.

- **Practise Your Prior Skills** provides exercises on relevant topics learned in earlier grades.
- An **Investigation**, or a series of Investigations, leads you to explore mathematical concepts.

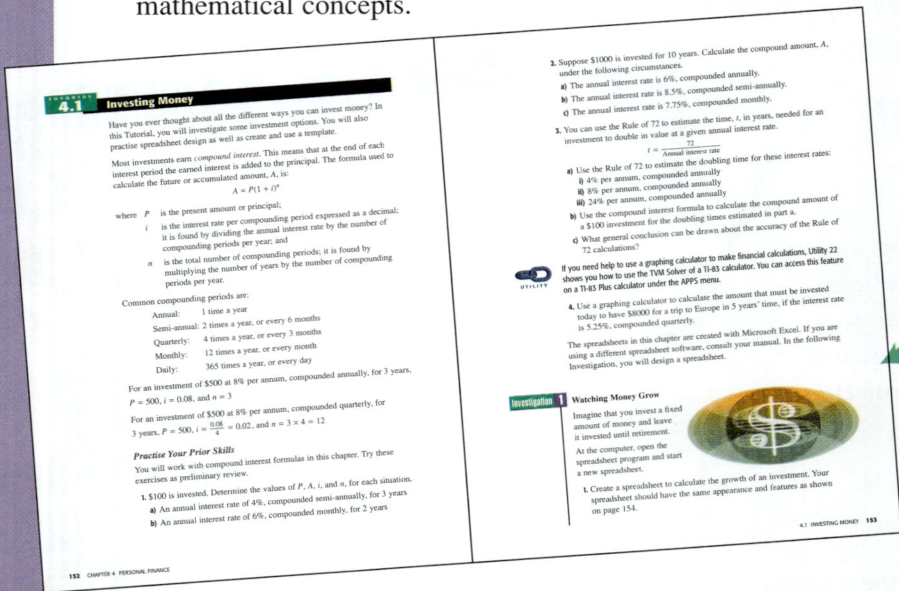

TUTORIAL 4.1 **Investing Money**

Have you ever thought about all the different ways you can invest money? In this Tutorial, you will investigate some investment options. You will also practise spreadsheet design as well as create and use a template.

Most investments earn *compound interest*. This means that at the end of each interest period the earned interest is added to the principal. The formula used to calculate the future or accumulated amount, A, is:

$$A = P(1 + i)^n$$

where P is the present amount or principal;
i is the interest rate per compounding period expressed as a decimal; it is found by dividing the annual interest rate by the number of compounding periods per year; and
n is the total number of compounding periods; it is found by multiplying the number of years by the number of compounding periods per year.

Common compounding periods are:

Annual: 1 time a year
Semi-annual: 2 times a year, or every 6 months
Quarterly: 4 times a year, or every 3 months
Monthly: 12 times a year, or every month
Daily: 365 times a year, or every day

For an investment of \$500 at 8% per annum, compounded annually, for 3 years, $P = 500$, $i = 0.08$, and $n = 3$

For an investment of \$500 at 8% per annum, compounded quarterly, for 3 years, $P = 500$, $i = \frac{0.08}{4} = 0.02$, and $n = 3 \times 4 = 12$

Practise Your Prior Skills

You will work with compound interest formulas in this chapter. Try these exercises as preliminary review.

1. \$100 is invested. Determine the values of P, A, i, and n, for each situation.
   a) An annual interest rate of 4%, compounded semi-annually, for 3 years
   b) An annual interest rate of 6%, compounded monthly, for 2 years

152 CHAPTER 4 PERSONAL FINANCE

2. Suppose \$1000 is invested for 10 years. Calculate the compound amount, A, under the following circumstances.
   a) The annual interest rate is 6%, compounded annually.
   b) The annual interest rate is 8.5%, compounded semi-annually.
   c) The annual interest rate is 7.75%, compounded monthly.
3. You can use the Rule of 72 to estimate the time, t, in years, needed for an investment to double in value at a given annual interest rate.
   $$t = \frac{72}{\text{Annual interest rate}}$$
   a) Use the Rule of 72 to estimate the doubling time for these interest rates:
      i) 4% per annum, compounded annually
      ii) 8% per annum, compounded annually
      iii) 24% per annum, compounded annually
   b) Use the compound interest formula to calculate the compound amount of a \$100 investment for the doubling times estimated in part a.
   c) What general conclusion can be drawn about the accuracy of the Rule of 72 calculations?

UTILITY If you need help to use a graphing calculator to make financial calculations, Utility 22 shows you how to use the TVM Solver of a TI-83 calculator. You can access this feature on a TI-83 Plus calculator under the APPS menu.

4. Use a graphing calculator to calculate the amount that must be invested today to have \$8000 for a trip to Europe in 5 years' time, if the interest rate is 5.25%, compounded quarterly.

The spreadsheets in this chapter are created with Microsoft Excel. If you are using a different spreadsheet software, consult your manual. In the following Investigation, you will design a spreadsheet.

Investigation 1 **Watching Money Grow**

Imagine that you invest a fixed amount of money and leave it invested until retirement.

At the computer, open the spreadsheet program and start a new spreadsheet.

1. Create a spreadsheet to calculate the growth of an investment. Your spreadsheet should have the same appearance and features as shown on page 154.

4.1 INVESTING MONEY 153

- **Examples** illustrate how to use new mathematical methods.

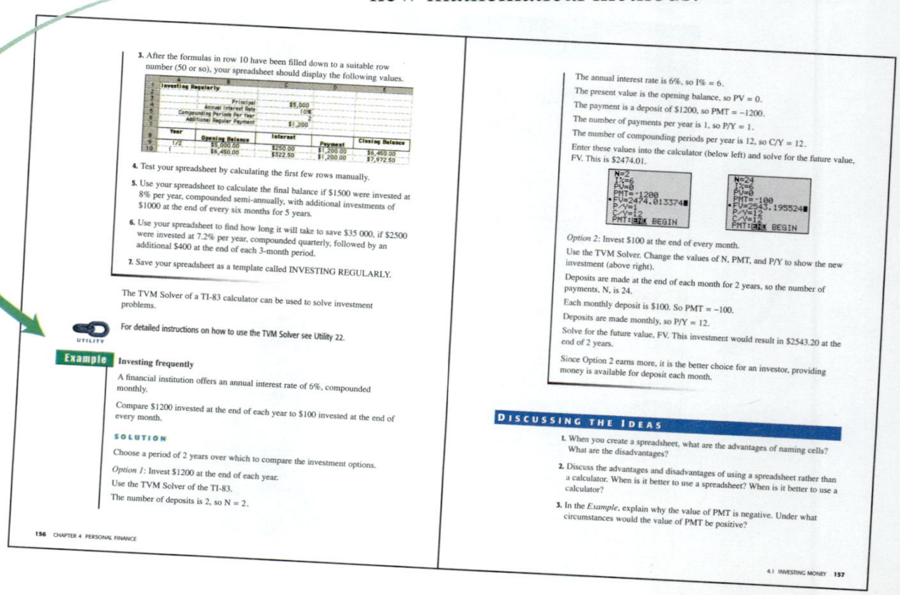

3. After the formulas in row 10 have been filled down to a suitable row number (50 or so), your spreadsheet should display the following values.

	A	B	C	D	E
1	Investing Regularly				
2					
3					
4		Principal	\$5,000		
5		Annual Interest Rate	10%		
6		Compounding Periods Per Year	2		
7		Additional Regular Payment	\$1,200		
8	Year	Opening Balance	Interest	Payment	Closing Balance
9	1/2	\$5,000.00	\$250.00	\$1,200.00	\$6,450.00
10	1	\$6,450.00	\$322.50	\$1,200.00	\$7,972.50

4. Test your spreadsheet by calculating the first few rows manually.
5. Use your spreadsheet to calculate the final balance if \$1500 were invested at 8% per year, compounded semi-annually, with additional investments of \$1000 at the end of every six months for 5 years.
6. Use your spreadsheet to find how long it will take to save \$35 000, if \$2500 were invested at 7.2% per year, compounded quarterly, followed by an additional \$400 at the end of each 3-month period.
7. Save your spreadsheet as a template called INVESTING REGULARLY.

The TVM Solver of a TI-83 calculator can be used to solve investment problems.

UTILITY For detailed instructions on how to use the TVM Solver see Utility 22.

Example **Investing frequently**

A financial institution offers an annual interest rate of 6%, compounded monthly.

Compare \$1200 invested at the end of each year to \$100 invested at the end of every month.

SOLUTION

Choose a period of 2 years over which to compare the investment options.

Option 1: Invest \$1200 at the end of each year.
Use the TVM Solver of the TI-83.
The number of deposits is 2, so N = 2.

156 CHAPTER 4 PERSONAL FINANCE

The annual interest rate is 6%, so I% = 6.
The present value is the opening balance, so PV = 0.
The payment is a deposit of \$1200, so PMT = −1200.
The number of payments per year is 1, so P/Y = 1.
The number of compounding periods per year is 12, so C/Y = 12.
Enter these values into the calculator (below left) and solve for the future value, FV. This is \$2474.01.

N=2
I%=6
PV=0
PMT=-1200
▪FV=2474.013374
P/Y=1
C/Y=12
PMT:END BEGIN

N=24
I%=6
PV=0
PMT=-100
▪FV=2543.195524
P/Y=12
C/Y=12
PMT:END BEGIN

Option 2: Invest \$100 at the end of every month.
Use the TVM Solver. Change the values of N, PMT, and P/Y to show the new investment (above right).
Deposits are made at the end of each month for 2 years, so the number of payments, N, is 24.
Each monthly deposit is \$100. So PMT = −100.
Deposits are made monthly, so P/Y = 12.
Solve for the future value, FV. This investment would result in \$2543.20 at the end of 2 years.

Since Option 2 earns more, it is the better choice for an investor, providing money is available for deposit each month.

DISCUSSING THE IDEAS

1. When you create a spreadsheet, what are the advantages of naming cells? What are the disadvantages?
2. Discuss the advantages and disadvantages of using a spreadsheet rather than a calculator. When is it better to use a spreadsheet? When is it better to use a calculator?
3. In the *Example*, explain why the value of PMT is negative. Under what circumstances would the value of PMT be positive?

4.1 INVESTING MONEY 157

- **Discussing the Ideas** provides suggestions for classroom discussion of concepts you worked with in Investigations or Examples.
- **Exercises** give you opportunities to work with the same mathematical concepts and skills in different contexts.

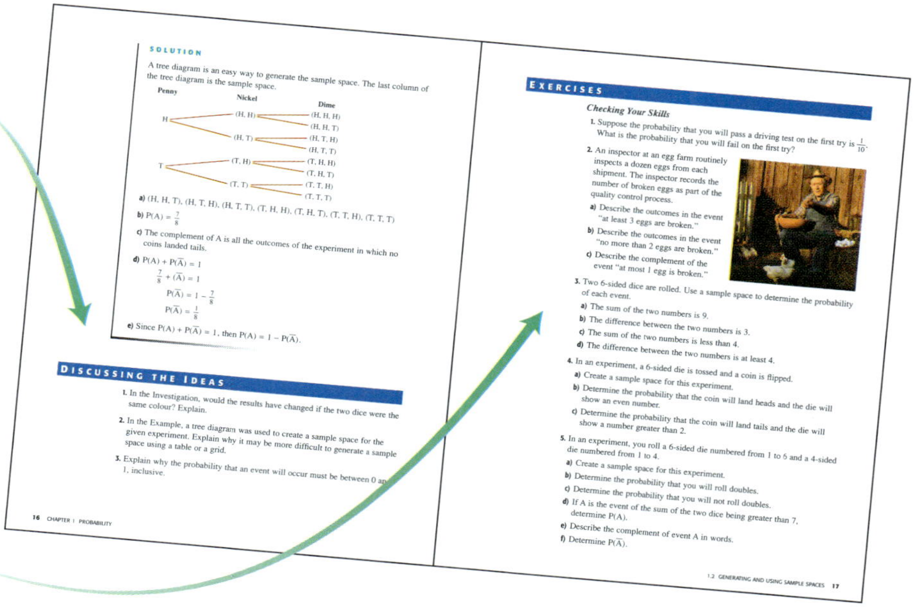

SOLUTION

A tree diagram is an easy way to generate the sample space. The last column of the tree diagram is the sample space.

Penny	Nickel	Dime
H	(H, H)	(H, H, H)
		(H, H, T)
	(H, T)	(H, T, H)
		(H, T, T)
T	(T, H)	(T, H, H)
		(T, H, T)
	(T, T)	(T, T, H)
		(T, T, T)

a) (H, H, T), (H, T, H), (H, T, T), (T, H, H), (T, H, T), (T, T, H), (T, T, T)

b) $P(A) = \frac{7}{8}$

c) The complement of A is all the outcomes of the experiment in which no coins landed tails.

d) $P(A) + P(\overline{A}) = 1$

$\frac{7}{8} + (\overline{A}) = 1$

$P(\overline{A}) = 1 - \frac{7}{8}$

$P(\overline{A}) = \frac{1}{8}$

e) Since $P(A) + P(\overline{A}) = 1$, then $P(A) = 1 - P(\overline{A})$.

DISCUSSING THE IDEAS

1. In the Investigation, would the results have changed if the two dice were the same colour? Explain.
2. In the Example, a tree diagram was used to create a sample space for the given experiment. Explain why it may be more difficult to generate a sample space using a table or a grid.
3. Explain why the probability that an event will occur must be between 0 and 1, inclusive.

16 CHAPTER 1 PROBABILITY

EXERCISES

Checking Your Skills

1. Suppose the probability that you will pass a driving test on the first try is $\frac{1}{10}$. What is the probability that you will fail on the first try?
2. An inspector at an egg farm routinely inspects a dozen eggs from each shipment. The inspector records the number of broken eggs as part of the quality control process.
   a) Describe the outcomes in the event "at least 3 eggs are broken."
   b) Describe the outcomes in the event "no more than 2 eggs are broken."
   c) Describe the complement of the event "at most 1 egg is broken."
3. Two 6-sided dice are rolled. Use a sample space to determine the probability of each event.
   a) The sum of the two numbers is 9.
   b) The difference between the two numbers is 3.
   c) The sum of the two numbers is less than 4.
   d) The difference between the two numbers is at least 4.
4. In an experiment, a 6-sided die is tossed and a coin is flipped.
   a) Create a sample space for this experiment.
   b) Determine the probability that the coin will land heads and the die will show an even number.
   c) Determine the probability that the coin will land tails and the die will show a number greater than 2.
5. In an experiment, you roll a 6-sided die numbered from 1 to 6 and a 4-sided die numbered from 1 to 4.
   a) Create a sample space for this experiment.
   b) Determine the probability that you will roll doubles.
   c) Determine the probability that you will not roll doubles.
   d) If A is the event of the sum of the two dice being greater than 7, determine P(A).
   e) Describe the complement of event A in words.
   f) Determine $P(\overline{A})$.

1.2 GENERATING AND USING SAMPLE SPACES 17

- **Project Link** emphasizes the focus of the Tutorial, and highlights the projects that involve the same concepts and skills.

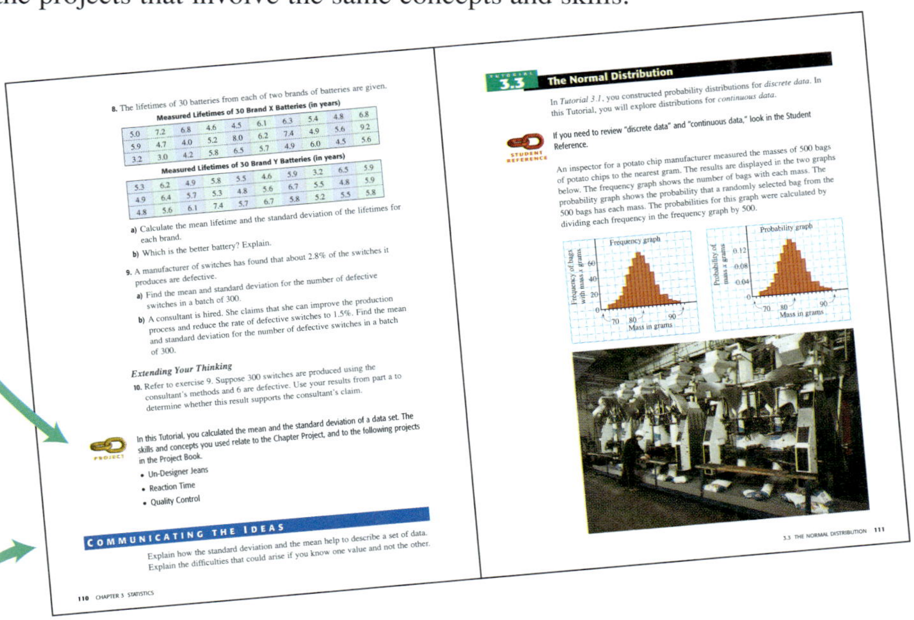

8. The lifetimes of 30 batteries from each of two brands of batteries are given.

Measured Lifetimes of 30 Brand X Batteries (in years)

5.0	7.2	6.8	4.6	4.5	6.1	6.3	5.4	4.8	6.8
5.9	4.7	4.0	5.2	8.0	6.2	7.4	4.9	5.6	9.2
3.2	3.0	4.2	5.8	6.5	5.7	4.9	6.0	4.5	5.6

Measured Lifetimes of 30 Brand Y Batteries (in years)

5.3	6.2	4.9	5.8	5.5	4.6	5.9	3.2	6.5	5.9
4.9	6.4	5.7	5.3	4.8	5.6	6.7	5.5	4.8	5.9
4.8	5.6	6.1	7.4	5.7	6.7	5.8	5.2	5.5	5.8

a) Calculate the mean lifetime and the standard deviation of the lifetimes for each brand.
b) Which is the better battery? Explain.

9. A manufacturer of switches has found that about 2.8% of the switches it produces are defective.
   a) Find the mean and standard deviation for the number of defective switches in a batch of 300.
   b) A consultant is hired. She claims that she can improve the production process and reduce the rate of defective switches to 1.5%. Find the mean and standard deviation for the number of defective switches in a batch of 300.

Extending Your Thinking

10. Refer to exercise 9. Suppose 300 switches are produced using the consultant's methods and 6 are defective. Use your results from part a to determine whether this result supports the consultant's claim.

PROJECT

In this Tutorial, you calculated the mean and the standard deviation of a data set. The skills and concepts you used relate to the Chapter Project, and to the following projects in the Project Book.
- Un-Designer Jeans
- Reaction Time
- Quality Control

COMMUNICATING THE IDEAS

Explain how the standard deviation and the mean help to describe a set of data. Explain the difficulties that could arise if you know one value and not the other.

110 CHAPTER 3 STATISTICS

TUTORIAL 3.3 The Normal Distribution

In *Tutorial 3.1*, you constructed probability distributions for *discrete data*. In this Tutorial, you will explore distributions for *continuous data*.

STUDENT REFERENCE

If you need to review "discrete data" and "continuous data," look in the Student Reference.

An inspector for a potato chip manufacturer measured the masses of 500 bags of potato chips to the nearest gram. The results are displayed in the two graphs below. The frequency graph shows the number of bags with each mass. The probability graph shows the probability that a randomly selected bag from the 500 bags has each mass. The probabilities for this graph were calculated by dividing each frequency in the frequency graph by 500.

3.3 THE NORMAL DISTRIBUTION 111

- **Communicating the Ideas** suggests a writing activity in which you summarize the key learning for the Tutorial.

Consolidating Your Skills

Each chapter ends with a **Consolidating Your Skills** section. This starts with a table that summarizes the work of the chapter.

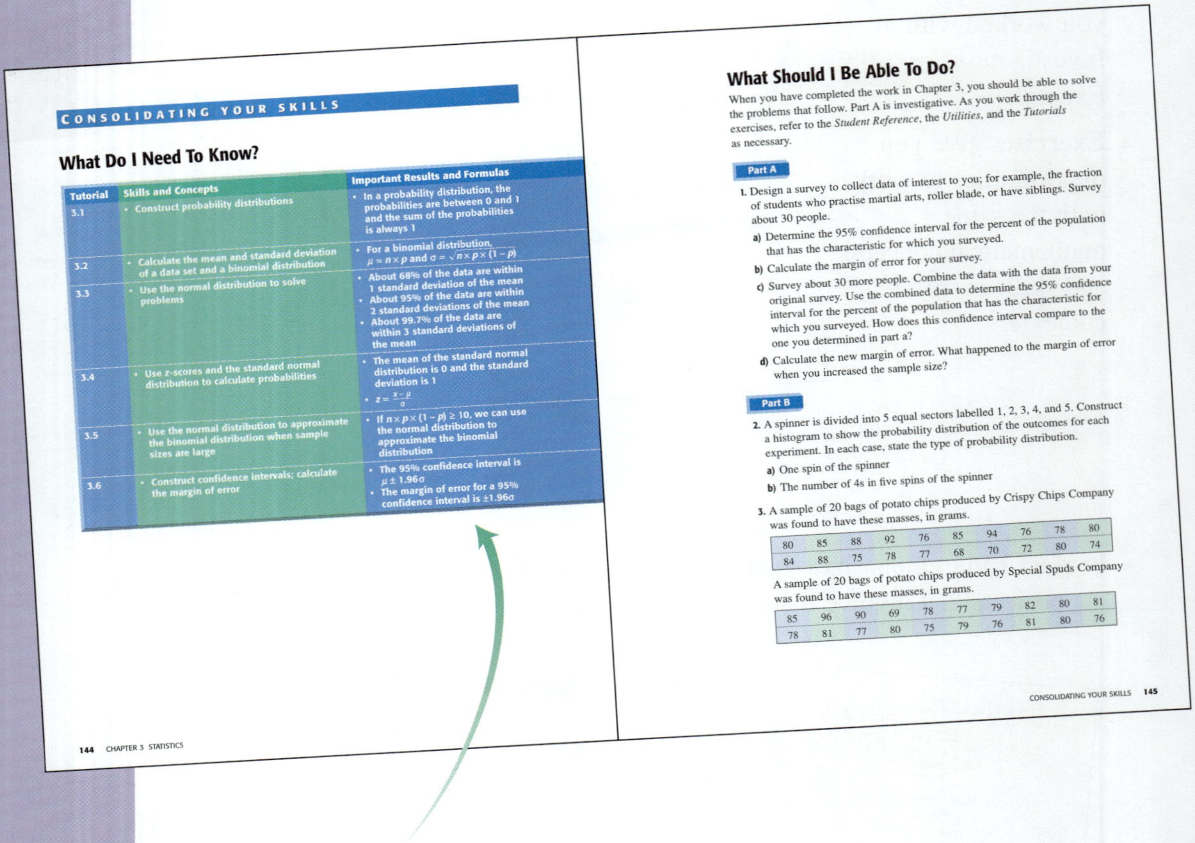

CONSOLIDATING YOUR SKILLS

What Do I Need To Know?

Tutorial	Skills and Concepts	Important Results and Formulas
3.1	• Construct probability distributions	• In a probability distribution, the probabilities are between 0 and 1 and the sum of the probabilities is always 1
3.2	• Calculate the mean and standard deviation of a data set and a binomial distribution	• For a binomial distribution, $\mu = n \times p$ and $\sigma = \sqrt{n \times p \times (1 - p)}$
3.3	• Use the normal distribution to solve problems	• About 68% of the data are within 1 standard deviation of the mean • About 95% of the data are within 2 standard deviations of the mean • About 99.7% of the data are within 3 standard deviations of the mean
3.4	• Use z-scores and the standard normal distribution to calculate probabilities	• The mean of the standard normal distribution is 0 and the standard deviation is 1 • $z = \frac{x - \mu}{\sigma}$
3.5	• Use the normal distribution to approximate the binomial distribution when sample sizes are large	• If $n \times p \times (1 - p) \geq 10$, we can use the normal distribution to approximate the binomial distribution
3.6	• Construct confidence intervals; calculate the margin of error	• The 95% confidence interval is $\mu \pm 1.96\sigma$ • The margin of error for a 95% confidence interval is $\pm 1.96\sigma$

144 CHAPTER 3 STATISTICS

What Should I Be Able To Do?

When you have completed the work in Chapter 3, you should be able to solve the problems that follow. Part A is investigative. As you work through the exercises, refer to the *Student Reference*, the *Utilities*, and the *Tutorials* as necessary.

Part A

1. Design a survey to collect data of interest to you; for example, the fraction of students who practise martial arts, roller blade, or have siblings. Survey about 30 people.
   a) Determine the 95% confidence interval for the percent of the population that has the characteristic for which you surveyed.
   b) Calculate the margin of error for your survey.
   c) Survey about 30 more people. Combine the data with the data from your original survey. Use the combined data to determine the 95% confidence interval for the percent of the population that has the characteristic for which you surveyed. How does this confidence interval compare to the one you determined in part a?
   d) Calculate the new margin of error. What happened to the margin of error when you increased the sample size?

Part B

2. A spinner is divided into 5 equal sectors labelled 1, 2, 3, 4, and 5. Construct a histogram to show the probability distribution of the outcomes for each experiment. In each case, state the type of probability distribution.
   a) One spin of the spinner
   b) The number of 4s in five spins of the spinner
3. A sample of 20 bags of potato chips produced by Crispy Chips Company was found to have these masses, in grams.

80	85	88	92	76	85	94	76	78	80
84	88	75	78	77	68	70	72	80	74

A sample of 20 bags of potato chips produced by Special Spuds Company was found to have these masses, in grams.

85	96	90	69	78	77	79	82	80	81
78	81	77	80	75	79	76	81	80	76

CONSOLIDATING YOUR SKILLS 145

Review this column to identify important results and formulas.

The remainder of the section gives a 3-part review of the chapter.

Part A involves an activity, just like the Investigations in the chapter.

Part B has some review problems, similar to the Examples and Exercises you did in the Tutorials.

Part C focuses on the Chapter Project.

These photographs show you some actual student work.

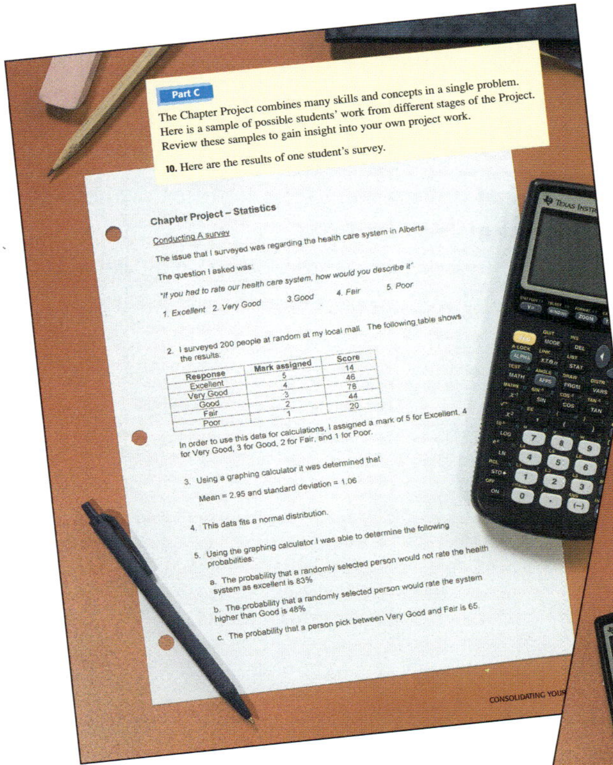

Follow-up questions develop your skills at assessing project work.

Use Part C for inspiration with your own project, and to learn how to examine your own work critically.

Reference Sections

People at work use many different references. For example, an editor refers to a dictionary regularly. A contractor needs to look up building codes. A bank worker refers to daily postings of approved interest rates.

Your Source Book has been constructed to develop your skills in using reference materials as the need arises. In addition to research you will do outside the text, you will be able to use three important reference sections within the Source Book: Utilities, Student Reference, and Answers.

In a Tutorial, you may see a comment that directs you to the Utilities or the Student Reference section at the back of the Source Book. Projects in the Project Book may also refer you to the Student Reference section. You decide whether you need to follow the references for a quick review or a mini-lesson.

Utilities

A link symbol indicates the Utilities section, highlighted by the blue colour across the top of each page.

Utilities 25–34 teach you how to use specific tools for this course.

Read the Utility while you work with the tool you are learning. If you are reading a Utility about the graphing calculator, key in the instructions as you read. Check that your results match the results shown in the Utility. Check with another student if you have a problem.

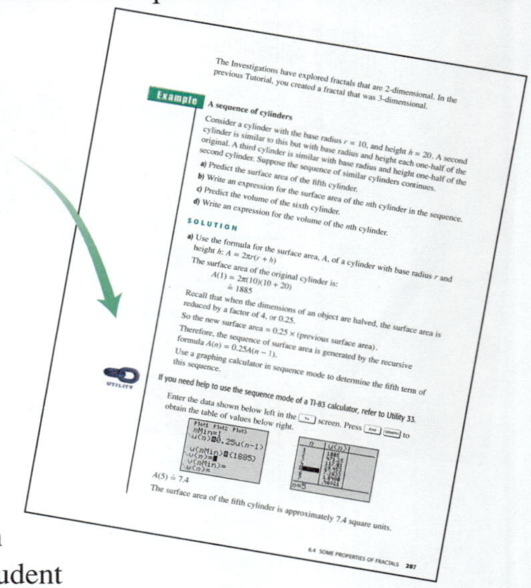

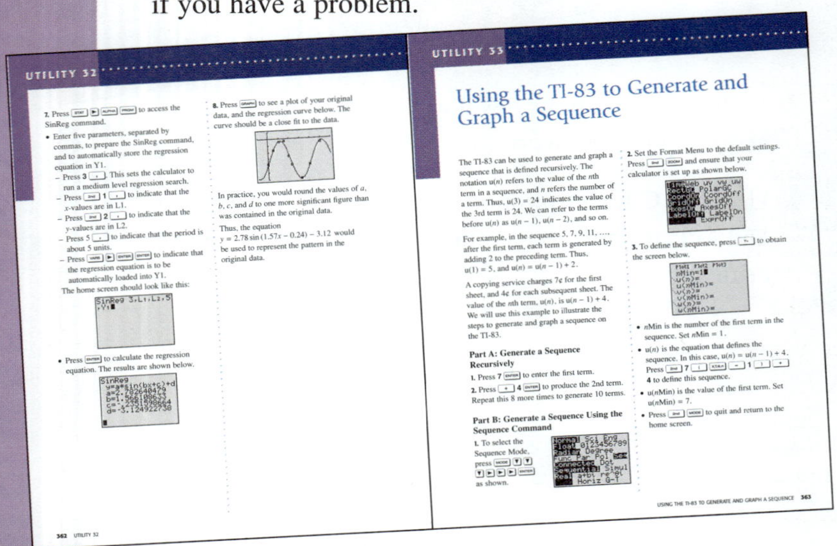

The Utilities 1–24 from *Applied Mathematics 10* and *Applied Mathematics 11* are available from your teacher, should you need them.

Student Reference

The link symbol indicates the Student Reference section with pages that have the red colour across the top.

6.1 Sequences

Many events produce data that form patterns. In this chapter, you will use technology to investigate patterns in numbers and in geometrical figures.

In 1705, the astronomer Edmund Halley noticed similarities in the records of major comets that had appeared in 1531, 1607, and 1682. He also noticed that those dates were separated by a little more than 75 years. He concluded that it was really the same comet returning on a regular basis. It last appeared in 1986 and records have been found for every appearance of it since 239 B.C.

The pattern described above is an arithmetic sequence. Any term in an arithmetic sequence, after the first, can be found by adding a constant to the preceding term.

STUDENT REFERENCE To review "arithmetic sequence," look in the Student Reference.

A potato stored in a warm, dark room starts to grow new shoots capable of producing whole new plants. Potato farmers typically divide 1 seed potato to start 6 new plants. Over several generations of potatoes, the number of potato plants descended from a single plant forms a sequence 1, 6, 36, 216, 1296, 7776, 46 656, 279 936, ... In each successive generation, the number of potato plants is 6 times greater than the number of potato plants in the preceding generation. These numbers form a *geometric sequence*.

In a geometric sequence, to calculate each term after the first, you multiply the preceding term by a constant number, called the *common ratio*.

In both arithmetic and geometric sequences, each term, after the first, is generated by using the terms that come before it. These are both *recursively-generated sequences*. The Fibonacci sequence 1, 1, 2, 3, 5, 8, 13, ... is a recursively generated sequence that is neither arithmetic nor geometric. It can be found in population growth, the number of petals of common flowers, or the shell of a Chambered Nautilus mollusk. To calculate a term of a Fibonacci sequence, after the second term, add the two preceding terms.

A spreadsheet program can be used to generate a sequence, particularly when the sequence is recursively generated.

260 CHAPTER 6 PATTERNS

This section of the Source Book is organized alphabetically, like a glossary. It does more than a glossary, though, because it gives a quick review for many prerequisite skills.

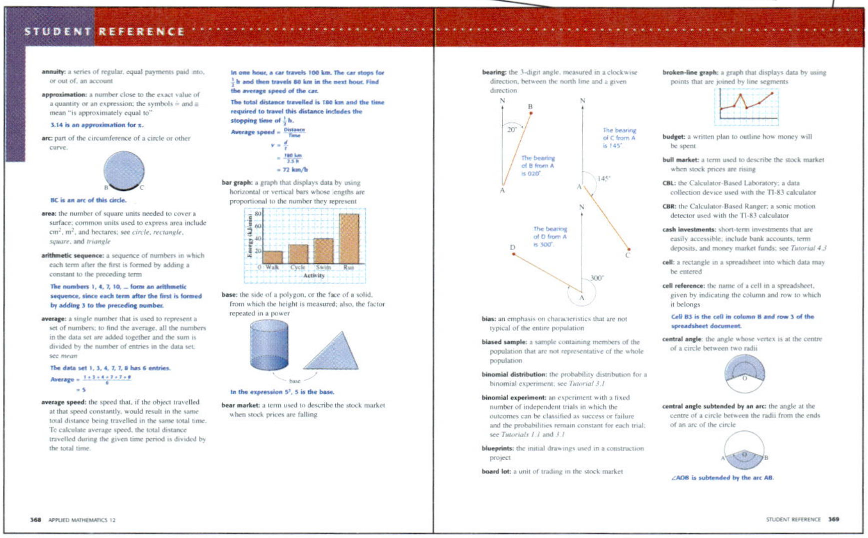

The Student Reference provides:

- definitions for key terms
- general results, such as the Pythagorean Theorem
- brief review of concepts from previous grades
- examples when relevant
- references to Source Book tutorials when relevant

Answers

Check your answers after completing a set of Investigations or Exercises to see how well you understood the material.

By using the Answers responsibly, you can improve your mathematical learning and develop good study habits.

Project Book

The Project Book is designed for regular use during the course. The Project Book provides 24 ideas for open-ended projects. Each project relates to a specific career area, and most tie to one or more chapters from the Source Book.

You do not have to complete all the projects, but you may be interested in pursuing several of them. To cover all of the grade 12 Applied Mathematics curriculum, you will need to complete projects in two compulsory areas. Within these areas, you will have some choice of project topics.

Here is how a typical project works.

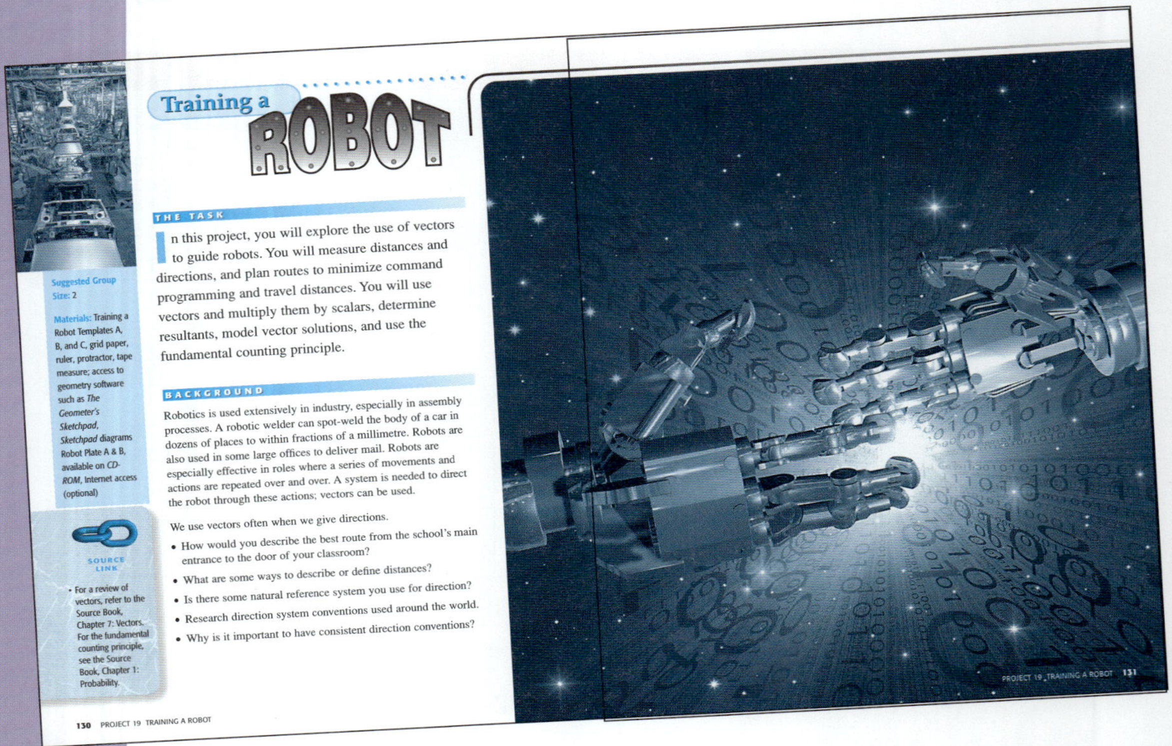
Training a ROBOT

THE TASK

In this project, you will explore the use of vectors to guide robots. You will measure distances and directions, and plan routes to minimize command programming and travel distances. You will use vectors and multiply them by scalars, determine resultants, model vector solutions, and use the fundamental counting principle.

Suggested Group Size: 2

Materials: Training a Robot Templates A, B, and C, grid paper, ruler, protractor, tape measure; access to geometry software such as *The Geometer's Sketchpad*, *Sketchpad* diagrams Robot Plate A & B, available on *CD-ROM*, Internet access (optional)

BACKGROUND

Robotics is used extensively in industry, especially in assembly processes. A robotic welder can spot-weld the body of a car in dozens of places to within fractions of a millimetre. Robots are also used in some large offices to deliver mail. Robots are especially effective in roles where a series of movements and actions are repeated over and over. A system is needed to direct the robot through these actions; vectors can be used.

We use vectors often when we give directions.

- How would you describe the best route from the school's main entrance to the door of your classroom?
- What are some ways to describe or define distances?
- Is there some natural reference system you use for direction?
- Research direction system conventions used around the world.
- Why is it important to have consistent direction conventions?

SOURCE LINK

- For a review of vectors, refer to the Source Book, Chapter 7: Vectors. For the fundamental counting principle, see the Source Book, Chapter 1: Probability.

130 PROJECT 19 TRAINING A ROBOT

PROJECT 19 TRAINING A ROBOT 131

- **The Task** explains the purpose of the project and describes a specific application you will investigate.
- **Materials** describes the supplies you need to complete the project.
- **Background** provides some interesting facts about the topic or sets up a possible scenario for your project.
- **Source Link** points you to required skills in the Source Book, if you need them.

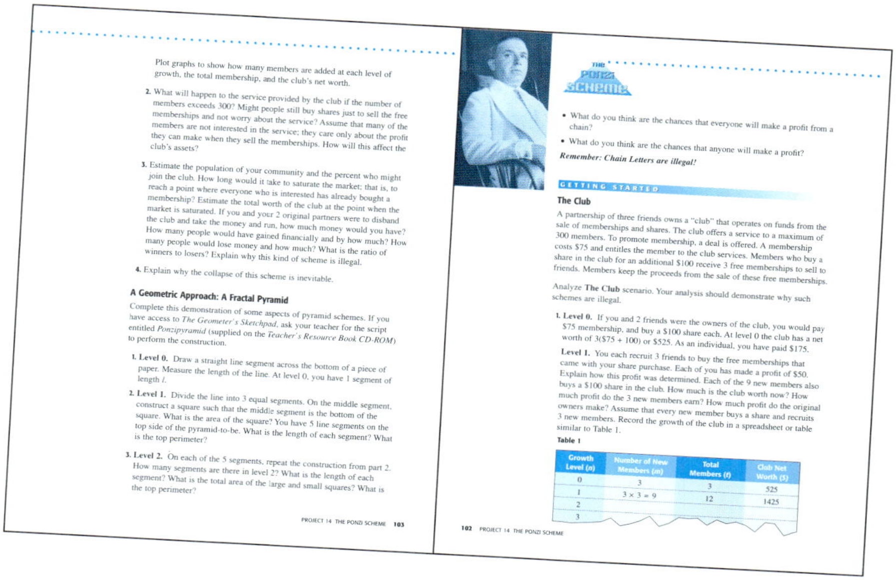

Plot graphs to show how many members are added at each level of growth, the total membership, and the club's net worth.

2. What will happen to the service provided by the club if the number of members exceeds 300? Might people still buy shares just to sell the free memberships and not worry about the service? Assume that many of the members are not interested in the service; they care only about the profit they can make when they sell the memberships. How will this affect the club's assets?

3. Estimate the population of your community and the percent who might join the club. How long would it take to saturate the market; that is, to reach a point where everyone who is interested has already bought a membership? Estimate the total worth of the club at the point when the market is saturated. If you and your 2 original partners were to disband the club and take the money and run, how much money would you have? How many people would have gained financially and by how much? How many people would lose money and how much? What is the ratio of winners to losers? Explain why this kind of scheme is illegal.

4. Explain why the collapse of this scheme is inevitable.

A Geometric Approach: A Fractal Pyramid

Complete this demonstration of some aspects of pyramid schemes. If you have access to *The Geometer's Sketchpad*, ask your teacher for the script entitled *Ponzipyramid* (supplied on the *Teacher's Resource Book CD-ROM*) to perform the construction.

1. **Level 0.** Draw a straight line segment across the bottom of a piece of paper. Measure the length of the line. At level 0, you have 1 segment of length l.

2. **Level 1.** Divide the line into 3 equal segments. On the middle segment, construct a square such that the middle segment is the bottom of the square. What is the area of the square? You have 5 line segments on the top side of the pyramid-to-be. What is the length of each segment? What is the top perimeter?

3. **Level 2.** On each of the 5 segments, repeat the construction from part 2. How many segments are there in level 2? What is the length of each segment? What is the total area of the large and small squares? What is the top perimeter?

PROJECT 14 THE PONZI SCHEME 103

THE PONZI SCHEME

- What do you think are the chances that everyone will make a profit from a chain?
- What do you think are the chances that anyone will make a profit?

Remember: Chain Letters are illegal!

GETTING STARTED

The Club

A partnership of three friends owns a "club" that operates on funds from the sale of memberships and shares. The club offers a service to a maximum of 300 members. To promote membership, a deal is offered. A membership costs $75 and entitles the member to the club services. Members who buy a share in the club for an additional $100 receive 3 free memberships to sell to friends. Members keep the proceeds from the sale of these free memberships.

Analyze **The Club** scenario. Your analysis should demonstrate why such schemes are illegal.

1. **Level 0.** If you and 2 friends were the owners of the club, you would pay $75 membership, and buy a $100 share each. At level 0 the club has a net worth of 3($75 + 100) or $525. As an individual, you have paid $175.

Level 1. You each recruit 3 friends to buy the free memberships that came with your share purchase. Each of you has made a profit of $50. Explain how this profit was determined. Each of the 9 new members also buys a $100 share in the club. How much is the club worth now? How much profit do the 3 new members earn? How much profit do the original owners make? Assume that every new member buys a share and recruits 3 new members. Record the growth of the club in a spreadsheet or table similar to Table 1.

Table 1

Growth Level (n)	Number of New Members (m)	Total Members (t)	Club Net Worth ($)
0	3	3	525
1	3 × 3 = 9	12	1425
2			
3			

102 PROJECT 14 THE PONZI SCHEME

- **Getting Started** leads you to explore some of the basic mathematical concepts underlying the application.
- **Project Presentation** lets you pursue a variety of extensions of the work done in Getting Started, and suggests ways to present your project report.

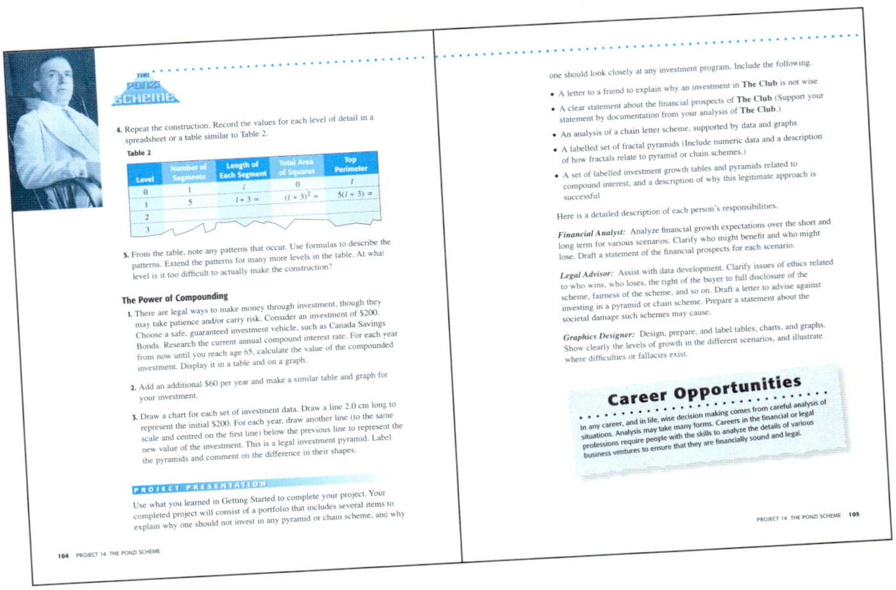

THE PONZI SCHEME

4. Repeat the construction. Record the values for each level of detail in a spreadsheet or a table similar to Table 2.

Table 2

Level	Number of Segments	Length of Each Segment	Total Area of Squares	Top Perimeter
0	1	l	0	l
1	5	$l \div 3 =$	$(l \div 3)^2 =$	$5(l \div 3) =$
2				
3				

5. From the table, note any patterns that occur. Use formulas to describe the patterns. Extend the patterns for many more levels in the table. At what level is it too difficult to actually make the construction?

The Power of Compounding

1. There are legal ways to make money through investment, though they may take patience and/or carry risk. Consider an investment of $200. Choose a safe, guaranteed investment vehicle, such as Canada Savings Bonds. Research the current annual compound interest rate. For each year from now until you reach age 65, calculate the value of the compounded investment. Display it in a table and on a graph.

2. Add an additional $60 per year and make a similar table and graph for your investment.

3. Draw a chart for each set of investment data. Draw a line 2.0 cm long to represent the initial $200. For each year, draw another line (to the same scale and centred on the first line) below the previous line to represent the new value of the investment. This is a legal investment pyramid. Label the pyramids and comment on the difference in their shapes.

PROJECT PRESENTATION

Use what you learned in Getting Started to complete your project. Your completed project will consist of a portfolio that includes several items to explain why one should not invest in any pyramid or chain scheme, and why one should look closely at any investment program. Include the following.

- A letter to a friend to explain why an investment in **The Club** is not wise
- A clear statement about the financial prospects of **The Club** (Support your statement by documentation from your analysis of **The Club**.)
- An analysis of a chain letter scheme, supported by data and graphs
- A labelled set of fractal pyramids (Include numeric data and a description of how fractals relate to pyramid or chain schemes.)
- A set of labelled investment growth tables and pyramids related to compound interest, and a description of why this legitimate approach is successful

Here is a detailed description of each person's responsibilities.

Financial Analyst: Analyze financial growth expectations over the short and long term for various scenarios. Clarify who might benefit and who might lose. Draft a statement of the financial prospects for each scenario.

Legal Advisor: Assist with data development. Clarify issues of ethics related to who wins, who loses, the right of the buyer to full disclosure of the scheme, fairness of the scheme, and so on. Draft a letter to advise against investing in a pyramid or chain scheme. Prepare a statement about the societal damage such schemes may cause.

Graphics Designer: Design, prepare, and label tables, charts, and graphs. Show clearly the levels of growth in the different scenarios, and illustrate where difficulties or fallacies exist.

Career Opportunities

In any career, and in life, wise decision making comes from careful analysis of situations. Analysis may take many forms. Careers in the financial or legal professions require people with the skills to analyze the details of various business ventures to ensure that they are financially sound and legal.

104 PROJECT 14 THE PONZI SCHEME

PROJECT 14 THE PONZI SCHEME 105

- **Career Opportunities** describes how the mathematical skills needed in the project relate to careers.

1 Probability

PROJECT

You should do at least one project that uses the skills you learn in this chapter. The recommended projects in the Project Book are:

- NiDi
- Genetic Disorders

Overview

Chapter Section	Skills and Concepts
Chapter Project: Voice-Mail Security, page 2	Analyze and develop a proposal for securing voice-mail
Tutorial 1.1: Experimental and Theoretical Probability, page 4	Calculate experimental and theoretical probabilities
Tutorial 1.2: Generating and Using Sample Spaces, page 13	Construct a sample space for two or three events; use sample spaces and complements to solve problems
Tutorial 1.3: The Fundamental Counting Principle, page 20	Use the Fundamental Counting Principle to determine the number of ways multi-step operations can be performed
Chapter Project: How Many Passwords Are Possible?, page 26	Use the Fundamental Counting Principle to examine proposed password systems
Tutorial 1.4: Independent and Dependent Events, page 28	Classify events as dependent or independent; determine the probability of independent events
Tutorial 1.5: Mutually Exclusive Events, page 34	Solve problems involving mutually exclusive events
Chapter Project: Recommending a Password System, page 40	Use probabilities to analyze proposed password systems; make a recommendation about which system to adopt
Consolidating Your Skills, page 42	Review the work of the chapter

CHAPTER PROJECT

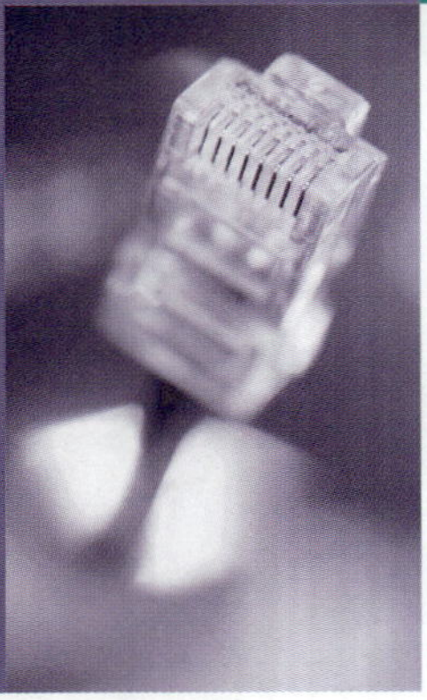

Voice-Mail Security

Today, we can easily access information by computer and by telephone. Although this is very convenient, the security of confidential personal or company information may be compromised if it is too easy to access. It is critical that a password system for confidential information be friendly but secure.

In this project, you will examine two security proposals for the voice-mail system of a telephone company and decide if the proposals are adequate for the anticipated number of users. You will also suggest improvements or alternative proposals.

In our daily life, we use many different types of passwords and identification numbers. Here are some questions to consider when you develop a password system.

- Describe the structures of some common identification numbers or passwords. Why are they structured in this way?
- What common problems are associated with the use of passwords?
- What provisions will the telephone company have to make to deal with these problems?
- What features might some customers look for in a voice-mail system?
- Since voice mail is accessed by telephone, what limitations are there on your system?

Conduct some research on password security. Sources of information include the Internet and the library.

CHAPTER PROJECT

To complete this Chapter Project, you will need to be able to determine the number of different passwords that would be available using different password systems. Complete *Tutorials 1.1* to *1.3* to develop an understanding of the necessary concepts. You will return to this Chapter Project on page 26.

FYI Visit: www.awl.ca/school/connections

You can start your research from our Internet site: Click on MATHLINKS, followed by the *Applied Mathematics 12* logo. Then select a topic under Voice-Mail Security.

TUTORIAL **1.1**

Experimental and Theoretical Probability

Probability is involved in such complex issues as the validity of DNA, fingerprinting, or the safety of various medical procedures. An understanding of chance is essential to work with *inferential statistics*, which are commonly used in business and science.

In the following Investigation, you will determine the probability of an event in two different ways.

Investigation

Finding Experimental and Theoretical Probabilities

When 3 coins are tossed, what is the probability that exactly 2 result in heads? Make a guess and record it.

Part A: Estimating the Probability

1. Work with a partner. One person tosses 3 coins and the other person records the results. Repeat this experiment 30 times. To determine the probability of getting exactly 2 heads, divide the number of results with exactly 2 heads by the total number of results. How does this answer compare to your guess?

2. In your notebook, make a chart like the one below.

Group	Number of results with exactly 2 heads	Total number of results
Totals		

Collect data from the other groups in your class and enter it in the table. Use the totals to calculate the probability of getting exactly 2 heads. How does this answer compare to your guess and your result from exercise 1?

Part B: Calculating the Probability

STUDENT REFERENCE

If you need help to draw a tree diagram, look in the Student Reference under "tree diagram."

3. Draw a tree diagram to represent the possible arrangements of heads and tails in the toss of 3 coins. Determine the number of branches that contain exactly 2 heads. Divide that number by the total number of branches to determine the probability that there are exactly 2 heads. How does this answer compare to your guess and to your estimates in exercises 1 and 2?

In the Investigation, you considered a probability problem in two different ways. In Part A, you used sampling to estimate the probability of an outcome. Probabilities determined this way are called *experimental probabilities.* In Part B, you determined the total number of successful outcomes and divided this by the total number of different possible outcomes to calculate the probability. Probabilities determined this way are called *theoretical probabilities.*

Simple examples of theoretical probability may involve tossing coins or rolling dice. For a fair coin, the theoretical probability of a head is $P(\text{head}) = \frac{1}{2}$. This does not mean that if we toss a coin 200 times that exactly 100 heads will result. Rather, it means that as the number of coin tosses increases, the probability of heads occurring gets closer and closer to $\frac{1}{2}$.

A *binomial experiment* is an experiment in which the outcomes can be classified as either a success or a failure. The experiment in the Investigation is a binomial experiment because the outcome "exactly 2 heads" is considered a success and all other outcomes are considered failures.

You can simulate a binomial experiment using the randBin command on a TI-83 graphing calculator. The command is

randBin(number of trials, probability of success on one trial, number of simulations).

UTILITY

If you need help to simulate a binomial experiment with a TI-83 calculator, Utility 25 shows you how. If you use a different graphing calculator, consult the user's manual that came with it.

Example 1 Determine the probability of rolling a 5.

Visualize tossing a 6-sided die.

a) What is the theoretical probability of getting a 5?

b) Toss a die 20 times. Record the number of times a 5 turns up. Calculate the experimental probability. Compare this probability to the theoretical probability.

c) Simulate tossing a die 300 times and record the number of times a 5 appears. From your simulation, what is the experimental probability of getting a 5? Compare this probability to the theoretical probability.

SOLUTION

a) There are 6 equally likely outcomes: 1, 2, 3, 4, 5, or 6.
$P(5) = \frac{1}{6}$, or approximately 0.167.
The theoretical probability of tossing a 5 is $\frac{1}{6}$ or about 0.167.

b) The results of one experiment are shown in the table.

The experimental probability of rolling a 5 is $\frac{4}{20}$ (or $\frac{1}{5}$) or 0.2. This is higher than the theoretical probability.

Number on Face of Die	Tally	Frequency
1	||||	4
2	|||	3
3	|||	3
4	||	2
5	||||	4
6	||||	4
Total		**20**

c) Enter randBin$(1, \frac{1}{6}, 300)$, store the results in list L1 and then add the data.

For this simulation, 1 appears 41 times. The experimental probability of tossing a 5 is $\frac{41}{300}$, or about 0.137. This is less than the theoretical probability of 0.167.

An experiment requires that a ball drop through the pin arrangement shown below. At each pin, the ball is equally likely to go left or right.

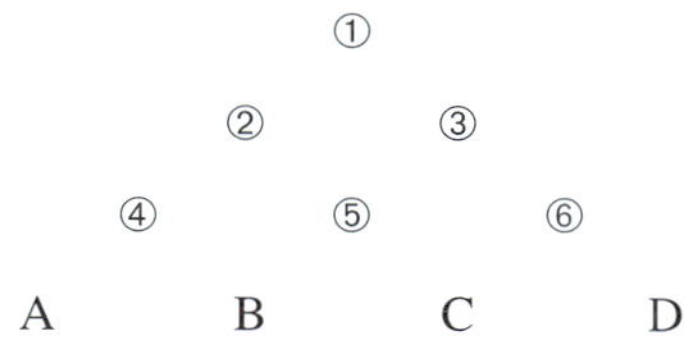

We can determine the number of possible routes to each exit. As shown in the diagram below, there is only 1 path the ball can take from pin ① to pin ②. Similarly, there is only 1 path the ball can take from pin ① to pin ③.

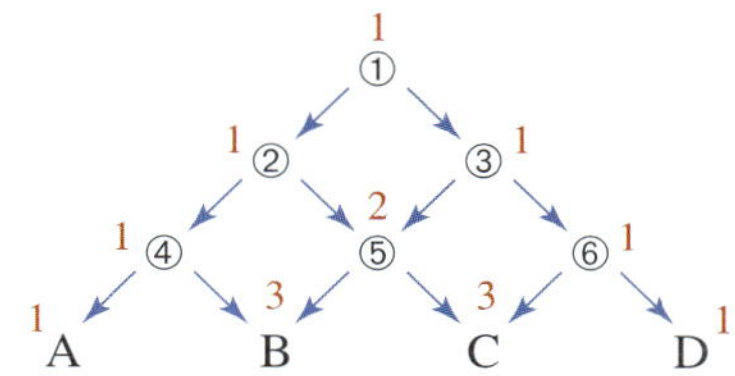

There is only 1 path from pin ① to pin ④ (via pin ②), and only 1 path from pin ① to pin ⑥ (via pin ③).

The ball can get to pin ⑤ from pin ① via pin ② or pin ③. Since there is 1 way to get to pin ② and 1 way to get to pin ③, there are $1 + 1 = 2$ different paths from pin ① to pin ⑤.

The ball can exit at A only from pin ④, so there is 1 path to exit A.

The ball can exit at D only from pin ⑥, so there is 1 path to exit D.

The ball can exit at B from pin ④ or pin ⑤. Since there is 1 path to pin ④, and 2 different paths to pin ⑤, there are $1 + 2 = 3$ different paths to exit B.

The ball can exit at C from pin ⑤ or pin ⑥. Since there are 2 different paths to ⑤ and 1 path to pin ⑥, there are $2 + 1 = 3$ different paths to exit C.

Example 2 **Determine the probability of a pathway.**

The pins in a pinball game are arranged as shown to the right. A ball is dropped in from above and exits at A, B, C, D, or E. At each pin, the ball is equally likely to go left or right.

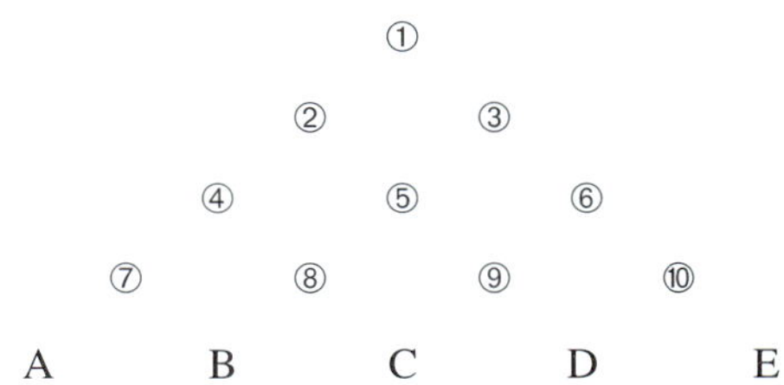

a) If 500 balls are dropped and 117 exit at D, what is the experimental probability that a ball will exit at D?

b) i) Determine the number of different pathways to each exit.

ii) Determine the total number of different pathways through the pins.

iii) What is the theoretical probability for exit D? How does this compare to the experimental probability?

SOLUTION

a) $P(D) = \frac{117}{500} = 0.234$

b) i)

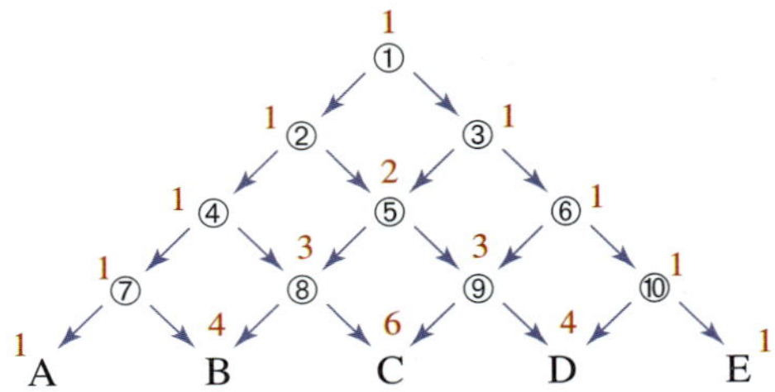

There is 1 pathway to exit A and 1 pathway to exit E.

There are 4 different pathways to exit B and 4 different pathways to exit D.

There are 6 different pathways to exit C.

ii) Total number of different pathways $= 1 + 4 + 6 + 4 + 1$
$= 16$

iii) $P(D) = \frac{4}{16} = 0.25$. This is close to the experimental probability.

An *event* is any outcome, or set of outcomes, of an experiment. Tossing a die and getting 5 is an event. Drawing a face card from a deck of 52 cards is another event.

Since 12 of the 52 cards in a deck are face cards, P(face card) $= \frac{12}{52}$, or $\frac{3}{13}$. This is the theoretical probability of drawing a face card. When we repeat the experiment and draw a card many times (replacing the card each time), the probability of drawing a face card (the experimental probability) should get closer and closer to $\frac{3}{13}$.

When you work with probability, it is inconvenient to use the adjective "theoretical" every time. Unless otherwise stated, the word "probability" means theoretical probability.

Probability

If an experiment has n equally likely outcomes of which r outcomes are favourable to event A, then the probability of event A is: $P(A) = \frac{r}{n}$.

Example 3 **Determine the probability of drawing a particular card.**

Euchre is played with the 9, 10, jack, queen, king, and ace of each suit. Suppose 1 card is drawn from those cards. What is the probability that it will be:

a) a face card?

b) a red ace?

SOLUTION

There are 6 cards in each of the 4 suits, or 24 cards.

a) There are 3 face cards in each suit, or 12 face cards in total.

$$\text{P(face card)} = \frac{12}{24} = \frac{1}{2}$$

b) There are 2 red aces. $\text{P(red aces)} = \frac{2}{24} = \frac{1}{12}$

DISCUSSING THE IDEAS

1. We frequently see or hear references to probability. Are these references more likely to theoretical probability or to experimental probability? Explain.

2. Describe a scenario in which experimental probabilities would have to be used to solve a problem.

3. The theoretical probability of winning a certain game is 30%. Your friend plays this game 10 times and wins 5 times. Based on these data, your friend argues that the probability of winning the game must be higher than 30%. Do you agree or disagree? Explain.

4. What advice would you offer to a person who wants to solve a problem involving experimental probability? Include reasons for this advice.

EXERCISES

Checking Your Skills

1. A fair coin lands heads 40 times in 60 tosses.

a) What is the experimental probability of a head?

b) What is the theoretical probability of a head?

2. In a raffle, 250 tickets are sold. Determine the probability of winning the raffle if you purchase:

a) 1 ticket **b)** 5 tickets **c)** 10 tickets

3. One card is drawn from a well-shuffled standard deck of 52 playing cards. Determine the following probabilities.

a) The card is a 7. **b)** The card is a diamond.

c) The card is red. **d)** The card is a black 10.

4. Simulate 300 trials of each event in exercise 3. Use the simulation results to estimate the experimental probability in each case.

5. At a traffic signal, the light is red for 30 s, amber for 5 s, and green for 45 s. What is the probability you will arrive when the light is red?

6. A roulette wheel consists of 38 numbers: 1 to 36, 0, and 00. What is the probability that a roulette ball will come to rest on an even number other than 0 or 00?

7. A pinball game has pins arranged as shown below.

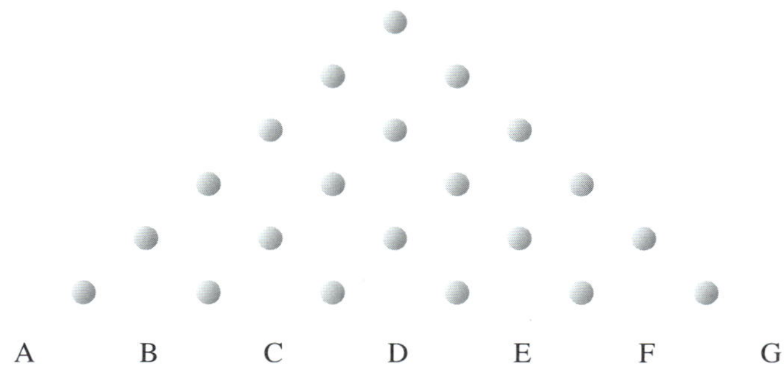

a) Determine the number of different pathways to each exit.

b) Determine the probability that a ball dropped through this arrangement will exit at D.

8. On the game board shown below, you can move the game piece diagonally forward.

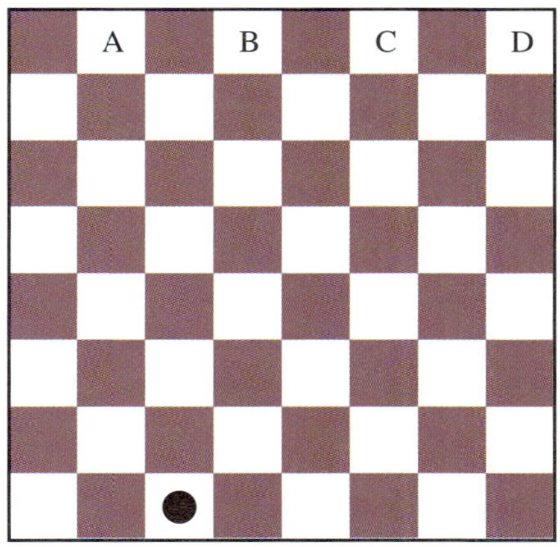

a) Determine the number of different pathways from the game piece to each of the squares on the other side of the board.

b) Assume that a player is equally likely to go left or right on each move. What is the probability he will land on:

i) square A? **ii)** square C?

9. Two coins are tossed. Consider the event "getting 2 heads."

a) Determine the experimental probability with 20 tosses of the 2 coins.

b) Determine the theoretical probability.

c) Compare the results in parts a and b. What do you notice?

10. In the game "In Between," 3 cards are dealt from a standard 52-card deck. To win, the value of the third card must be in between the first two.

a) Determine the probability of winning, given that the first 2 cards dealt are:

i) a 2 and a 6

ii) a 5 and a queen

iii) a 7 and an 8

iv) a jack and a king

b) Identify any assumptions you made in your answers to part a.

Extending Your Thinking

11. When we write P(heads) $= \frac{1}{2}$, we are modelling a coin-toss experiment. The model assumes that when a coin is tossed, only 2 equally likely outcomes can occur.

a) Although cylindrical in shape, a coin is not a perfect cylinder. Researchers once analyzed the distribution of mass in the design on the two sides of a coin and concluded that one side was slightly heavier than the other. Which side is slightly more likely to appear when a coin is tossed? Explain.

b) Do you think the assumption that there are only 2 equally likely outcomes is reasonable? Explain.

PROJECT

In this Tutorial, you determined the relationship between theoretical and experimental probabilities. The skills and concepts you used relate to the Chapter Project, and to the following projects in the Project Book.

- NiDi
- Genetic Disorders

COMMUNICATING THE IDEAS

A friend asks: "Since the probability of tossing heads on a coin is $\frac{1}{2}$, does this mean that it should land heads once for every two flips of a coin?" Write a brief response to your friend.

TUTORIAL 1.2

Generating and Using Sample Spaces

To calculate theoretical probabilities, you need to know the total number of possible outcomes. In simple experiments, such as rolling a single die, it is easy to list all of the outcomes. In this Tutorial, you will explore different methods to generate the list of all possible outcomes, the *sample space*.

The sample space is the list of all the possible outcomes of an experiment.

Investigation

Rolling Two Dice

Visualize rolling a black die and a green die at the same time. When the black die shows a 2 and the green die shows a 4, we can record the result as an ordered pair, in this case (2, 4). Since each of the events for this experiment are equally likely, their probabilities can be determined by listing all the possible outcomes.

List the Sample Space Using a Table.

1. Copy and complete this table.

Black Die

Green Die		1	2	3	4	5	6
	1						
	2						
	3						
	4		(2, 4)				
	5						
	6						

2. How many different possible outcomes are there for this experiment?

List the Sample Space Using a Grid.

3. The outcomes for this experiment can be represented with points on a grid. To create this grid, list all the possible outcomes for the black die on the horizontal axis and all the possible outcomes for the green die on the vertical axis. The points, or ordered pairs, represent all of the possible outcomes for this experiment. Plot these points on the grid.

Determine Probabilities.

4. Let B represent the following event.

B: Roll the same number on each die (doubles).

a) Place a triangle around each ordered pair in the table from exercise 1 that represents a favourable outcome for event B.

b) What is the probability that event B will occur?

c) What is the probability that event B will not occur?

5. Let C represent the following event.

C: The sum of the two dice is 5.

a) Place a square around each ordered pair in the table from exercise 1 that represents a favourable outcome for event C.

b) What is the probability that event C will occur?

c) What is the probability that event C will not occur?

6. Let D represent the following event.

D: The sum of the two dice is greater than 9.

a) Circle each ordered pair in the table from exercise 1 that represents a favourable outcome for event D.

b) What is the probability that event D will occur?

c) What is the probability that event D will not occur?

7. Use the grid you created in exercise 3 and repeat exercises 4 to 6. Do you get the same results?

8. What is the relationship between the probability that an event will occur and the probability that the event will not occur?

If an event, A, describes a set of outcomes that satisfy a given condition, then the *complement* of event A is all of the outcomes that do not satisfy this condition. The complement of event A is denoted by $\overline{A}$. Let A represent the following event.

A: The sum of numbers on the two dice in the Investigation is greater than 7.

This event can be illustrated on a grid.

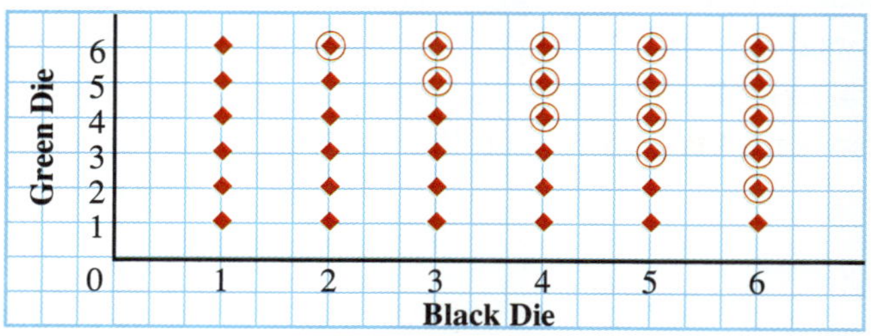

There are 15 outcomes favourable to A, so $P(A) = \frac{15}{36}$ or $\frac{5}{12}$.

This leaves 21 outcomes that are not favourable to A. These outcomes make up the complement, $\overline{A}$.

$P(\overline{A}) = \frac{21}{36}$ or $\frac{7}{12}$.

In any experiment, the outcomes of any event, A, plus the outcomes of its complement, $\overline{A}$, form the sample space. This is illustrated in the following *Venn diagram*.

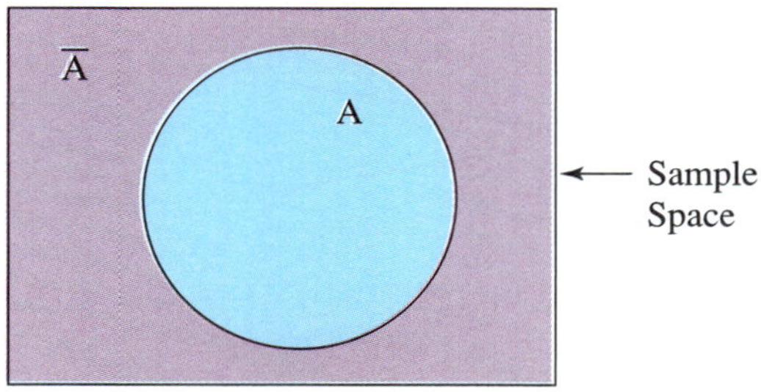

When an experiment is conducted, it is certain that either event A or its complement, $\overline{A}$, will occur.

For any event A, $P(A) + P(\overline{A}) = 1$.

In some situations, it is easier to determine $P(\overline{A})$ directly and then use the equation to calculate P(A).

Example

Tossing 3 coins

List the sample space for the experiment of tossing a penny, a nickel, and a dime. Event A is at least one coin landing tails.

a) List all the outcomes that belong to event A.

b) Determine P(A).

c) Describe the complement of event A in writing.

d) Determine $P(\overline{A})$.

e) Explain how $P(\overline{A})$ could be used to find P(A).

SOLUTION

A tree diagram is an easy way to generate the sample space. The last column of the tree diagram is the sample space.

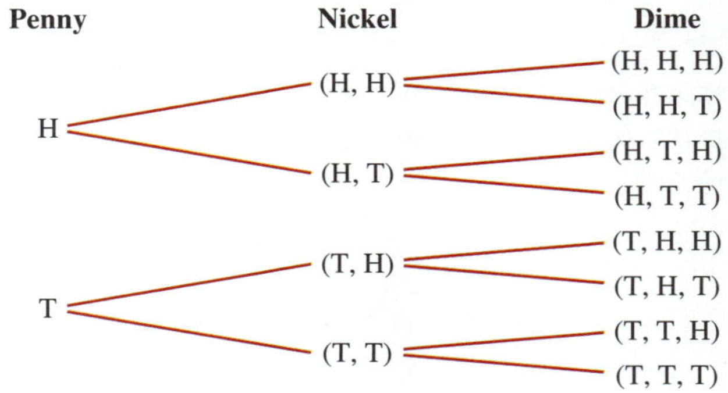

a) (H, H, T), (H, T, H), (H, T, T), (T, H, H), (T, H, T), (T, T, H), (T, T, T)

b) $P(A) = \frac{7}{8}$

c) The complement of A is all the outcomes of the experiment in which no coins landed tails.

d) $P(A) + P(\overline{A}) = 1$

$$\frac{7}{8} + (\overline{A}) = 1$$

$$P(\overline{A}) = 1 - \frac{7}{8}$$

$$P(\overline{A}) = \frac{1}{8}$$

e) Since $P(A) + P(\overline{A}) = 1$, then $P(A) = 1 - P(\overline{A})$.

DISCUSSING THE IDEAS

1. In the Investigation, would the results have changed if the two dice were the same colour? Explain.

2. In the Example, a tree diagram was used to create a sample space for the given experiment. Explain why it may be more difficult to generate a sample space using a table or a grid.

3. Explain why the probability that an event will occur must be between 0 and 1, inclusive.

EXERCISES

Checking Your Skills

1. Suppose the probability that you will pass a driving test on the first try is $\frac{1}{10}$. What is the probability that you will fail on the first try?

2. An inspector at an egg farm routinely inspects a dozen eggs from each shipment. The inspector records the number of broken eggs as part of the quality control process.

a) Describe the outcomes in the event "at least 3 eggs are broken."

b) Describe the outcomes in the event "no more than 2 eggs are broken."

c) Describe the complement of the event "at most 1 egg is broken."

3. Two 6-sided dice are rolled. Use a sample space to determine the probability of each event.

a) The sum of the two numbers is 9.

b) The difference between the two numbers is 3.

c) The sum of the two numbers is less than 4.

d) The difference between the two numbers is at least 4.

4. In an experiment, a 6-sided die is tossed and a coin is flipped.

a) Create a sample space for this experiment.

b) Determine the probability that the coin will land heads and the die will show an even number.

c) Determine the probability that the coin will land tails and the die will show a number greater than 2.

5. In an experiment, you roll a 6-sided die numbered from 1 to 6 and a 4-sided die numbered from 1 to 4.

a) Create a sample space for this experiment.

b) Determine the probability that you will roll doubles.

c) Determine the probability that you will not roll doubles.

d) If A is the event of the sum of the two dice being greater than 7, determine P(A).

e) Describe the complement of event A in words.

f) Determine $P(\overline{A})$.

6. A bus is scheduled to arrive at a train station at any time between 07:08 and 07:13, inclusively. A train is scheduled to arrive between 07:11 and 07:15, inclusively. The ordered pair (8, 14) can represent the arrival of a bus at 07:08 and a train at 07:14. Assume that all of the arrival times are equally likely and are expressed in whole minutes.

a) Create a sample space for this situation.

b) How many outcomes are there in this sample space?

c) For how many outcomes do the bus and train arrive at the same time?

d) For how many outcomes does the bus arrive after the train?

e) What is the probability that the bus will arrive after the train?

7. A homework assignment is given to Alice, Bob, and Carrie on Friday and must be handed in Monday morning. All three students were able to finish the assignment in one evening.

a) Create the sample space for this situation. Use ordered triples to show the outcomes. The ordered triple (F, Sa, Su) means that Alice completed her homework on Friday, Bob on Saturday, and Carrie on Sunday.

b) How many outcomes are there in this sample space?

c) Assume that each of the outcomes in the sample space is equally likely. Determine the probability that all three students completed the assignment on the same evening.

d) Determine the probability that at least one of the students did not complete the assignment until Sunday evening.

8. A couple has decided to have 4 children. Assume that each child is equally likely to be a boy or a girl.

a) Create a sample space for a family with 4 children. The notation (G, B, G, B) would represent the first child being a girl, the second a boy, the third a girl, and the fourth a boy.

b) Determine the probability that all 4 children will be girls.

c) Determine the probability that the couple will have at least 1 girl.

d) Determine the probability that the couple will have 2 boys and 2 girls.

9. Human blood is classified by the presence of 3 main antigens: A, B, and Rh. When the A or B antigens are present, the blood type contains the letter(s) and a positive or negative sign for the Rh antigen. For example, A+ means A and Rh antigens are present. AB– means A and B are present, but Rh is not. When neither A nor B is present, the letter O is used. O+ and O– are the two blood types with Rh present and absent, respectively. Determine the sample space for the different blood types.

Extending Your Thinking

10. A 5-question true/false quiz is given. Assume that a student randomly guesses at each question. This makes the probability that each question is correct the same as the probability that each question is incorrect.

a) Without creating the sample space, how many outcomes will be in this sample space? Explain how you know this.

b) Determine the probability that the student will score 100% on this quiz. Explain how you determined this.

c) Determine the probability that this student will get at least 1 of the 5 questions correct. Explain your strategy.

PROJECT

In this Tutorial, you learned how to construct a sample space for two or three events, and used sample spaces and complements to solve problems. The skills and concepts you used relate to the Chapter Project, and to the following projects in the Project Book.

- NiDi
- Genetic Disorders

COMMUNICATING THE IDEAS

A friend tells you that a tree diagram can be used to create the sample space for any experiment, but a table or grid cannot. Is this true? Use examples to explain your response.

TUTORIAL 1.3

The Fundamental Counting Principle

Sample spaces are very useful for calculating probabilities, but often it is not practical to create a complete sample space. In this Tutorial, we will investigate another method for counting outcomes.

Investigation

Counting without Counting

1. At a local college, first-year students must take one of 3 different literature classes (L1, L2, L3) and one of 5 different science classes (S1, S2, S3, S4, S5).

a) Create a sample space that lists the different ways that students can satisfy these requirements.

b) In how many different ways can these requirements be satisfied?

c) In how many different ways can a student choose a literature class?

d) In how many different ways can a student choose a science class?

e) How can you determine the number of ways to select 1 literature class and 1 science class without listing them all?

2. Joe has 3 different shirts (S1, S2, S3), 2 different pairs of pants (P1 and P2), and 3 different pairs of shoes (H1, H2, H3).

a) Create a sample space that lists the different ways Joe could get dressed in the morning.

b) In how many different ways could Joe get dressed in the morning?

c) How can you determine the number of different outfits Joe has without listing them all?

3. There are 5 different choices for pizza (P1, P2, P3, P4, P5) and 3 different choices for a drink (D1, D2, D3).

a) Predict the number of different selections there are of a pizza and a drink.

b) Create a sample space to verify your prediction.

4. Suppose one item can be selected in m ways, and a second item can be selected in n ways. In how many ways can the 2 items be selected? This result is known as the *fundamental counting principle*.

When two or more choices must be made together, the total number of possibilities can be determined without listing them all. The results of the Investigation illustrate the following principle.

The Fundamental Counting Principle

If one item can be selected in m different ways and a second item can be selected in n different ways, then the two items can be selected in $m \times n$ ways.

Example 1 **Purchasing a computer**

A computer store sells 6 different computers, 4 different monitors, 5 different printers, and 3 different multimedia packages. How many different computer systems are available?

SOLUTION

There are 6 different computers.	For each computer, there are 4 monitors.	For each monitor, there are 5 printers.	For each printer, there are 3 multimedia packages.
6 ×	**4** ×	**5** ×	**3**

There are $6 \times 4 \times 5 \times 3 = 360$ different systems.

When there are restrictions on the choices, the restrictions should always be considered first.

Example 2 **How many digits are possible?**

How many 4-digit odd numbers are there in which no digit is repeated?

SOLUTION

The number has 4 digits so we have four choices to make.

________	________	________	________
Thousands Digit	Hundreds Digit	Tens Digit	Ones Digit

To ensure the number is odd, the ones digit must be 1, 3, 5, 7, or 9. There are 5 ways to select the ones digit.

To ensure the number has 4 digits, the thousands digit can be any digit except 0. Since we are not allowed to repeat any of the digits, we cannot choose the digit that was selected as the ones digit. There are 8 ways to select the thousands digit.

The hundreds digit can be any of the digits from 0 to 9, except the digits already used in the ones or thousands places. There are 8 ways to select the hundreds digit.

The tens digit can be any of the digits from 0 to 9, except those already used in the ones, hundreds, or thousands places. There are 7 ways to select the tens digit.

There are $8 \times 8 \times 7 \times 5$, or 2240, 4-digit odd numbers in which no digit is repeated.

Example 3 **Calculating probabilities**

A 3-digit number is to be randomly generated.

a) How many different 3-digit numbers are possible?

b) How many different 3-digit numbers are there with no digits repeated?

c) What is the probability that a 3-digit number has at least 2 digits that are the same?

SOLUTION

a) The only restriction in this case is that the hundreds digit cannot be zero. Thus, there will be $9 \times 10 \times 10$, or 900, 3-digit numbers.

b) Since the digits cannot be repeated there will be 9 ways to select the hundreds digit (since it cannot be zero), 9 ways to select the tens digit, and 8 ways to select the ones digit. Thus, there are $9 \times 9 \times 8$, or 648, 3-digit numbers with no digits being repeated.

c) There are 900 different 3-digit numbers and 648 of these have no digits repeated. This means $900 - 648$, or 252, of these numbers have at least 2 digits that are the same. Thus, the probability that at least 2 digits are the same is $\frac{252}{900} = \frac{7}{25}$, or 0.28.

DISCUSSING THE IDEAS

1. Consider the fundamental counting principle. Use an example to explain why the number of ways the items can be selected are multiplied and not added.

2. Suppose numbers such as 0135 were considered 4-digit numbers. How would the solution to *Example 2* change?

3. You have just completed a true/false test with 10 questions. Since there are 2 possible choices for each question and 10 questions, a friend states that there are 20 different ways of completing the exam. Explain the error your friend has made.

EXERCISES

Checking Your Skills

1. Three candidates (P1, P2, and P3) are running for class president and 5 candidates (T1, T2, T3, T4, and T5) are running for class treasurer.
   a) Create a sample space that lists all the possible outcomes for this election. How many outcomes are possible?
   b) Use the fundamental counting principle to determine the number of possible outcomes for this election. Does your answer match part a?

2. A person must choose 1 card from a well-shuffled deck of 52 playing cards, roll a 6-sided die, and then toss a coin. How many different results are possible?

3. Marina has 5 blouses, 4 skirts, and 4 sweaters. Determine the number of ways she can select an outfit given the following conditions.
   a) The outfit is comprised of 1 blouse, 1 skirt, and 1 sweater.
   b) The outfit may or may not include a sweater, but must include a blouse and skirt.

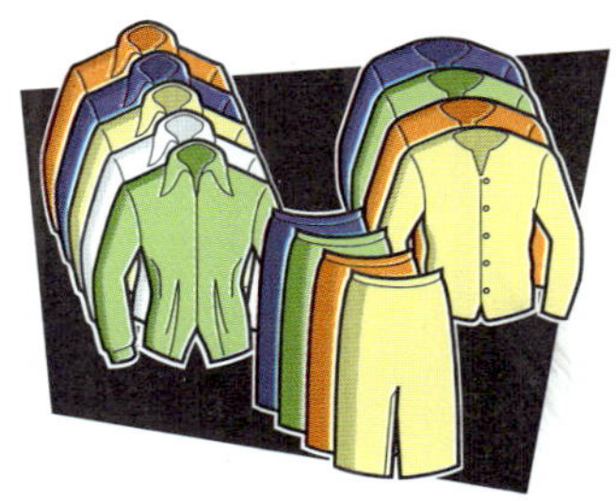

4. Use the digits 1, 3, 5, 7, and 9.
   a) How many 2-digit numbers can be formed?
   b) How many 2-digit numbers can be formed if no digit is repeated?
   c) How many 3-digit numbers can be formed?
   d) How many 3-digit numbers can be formed if no digit is repeated?

5. To get from his home to the downtown train station, Jai can take one of 3 different buses or 2 different trains. To get from the downtown station to work he can take one of 2 different buses or 4 different trains. Determine the number of different ways Jai can get from home to work, given the following conditions.

a) He uses only buses to get to work.

b) He uses only trains to get to work.

c) There are no restrictions.

6. In Saskatchewan, license plates are made up of 3 letters followed by 3 digits.

a) How many different license plates are possible? State any assumptions you made.

b) How many license plates are there where no letter is repeated?

c) How many license plates are there where no digit is repeated?

d) What is the probability that you will receive a license plate with at least one letter repeated?

e) What is the probability that you will receive a license plate with at least one digit repeated?

7. The final score at the end of a hockey game is 5 to 2. How many different scores are possible at the end of the second period, given the following conditions?

a) There are no restrictions.

b) Both teams had each scored at least 1 goal.

c) The winning team was in the lead at the end of the second period.

8. The dial on a 3-number combination lock contains markings to represent the numbers from 0 to 59.

a) How many combinations are possible with no restrictions?

b) How many combinations are possible if no number is repeated?

c) What is the probability that all 3 numbers of the combination are odd if repetitions are allowed?

d) Repetitions are allowed. What is the probability that all 3 numbers of the combination are prime numbers?

9. A multiple-choice test has 5 questions, with 4 possible answers for each question. Suppose that a student answers each question by guessing.

a) In how many ways can each question be answered?

b) Assume each question must be answered. How many different ways are there of completing the test?

c) What is the probability that all 5 questions will be answered correctly?

10. In a hotel in Hong Kong, the room keys are cards. Each card has positions for holes that form a 5 by 10 array. Each position in the array is either punched with a hole or left blank.

a) How many different keys are possible?

b) Assume Earth's population is 6 billion. Suppose all the possible keys were distributed equally among all the people on Earth. How many keys would there be for each person?

Extending Your Thinking

11. In the Pick 3 lottery, participants pick 3 digits from 0 to 9 and choose Straight Play or Box Play. The lottery corporation randomly selects 3 winning digits from 0 to 9, which can be the same. Straight Play wins if the 3 digits picked match the 3 winning digits in the same order. Box Play wins if the 3 digits picked match the 3 winning digits in any order. According to the lottery corporation, the odds that you will win are 1 in 1000 for Straight Play and 1 in 167 for Box Play. Write to explain how these odds are determined for both games.

12. In a contest at a local music store, the winner will be permitted to choose two CDs from the current top 10.

a) Create a sample space that lists the number of ways you could choose 2 CDs from:

i) 3 CDs (A, B, and C).

ii) 4 CDs (A, B, C, and D).

iii) 5 CDs (A, B, C, D, and E).

b) Use the fundamental counting principle to redo part a. How does your answer compare with part a? Why do you think this is so?

c) Use your results from parts a and b. In how many ways could a person choose 2 CDs from the current top 10?

PROJECT

In this Tutorial, you learned how to determine the number of ways multi-step operations can be performed using the fundamental counting principle. The skills and concepts you used relate to the Chapter Project, and to the following project in the Project Book.

- NiDi

COMMUNICATING THE IDEAS

Use an example to explain to a friend how the fundamental counting principle can be used to calculate probabilities.

How Many Passwords Are Possible?

A password should be easy to remember but difficult for others to guess. To ensure the security of users' voice-mail, the telephone company requires that a randomly generated and unique password be assigned to each customer. To access their voice-mail, users input their telephone number and their password.

The telephone company estimates that after 1 year they will have up to 100 000 users, and after 5 years, there will be 250 000 users. The voice-mail system allows passwords up to 15 digits long. The asterisk (*) and pound (#) symbols are used to navigate within the voice-mail system. Most people can remember 5 to 9 bits of information. In keeping with these constraints, the following initial proposals have been made.

CHAPTER PROJECT

You will return to the Chapter Project on page 40 of this chapter. *Tutorials 1.4* and *1.5* will help you develop the relevant skills for the last stage of the Chapter Project.

Proposal A

Each user is assigned a randomly generated 3-, 4-, or 5-digit number. To ensure the password is difficult to guess, the digits cannot all be the same (that is, passwords such as 111 or 2222 will not be used). Complete these exercises to determine if this proposal will provide enough user passwords.

1. How many different passwords are possible that have:

a) 3-digits? **b)** 4-digits? **c)** 5-digits?

2. What is the total number of passwords that are available if you use this system?

3. Are there any other numbers that should not be used? Why?

4. Will this proposal provide enough passwords for the number of anticipated users? Explain.

Career Skills

People who work for telephone companies, Internet-service providers, and banks must provide customers with secure, user-friendly access to their accounts.

Proposal B

Some phones do not have the letters Q and Z on the telephone keypad. The letters A, B, and C are on the key with the number 2. With this system, each user is assigned a randomly generated 4-letter password. To input the letter A, press 2 once. To input B, press 2 twice, and to input C, press 2 three times. Passwords cannot have two consecutive letters from the same number key. For example, BCPU is not permitted since the letters B and C are both on the number 2. Complete these exercises to determine if this proposal will provide enough passwords.

1. List 5 possible passwords and their corresponding number sequences.
2. What is the minimum number of digits that a password could have? Give one example of such a password.
3. What is the maximum number of digits that a password could have? Give one example of such a password.
4. How many different passwords could be generated?
5. Will this proposal provide enough passwords for the anticipated number of users? Explain.

You have determined the number of passwords generated by each proposal. Keep your answers to these exercises. You will include them with your final report. Think about ways you could improve the security of a voice-mail system without making it too difficult to use. Include these suggestions in your final report.

TUTORIAL 1.4

Independent and Dependent Events

In this Tutorial, we will calculate the probability that an event A and a second event B both occur.

Investigation 1 "AND" Probabilities

Visualize tossing a coin and rolling a 6-sided die.

1. Use a grid to create a sample space for this experiment.

2. Let A and B represent the following events.

   A: tossing a head
   B: rolling a 6
   Find the following probabilities.

   a) P(A) **b)** P(B) **c)** P(A and B)

   d) How is your result from part c related to your results from parts a and b?

3. Let M and N represent the following events.

   M: tossing a tail
   N: rolling an odd number
   Find the following probabilities.

   a) P(M) **b)** P(N) **c)** P(M and N)

   d) How is your result from part c related to your results from parts a and b?

4. Let S and T represent the following events.

   S: tossing a head
   T: rolling a number less than 5
   Find the following probabilities.

   a) P(S) **b)** P(T) **c)** P(S and T)

   d) How is your result from part c related to your results from parts a and b?

To toss heads on a coin and to roll a 6 on a die are *independent events*. The occurrence of one has no effect on the occurrence of the other. Not all events are independent. For example, a spade is drawn from a deck of 52 cards and then a second card is drawn from the same deck, without replacing the first card. Once the first spade is drawn, there are only 51 cards left, and only 12 spades. These are *dependent events* because the occurrence of the first event affects the probability of the occurrence of the second event.

Investigation 2 More "AND" Probabilities

Three marbles (X, Y, and Z) are placed in a bag. One marble is selected and not placed back in the bag. Then, a second marble is selected.

1. Would the selection of these 2 marbles be considered independent events? Explain your response.
2. Use the fundamental counting principle to determine the number of different ways these 2 marbles can be selected.
3. Use a tree diagram to create the sample space for this experiment.
4. Let A and B represent the following events.

   A: drawing marble X in either of the two selections
   B: drawing marble Y in either of the two selections
   Find the following probabilities.

   a) P(A) **b)** P(B)

   c) Use your sample space to determine P(A and B).

   d) Does $P(A \text{ and } B) = P(A) \times P(B)$ for these events?

If the occurrence of event A does not affect the probability that event B occurs, then events A and B are independent.

For independent events: $P(A \text{ and } B) = P(A) \times P(B)$.

Example 1 Determine the probability of tossing 4 heads in a row.

What is the probability that a coin will land heads in each of 4 consecutive tosses?

SOLUTION

These events are independent, since the result of the first toss does not affect the result of the second toss, and so on.

$$\begin{aligned} P(4 \text{ Heads in a Row}) &= P(\text{Heads}) \times P(\text{Heads}) \times P(\text{Heads}) \times P(\text{Heads}) \\ &= \frac{1}{2} \times \frac{1}{2} \times \frac{1}{2} \times \frac{1}{2} \\ &= \frac{1}{16} \end{aligned}$$

The probability of tossing 4 successive heads is $\frac{1}{16}$.

Example 2 Determine "at least" probabilities.

A quiz is made up of 5 multiple-choice questions, each with 4 possible answers. Determine the probability that you will get at least 1 answer correct if you guess the answers to all 5 questions.

SOLUTION

These events are independent, since a success on 1 question does not affect the probability of success on subsequent questions. Getting at least one correct response means there can be 1, 2, 3, 4, or 5 correct responses. The only outcome that is not included in this event is obtaining no correct responses. We can use complements to calculate this probability.

$$P(\text{no correct responses}) = \frac{3}{4} \times \frac{3}{4} \times \frac{3}{4} \times \frac{3}{4} \times \frac{3}{4}$$
$$= \left(\frac{3}{4}\right)^5$$

$$P(\text{at least one correct response}) = 1 - P(\text{no correct responses})$$
$$= 1 - \left(\frac{3}{4}\right)^5$$
$$\doteq 0.763$$

The probability of obtaining at least one correct response is approximately 0.763.

Example 3 A test for independence

Thirty percent of seniors catch the flu every year. Fifty percent of seniors get a flu shot annually. Ten percent of seniors who get the flu shot also get the flu. Are getting a flu shot and getting the flu independent events?

SOLUTION

Let S and F represent these events.

S: A senior gets a flu shot. F: A senior gets the flu.

Then $P(S) = 0.5$ and $P(F) = 0.3$.

These events will be independent if $P(F \text{ and } S) = P(F) \times P(S)$.

From the given information:

$$P(F \text{ and } S) = 0.1$$
$$P(F) \times P(S) = 0.3 \times 0.5$$
$$= 0.15$$

Since $P(F \text{ and } S) \neq P(F) \times P(S)$, getting the flu and getting a flu shot are not independent events.

Discussing the Ideas

1. The events described in *Investigation 2* were dependent. How could the experiment be changed so that the events would be independent?

2. In *Example 2*, how would the probability of getting at least one answer correct change if there were more than 5 questions? Explain.

3. A student tosses a coin 3 times and gets a head each time. Her partner states: "The probability of getting 4 heads in a row is very small, so I think the next toss will be tails." Do you agree or disagree with this statement? Explain.

Exercises

Checking Your Skills

1. Classify the following events as independent or dependent.
   a) Tossing a head in a coin toss and rolling a 6 on a 6-sided die.
   b) Drawing an ace for the first card and another ace for the second card, if the first card is not replaced.
   c) Drawing a king for the first card and a queen for the second card, if the first card is replaced.
   d) Rolling a 6-sided die and getting a 6 then rolling the die a second time and getting a 1.
   e) Being chosen as 1 of 3 students to represent your class if your friend has already been chosen.
   f) Winning at solitaire twice in a row.

2. Sixty percent of young drivers take driver training, and 25% of young drivers have an accident in their first year of driving. Statistics show that 10% of those who take driver training have an accident in their first year. Are taking driver training and having an accident in the first year independent events? Support your answer with the appropriate mathematical calculations.

3. Forty percent of grade 12 students enroll in a tutorial program. Fifty percent of grade 12 students find the academic demands tougher than in grade 11. Twenty percent of grade 12 students are in the tutorial program and feel that the academic demands of grade 12 are tougher than in grade 11. Are being enrolled in a tutorial program, and finding the academic demands tougher independent events? Support your answer with the appropriate mathematical calculations.

4. A card is drawn from a well-shuffled deck of 52 playing cards and the card is replaced. The deck is shuffled again, and then a second card is drawn. Determine the following probabilities.

a) Both cards are diamonds.
b) Both cards are red.
c) Both cards are face cards.
d) Both cards are 5s.
e) Neither card is a club.

5. A couple has decided to have 4 children. For each child, assume there is an equal probability that it will be a boy or a girl. Determine the following probabilities.

a) The couple has 4 girls.
b) The couple has 2 boys and then 2 girls.
c) The couple has at least 1 girl.

6. A bag contains 3 white balls and 2 black balls. The first draw produces a white ball, which is not replaced. What is the probability that the second draw will produce another white ball?

7. The probability that a student completes her mathematics assignment is 0.6. The probability that she completes her French assignment is 0.4. Assume these events are independent. What is the probability of the following?

a) She completes both assignments.
b) She completes neither assignment.

8. A coin is tossed many times in succession.

a) What is the probability that the coin lands tails on 5 successive tosses?
b) What is the probability that the coin lands heads at least once if the coin is tossed 5 times?
c) How many times does the coin have to be tossed so that the chance that it lands heads at least once is greater than or equal to 99%?

9. A lottery is advertising a house as its grand prize as well as 499 additional prizes. Only 10 000 tickets will be sold. Once a ticket is drawn, it is re-entered into the draw and is still eligible for each of the remaining prizes.

a) If you buy 1 ticket, determine the probability that you will win the grand prize.
b) If you buy one ticket, determine the probability that you will win at least 1 prize.
c) If you buy 10 tickets, determine the probability that you will win the grand prize.
d) If you buy 10 tickets, determine the probability that you will win at least 1 prize.

Extending Your Thinking

10. Three marbles (X, Y, and Z) are placed in a bag. One marble is selected and then a second marble is selected without replacing the first. Let A and B represent these events.

A: drawing marble X in either of the two selections
B: drawing marble Y in either of the two selections

a) Find P(A).

b) *Conditional probability* is the probability that an event B will occur given that another event A has already occurred. It is denoted as $P(B|A)$. Determine the conditional probability of selecting marble Y if marble X has already been selected.

c) Use the sample space from *Investigation 2* to determine P(A and B).

d) How is P(A and B) related to P(A) and $P(B|A)$?

11. Two cards are drawn, without replacement, from a standard deck of 52 playing cards. Use the result from exercise 10 to determine the following probabilities.

a) Both cards are diamonds.
b) Both cards are red.
c) Both cards are face cards.
d) Both cards are 5s.
e) Neither card is a club.

PROJECT

In this Tutorial, you learned how to classify events as dependent or independent and how to determine the probability of independent events. The skills and concepts you used relate to the Chapter Project, and to the following projects in the Project Book.

- NiDi
- Genetic Disorders

COMMUNICATING THE IDEAS

Write a paragraph to explain the difference between dependent and independent events.

TUTORIAL 1.5

Mutually Exclusive Events

In this Tutorial, we will develop a method to calculate the probability that event A occurs, or event B occurs, or both.

Investigation

"OR" Probabilities

Visualize tossing a pair of 6-sided dice. Consider these events.

X: The sum of the two dice is 5.

Y: The sum of the two dice is 8.

Z: Doubles are rolled.

1. Create a sample space for this experiment.

2. Use your sample space to determine the following probabilities.

a) P(X) **b)** P(Y)

c) P(X and Y) **d)** P(X or Y)

e) How is P(X or Y) related to P(X), P(Y), and P(X and Y)?

3. Use your sample space to determine the following probabilities.

a) P(X) **b)** P(Z)

c) P(X and Z) **d)** P(X or Z)

e) How is P(X or Z) related to P(X), P(Z), and P(X and Z)?

4. Use your sample space to determine the following probabilities.

a) P(Y). Circle all the outcomes that are part of event Y.

b) P(Z). Place a square around all the outcomes that are part of event Z.

c) P(Y and Z). Place a triangle around all the outcomes that are part of event Y and Z.

d) P(Y or Z)

e) How is P(Y or Z) related to the P(Y), P(Z), and P(Y and Z)? Use the outcomes in squares, circles, and triangles to explain your results.

5. How are the events in exercise 4 different from the events considered in exercises 2 and 3?

Two events that do not share any common outcomes are called *mutually exclusive* events. In the Investigation, events X and Y are mutually exclusive (it is not possible to get a sum of 5 and a sum of 8 in one roll of two dice). Events X and Z are also mutually exclusive.

If events A and B are mutually exclusive, they cannot happen at the same time, so P(A and B) will be zero and the probability that A or B will occur will be the sum of P(A) and P(B).

Two events A and B that cannot occur at the same time are mutually exclusive events. They have no common outcomes.

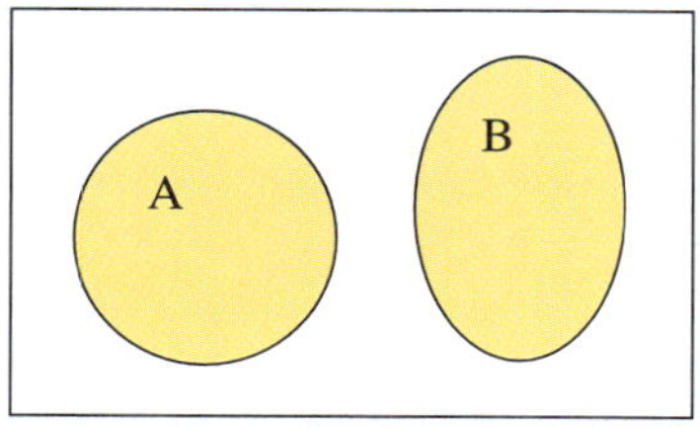

$$P(\text{A and B}) = 0$$
$$P(\text{A or B}) = P(\text{A}) + P(\text{B})$$

If events A and B are not mutually exclusive, they will have some common outcomes as in exercise 4 of the Investigation. Notice in the diagram below that the region (A and B) is contained within A and within B. When we find the sum P(A) + P(B) we add P(A and B) twice. To ensure common outcomes are not counted twice, P(A and B) is subtracted from the sum of P(A) and P(B).

Two events A and B that are not mutually exclusive have some common outcomes.

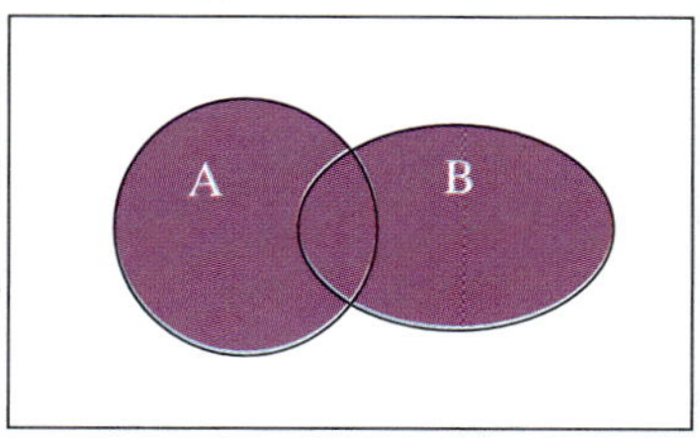

A or B

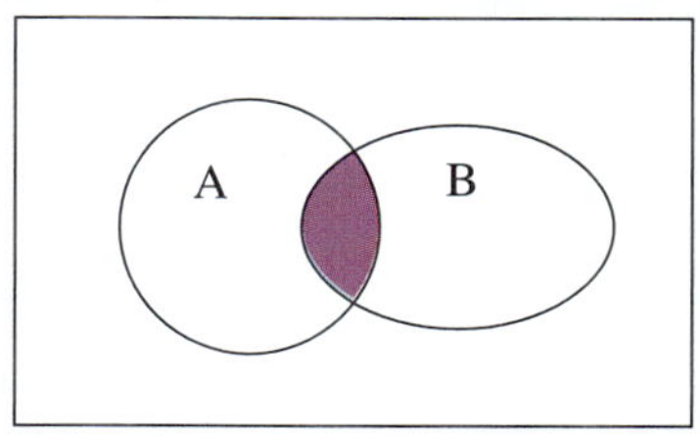

A and B

$$P(\text{A or B}) = P(\text{A}) + P(\text{B}) - P(\text{A and B})$$

Events that can occur at the same time are not mutually exclusive; they are either dependent or independent. Events that cannot occur at the same time are mutually exclusive.

Example 1 **Using a computer**

There are 30 students in a class. Sixteen students surf the Internet and 10 students use e-mail. Of these students, 6 do both. What is the probability that a randomly selected student in the class surfs the Internet or uses e-mail?

SOLUTION

Suppose a student is randomly selected. Let S and E represent these events:

S: The student surfs the Internet.
E: The student uses e-mail.

The events S and E are not mutually exclusive.

$$\begin{aligned} P(S \text{ or } E) &= P(S) + P(E) - P(S \text{ and } E) \\ &= \frac{16}{30} + \frac{10}{30} - \frac{6}{30} \\ &= \frac{20}{30} \text{ or } \frac{2}{3} \end{aligned}$$

The probability that a randomly selected student surfs the Internet or uses e-mail is $\frac{2}{3}$.

Example 2 **Combining "and" and "or" probabilities**

A card is drawn from a well-shuffled standard deck of 52 playing cards. The card is returned to the deck, the deck is shuffled, and a second card is drawn. Determine the probability that both cards are red or both cards are aces.

SOLUTION

Let R and A represent these events:

R: Both cards are red. A: Both cards are aces.

These two events are not mutually exclusive since both cards could be red aces.

We must determine P(R), P(A), and P(R and A).

$$\begin{aligned} P(R) &= P(1^{st} \text{ card is red}) \times P(2^{nd} \text{ card is red}) \\ &= \frac{26}{52} \times \frac{26}{52} \\ &= \frac{676}{2704} \end{aligned}$$

$$\begin{aligned} P(A) &= P(1^{st} \text{ card is an ace}) \times P(2^{nd} \text{ card is an ace}) \\ &= \frac{4}{52} \times \frac{4}{52} \\ &= \frac{16}{2704} \end{aligned}$$

$$
\begin{aligned}
P(\text{R and A}) &= P(1^{\text{st}} \text{ card is a red ace}) \times P(2^{\text{nd}} \text{ card is a red ace}) \\
&= \frac{2}{52} \times \frac{2}{52} \\
&= \frac{4}{2704}
\end{aligned}
$$

$$
\begin{aligned}
P(\text{R or A}) &= P(\text{R}) + P(\text{A}) - P(\text{R and A}) \\
&= \frac{676}{2704} + \frac{16}{2704} - \frac{4}{2704} \\
&= \frac{688}{2704} \text{ or } \frac{43}{169} \\
&\doteq 0.254
\end{aligned}
$$

The probability of both cards being red or both being aces is approximately 0.254.

DISCUSSING THE IDEAS

1. In *Examples 1* and *2*, explain why the fractions were not reduced until the final probability was calculated.

2. A and B represent two mutually exclusive events. Explain how we know P(A and B) = 0 and P(A or B) = P(A) + P(B).

3. Is it possible for mutually exclusive events to be independent? Explain.

EXERCISES

Checking Your Skills

1. For each experiment, are the events mutually exclusive? If they are not, identify one common outcome in the sample space.

	Experiment	Event
a)	Rolling 2 six-sided dice	Doubles or a sum of 11
b)	Drawing a card from a standard deck	The card is a club or a 7.
c)	Randomly selecting a number from 1 to 100	The number is odd or greater than 90.
d)	Tossing a coin twice	2 heads or 2 tails
e)	Drawing a card from a standard deck	The card is black or a diamond.

2. Let A represent any event in an experiment. Determine each probability and explain what it represents.

a) $P(A \text{ or } \overline{A})$

b) $P(A \text{ and } \overline{A})$

3. A number is randomly picked from the numbers 1 to 10 inclusive. Determine the probability of the following events.

a) The number is greater than 7 or less than 4.

b) The number is odd or a perfect square.

c) The number is odd or a multiple of 5.

d) The number is even or a multiple of 3.

e) The number is odd or greater than 5.

4. A card is drawn from a shuffled deck of 52 playing cards. Determine each probability.

a) The card is a heart or a spade.

b) The card is an ace or a face card.

c) The card is a heart or a face card.

d) The card is an ace or a spade.

5. Use a sample space that represents rolling two dice and determine the probability of each event.

a) The sum of the dice is less than or equal to 7.

b) The sum of the dice is less than 5 or greater than 9.

c) The sum of the dice is less than 9 or greater than 5.

d) The sum of the dice is 10 or doubles are rolled.

e) The sum of the dice is 7 or doubles are rolled.

6. A computer store advertised its annual half-price sale in the newspaper and on TV. A survey of 200 customers indicated that 60 learned about the sale from the newspaper and 50 from TV. Of these 110 customers, 30 learned about the sale from both sources. What is the probability of each event?

a) A randomly selected customer saw the advertisement in the newspaper and on TV.

b) A randomly selected customer saw the advertisement in at least one of the two media.

c) A randomly selected customer saw the advertisement on TV, but did not see it in the newspaper.

7. Two teams are playing a best of seven games playoff series. The probability that team A will win any one game is 0.55. Determine the probability that either team A or team B wins the series in the first 4 games.

8. A bag contains 3 white and 5 green marbles. A marble is selected, replaced, and then a second marble is selected. Determine the probability that both marbles are white or both are green.

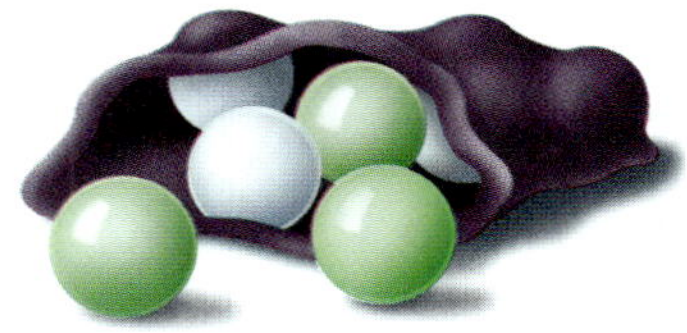

9. A card is drawn from a standard deck of playing cards, returned to the deck, and then a second card is drawn. Determine the following probabilities.

a) Both cards are clubs or both cards are diamonds.

b) Both cards are black or both cards are 2s.

c) Both cards are face cards or both cards are 7s.

d) Both cards are diamonds or both cards are face cards.

10. A pair of dice is rolled twice. Determine the following probabilities.

a) Both rolls result in doubles or both result in a sum of 7.

b) Both rolls result in a sum that is odd or both rolls result in a sum that is even.

c) Both rolls result in doubles or both rolls result in a sum of 10.

d) Both rolls result in a sum that is even or both rolls result in a sum that is greater than 9.

Extending Your Thinking

11. In the Venn diagram, what regions are included in each section?

a) A and B　　**b)** A and C

c) B and C　　**d)** A or B

e) B or C　　**f)** A or C

g) A and B and C　　**h)** A or B or C

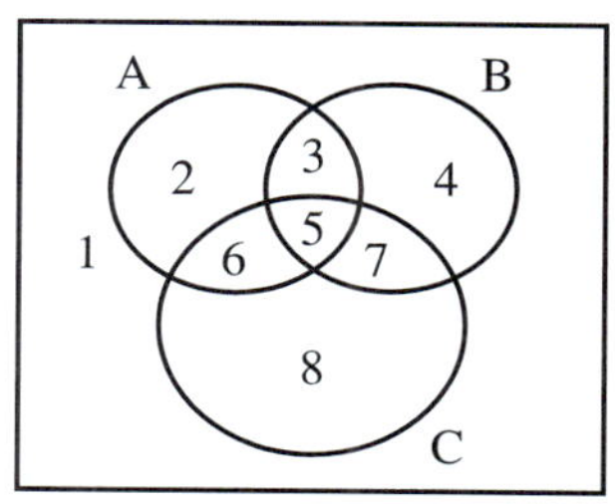

PROJECT

In this Tutorial, you learned how to solve problems that involve mutually exclusive events. The skills and concepts you used relate to the Chapter Project, and to the following projects in the Project Book.

- NiDi
- Genetic Disorders

COMMUNICATING THE IDEAS

Your friend missed class today. Write to her to explain what it means when events are mutually exclusive and how the probabilities of events A and B are affected when they are not mutually exclusive.

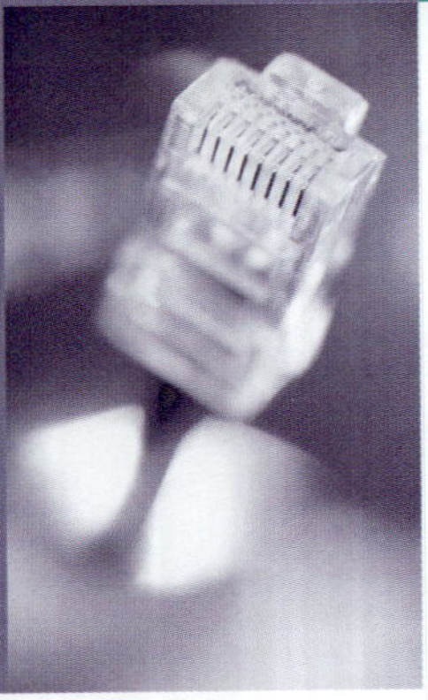

Recommending a Password System

On pages 26 and 27, you considered two proposals for generating passwords for voice-mail customers. In this part of the project, you will consider the security of each system and the likelihood that users will forget their password. You will then write reports to summarize each proposal and explain your recommendation to the company.

The company has done an initial study to determine the probability that users will forget their password. The experimental probabilities are given in the table.

Number of Characters	Probability of Forgetting
3	0.0001
4	0.000 15
5	0.0002

1. On any given day, 10 000 of your customers are expected to use this voice-mail system. Determine the probability that at least 1 customer will forget their password if it has

a) 3 characters **b)** 4 characters **c)** 5 characters

2. Which proposal will minimize this problem? Explain your response.

Let A and B represent these events.

A: A person randomly guesses a password that was generated under proposal A.

B: A person randomly guesses a password that was generated under proposal B.

3. Determine the following.

a) P(A) **b)** P(B)

c) Which event is more likely to occur?

Career Skills

Many companies focus on providing information services to their customers. The problems you have considered are an important part of providing these services. Contact some local companies that provide e-mail or voice-mail systems and inquire about their security systems and any problems they have encountered.

4. To gain access to your voice-mail account, you need to enter both your telephone number and your password. A colleague suggests that this makes it unlikely for someone else to access your voice mail. Do you agree or disagree? Explain your response.

5. To improve security, users could be assigned 2 passwords, one generated by each proposal. Determine P(A and B). Would you suggest this system? Explain your response.

6. What are some other ways the telephone company could improve the security of the voice-mail system?

You have determined the probability that on any given day at least one user will forget her or his password. You have also considered the level of security offered by both proposals. Write a complete report on each proposal including:

- the number of possible passwords
- the probability of users forgetting their passwords
- the level of security offered
- other considerations
- suggestions for improvement
- all relevant calculations

Write a separate report to outline your recommendations to the company. You may include an alternative proposal as part of this report. If you do, be sure to discuss the advantages and disadvantages of your proposal.

Your completed project will include all three reports.

CONSOLIDATING YOUR SKILLS

What Do I Need To Know?

Tutorial	Skills and Concepts	Important Results and Formulas
1.1	• Calculate theoretical probabilities • Calculate experimental probabilities	• The probability that event A will occur is given by $P(A) = \frac{r}{n}$, where r is the number of favourable outcomes and n is the total number of possible outcomes
1.2	• Construct a sample space for two or three events • Use sample spaces and complements to solve problems	• A sample space can be created using a grid, a table, or a tree diagram • For any event A, $P(A) + P(\overline{A}) = 1$
1.3	• Use the fundamental counting principle to determine the number of ways multi-step events can occur	• If one item can be selected in m ways and a second item can be selected in n ways, the two items can be selected in $m \times n$ ways
1.4	• Classify events as dependent or independent • Determine the probability of independent events	• Two events, A and B, are independent if the occurrence of A does not affect the probability of B occurring • For independent events A and B, P(A and B) = P(A) × P(B)
1.5	• Solve problems involving mutually exclusive events	• Two events are mutually exclusive if they cannot occur at the same time • If events A and B are mutually exclusive, P(A and B) = 0, P(A or B) = P(A) + P(B) • If events A and B are not mutually exclusive, P(A or B) = P(A) + P(B) − P(A and B)

What Should I Be Able To Do?

After you have finished the work in Chapter 1, you should be able to solve the problems that follow. Part A is investigative in nature.

As you work through the exercises, refer to the Student Reference, the Utilities, and the Tutorials as necessary.

Part A

1. A spinner is divided into 3 equal sections. Two sections are labelled "win" and the third is marked "lose." You spin the spinner twice.

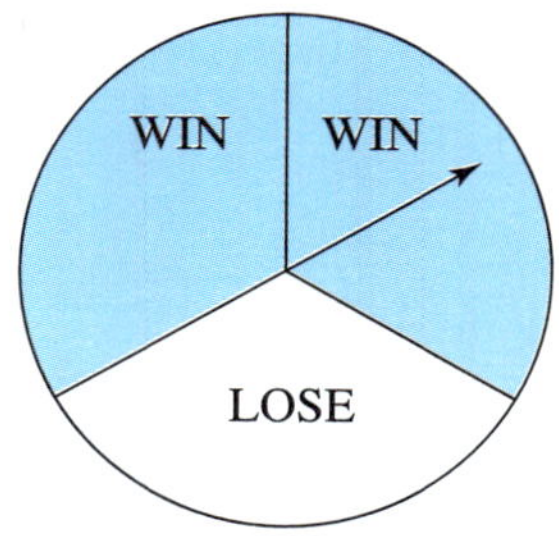

a) Use a table to create a sample space for this experiment.

b) Use your sample space to determine these probabilities.

i) The spinner lands on "win" on the first and second spins.

ii) The spinner lands on "win" on the first or second spin.

c) Let A and B represent these events.

A: The spinner lands on "win" on the first spin.

B: The spinner lands on "win" on the second spin.

Determine these probabilities.

i) P(A) **ii)** P(B)

iii) P(A and B) **iv)** P(A or B)

d) Do your results for part c agree with the probabilities determined in part b?

e) Enter the command randBin(1, $\frac{2}{3}$, 2) into the TI-83 calculator to simulate spinning this spinner twice. Repeat this experiment 15 times. Record the results of each experiment. Determine these experimental probabilities.

i) The spinner lands on "win" on the first and second spins.

ii) The spinner lands on "win" on the first or second spin.

f) Compare the experimental probabilities to the theoretical probabilities you have calculated.

g) Repeat part e. This time conduct the experiment 25, 50, and then 75 times. What happens to the experimental probabilities as you increase the number of trials?

Part B

2. You spin the spinner described in exercise 1 three times.

a) Use a tree diagram to create a sample space for this experiment.

b) Let A represent this event.
A: the spinner lands on "win" all 3 times.
Determine P(A).

c) Describe the complement of event A.

d) Determine $P(\overline{A})$.

3. Every street in Canada is assigned a postal code. A postal code is made up of 3 letters and 3 digits that alternate. How many different postal codes can be created given the following conditions?

a) There are no restrictions.

b) The first letter must be S.

c) The first letter must be S and the first digit must be 7.

4. A 4-digit number is to be generated using the digits 1 to 9.

a) How many different 4-digit numbers can be formed?

b) How many different 4-digit numbers can be formed if no digit is repeated?

c) If a 4-digit number is chosen at random, what is the probability that at least 1 digit is repeated?

5. A card is drawn from a well-shuffled deck of 52 playing cards and then replaced. This is repeated 2 more times. Determine the following probabilities.

a) All 3 cards are clubs.

b) All 3 cards are red.

c) All 3 cards are face cards.

d) At least 1 card is a heart.

6. Twenty percent of students take vitamin C daily. Sixty percent of students catch a cold every year. Eight percent of students who take vitamin C every day caught a cold last year. Are taking vitamin C and catching a cold independent events? Support your answer with the appropriate mathematical calculations.

7. In a certain card game, all 12 face cards are removed from a standard deck of cards. One card is drawn from the 40 remaining cards. Determine these probabilities.

a) The card is a spade or a red card.

b) The card is a 4 or a club.

c) The card is diamond or a red 8.

8. The probability that José beats Larry in a tennis match is 0.75. If they play 2 tennis matches, what is the probability that José wins the first or second match but not both?

Part C

The Chapter Project combines many skills and concepts covered in this Chapter. Here is a sampling of student work from different stages of the project. Review these samples to gain insights into your own project work.

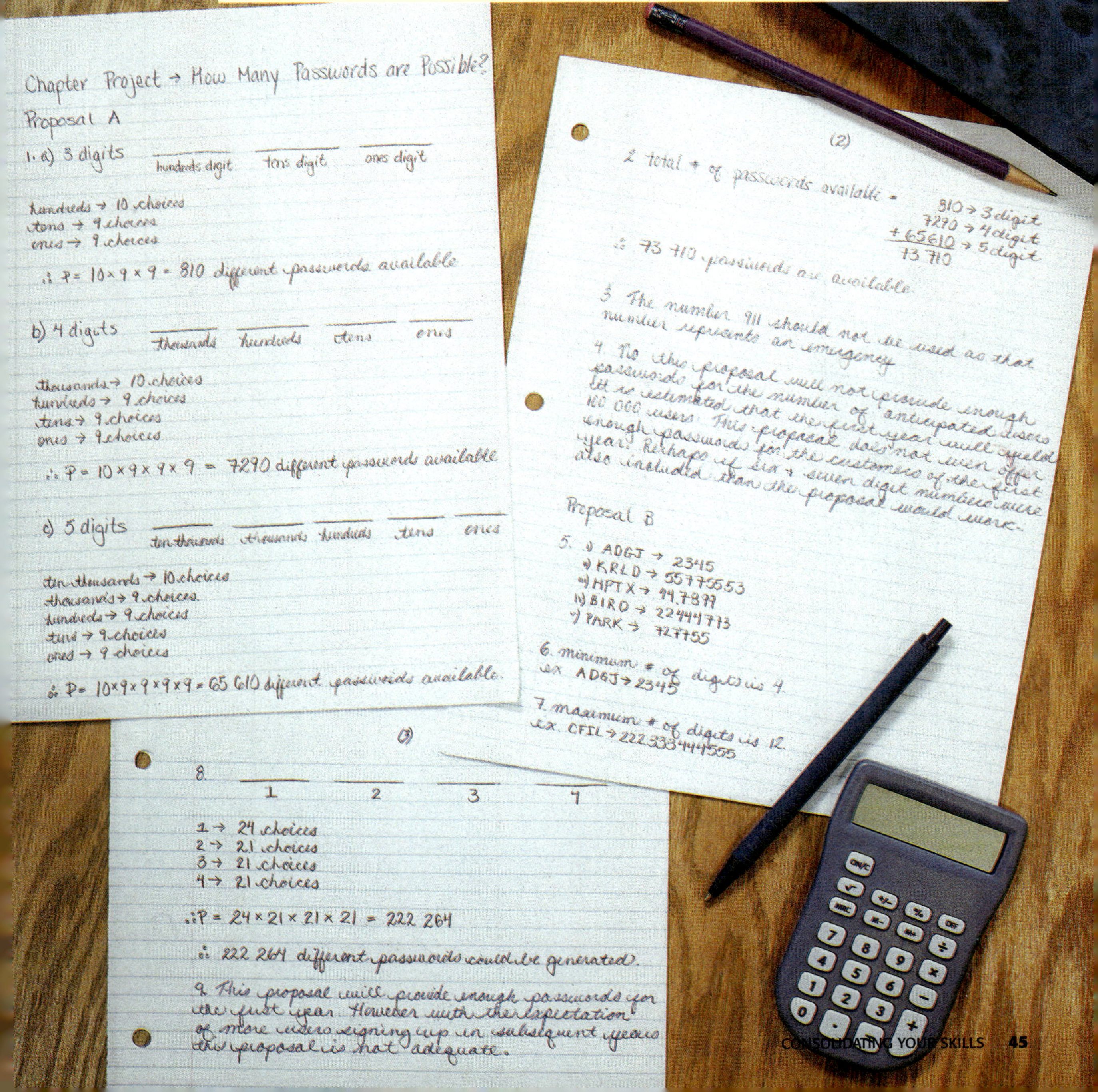

9. Here are the results of a student's analysis of proposal A.

- Were the calculations completed correctly? Explain.
- Are the advantages and disadvantages clearly explained?
- Are there other factors the student should have considered when analyzing this proposal?

10. Below and on the following page are reports and recommendations for the password system developed by a student.

- Are they clearly explained?
- Are the advantages and disadvantages clear?
- Are there other factors the student should have considered when analyzing this situation?

Write a summary of your assessment of this student's work.

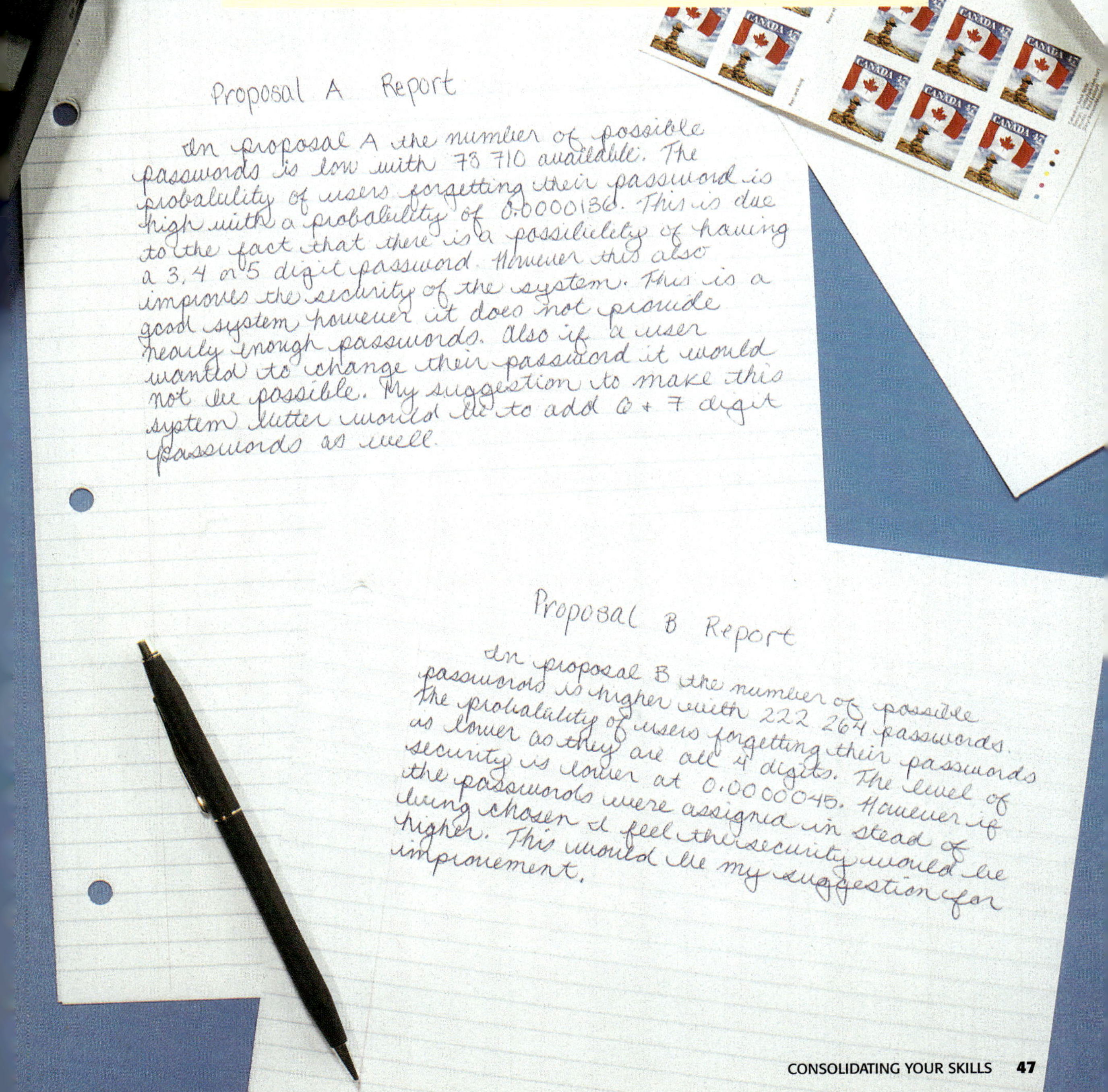

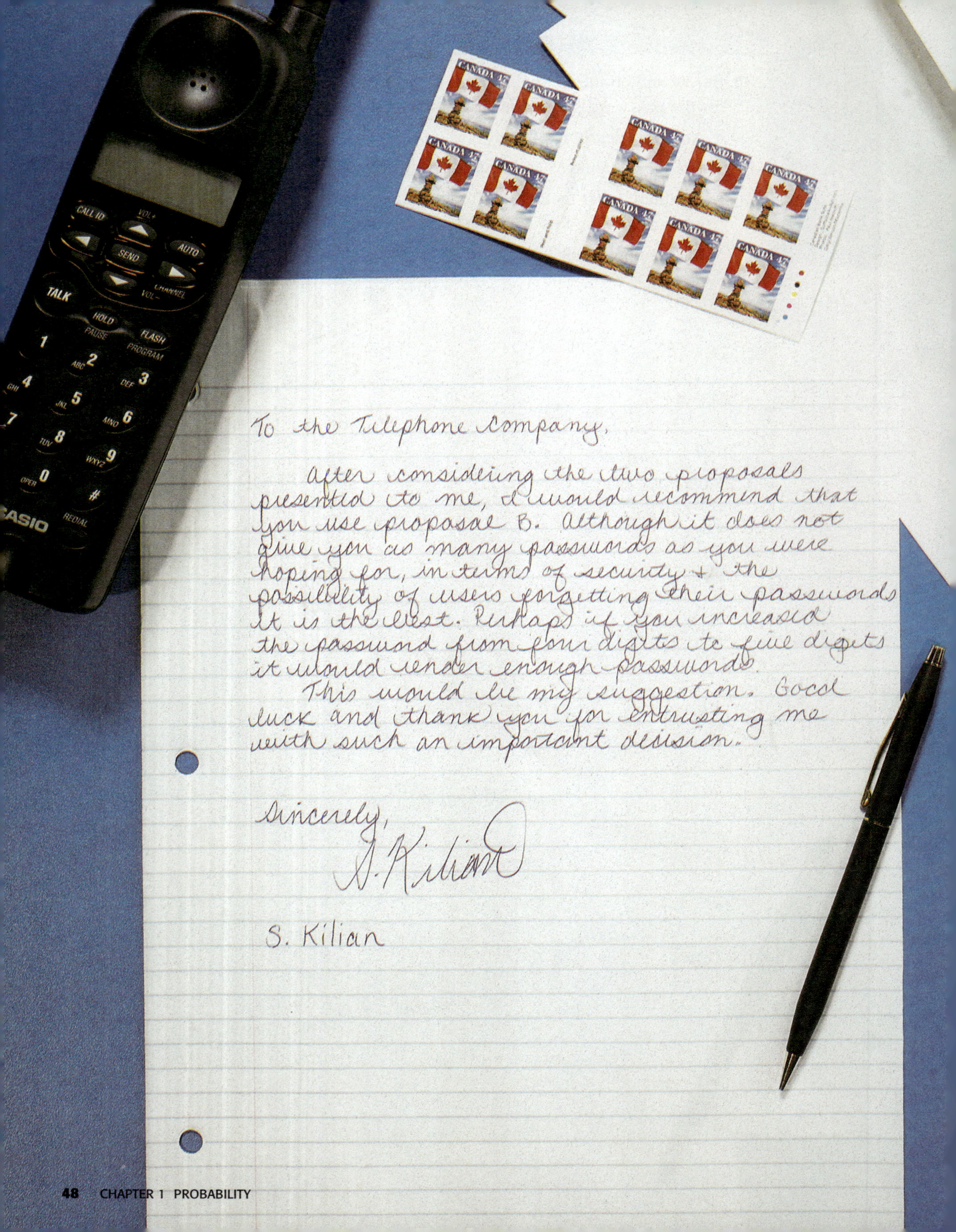
To the Telephone Company,
After considering the two proposals presented to me, I would recommend that you use proposal B. Although it does not give you as many passwords as you were hoping for, in terms of security + the possibility of users forgetting their passwords it is the best. Perhaps if you increased the password from four digits to five digits it would render enough passwords.
This would be my suggestion. Good luck and thank you for entrusting me with such an important decision.
Sincerely,
S. Kilian
S. Kilian

2

Matrices

Overview

Chapter Section	Skills and Concepts
Chapter Project: Creating a Long-Distance Telephone Plan, page 50	Research competitive long-distance telephone plans
Tutorial 2.1: Matrix Operations, page 52	Model and solve problems using technology to perform addition, subtraction, and scalar multiplication of matrices
Tutorial 2.2: Multiplying Matrices, page 60	Model and solve problems that involve matrix multiplication
Chapter Project: Comparing the Competition, page 68	Use matrices to analyze and compare existing long-distance telephone plans; create your own long-distance telephone plan
Tutorial 2.3: Solving Network Problems with Matrices, page 72	Model and solve network problems; use the network matrix A^2
Tutorial 2.4: Solving Transition Problems with Matrices, page 80	Model and solve transition problems
Chapter Project: Determining Your Competitiveness, page 86	Use sampling and transition matrices to predict how many customers would switch to your long-distance telephone plan
Consolidating Your Skills, page 88	Review the work of the chapter

PROJECT

You should do at least one project that uses the skills you develop in this chapter. Recommended projects in the Project Book are:

- Creating and Selling Rustic Twig Furniture
- Organizing a Meeting

CHAPTER PROJECT

Creating a Long-Distance Telephone Plan

In 1992 the Canadian government began to allow competition in the long-distance telephone market. Since then, the market for long-distance telephone service has become extremely competitive. In this project, you will research existing long-distance telephone plans, develop your own plan, and then conduct a survey to determine the popularity of your plan. Here are some initial questions to consider.

- How many long-distance providers are there in your area?
- How many of them are Canadian?
- What different incentives do they offer to be competitive?
- Why has your family chosen your present long-distance carrier?
- What services could your present carrier offer to make it more competitive?

Research long-distance telephone plans on the Internet or call companies to compare rates.

CHAPTER PROJECT

Complete *Tutorials* 2.1 to 2.2 to learn about matrices and matrix operations. You will return to this Chapter Project on page 68.

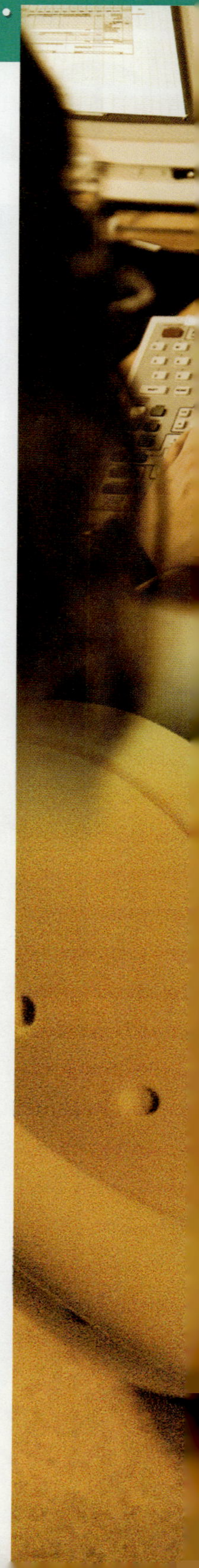

FYI Visit: www.awl.ca/school/connections

You can start your research from our Internet site: Click on MATHLINKS, followed by the *Applied Mathematics 12* logo. Then select a topic under Long Distance Plans.

TUTORIAL 2.1

Matrix Operations

In previous grades, you solved problems using tables and systems of equations. In this chapter, you will use *matrices* to solve problems.

A *matrix* is a rectangular array of numbers, known as the *elements* of the matrix. The array is enclosed within square brackets.

$$X = \begin{bmatrix} 3 & -5 & 7 & 9 & 10 \\ 18 & 2 & 1 & 6 & 4 \\ 2 & 4 & 3 & 8 & 0 \end{bmatrix}$$

Matrix X has 3 rows and 5 columns, and is called a 3×5 (read as "3-by-5") matrix. The number of rows and columns determine its *dimensions*. The matrix itself does not have a numerical value.

We refer to elements by their position in the matrix. For example, x_{12} refers to the element in row 1, column 2. In this case, $x_{12} = -5$.

Some matrices are given special names. A *column matrix* has only one column, a *row matrix* has only one row, and a *square matrix* has the same number of rows as columns.

Column Matrix

$$\begin{bmatrix} 3 \\ 8 \\ 2 \end{bmatrix}$$

Row Matrix

$$\begin{bmatrix} 18 & 32 & 21 \end{bmatrix}$$

Square Matrix

$$\begin{bmatrix} -1 & 5 & 11 \\ 14 & 3 & 0 \\ 7 & -9 & 8 \end{bmatrix}$$

Example 1

Create a matrix from a table.

This table gives the average unit test marks for 4 math classes.

	Unit 1 Test	Unit 2 Test	Unit 3 Test
Class A	86	59	45
Class B	73	65	50
Class C	76	72	56
Class D	71	68	52

a) Represent this information in a matrix, A.

b) State the dimensions of matrix A.

c) How many elements does matrix A have?

d) Which element is in position a_{42}?

e) What is the position of the element 50?

SOLUTION

a)

$$A = \begin{array}{c} \\ \\ \end{array}\begin{matrix} \textbf{T1} & \textbf{T2} & \textbf{T3} \\ \end{matrix}$$

$$A = \begin{bmatrix} 86 & 59 & 45 \\ 73 & 65 & 50 \\ 76 & 72 & 56 \\ 71 & 68 & 52 \end{bmatrix} \begin{matrix} \textbf{A} \\ \textbf{B} \\ \textbf{C} \\ \textbf{D} \end{matrix}$$

b) The dimensions of matrix A are 4×3.

c) There are 12 elements in matrix A.

d) Locate the element in the 4th row and 2nd column. Thus, $a_{42} = 68$.

e) The element 50 is in the 2nd row, 3rd column. Thus, $a_{23} = 50$.

We can add or subtract matrices if they have the same dimensions. We add or subtract the corresponding elements.

UTILITY

If you need help to perform operations with matrices on a TI-83, see Utility 26. If you use a different graphing calculator, consult the user's guide that came with it.

Example 2 Add matrices.

Long-distance rates for four telephone companies are given in the table below.

	Rates per Minute from 8 a.m. to 6 p.m.	Rates per Minute from 6 p.m. to 8 a.m.
Company I	15 cents	10 cents
Company II	14 cents	14 cents
Company III	18 cents	8 cents
Company IV	14 cents	12 cents

Due to competition, the companies adjust their rates as shown in the following table.

	Adjustment to Daytime Rates (per minute)	Adjustment to Evening Rates (per minute)
Company I	− 1 cent	+ 1 cent
Company II	0 cents	0 cents
Company III	− 2 cents	+ 1 cent
Company IV	− 1 cent	− 1 cent

a) Display the long-distance rates in a matrix, *A*, and the adjustments in a matrix, *B*.

b) Use *A* and *B* to determine the companies' new long-distance rates.

SOLUTION

a) Use the MATRIX function to define matrices *A* and *B*.

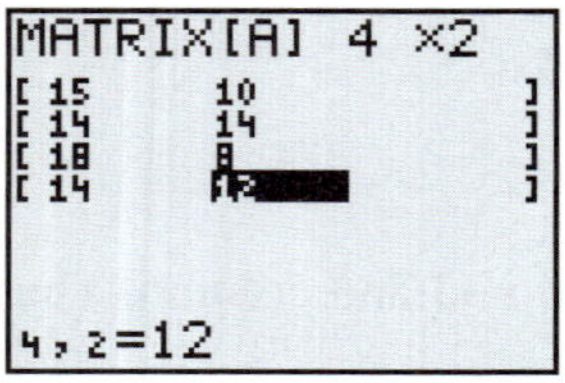

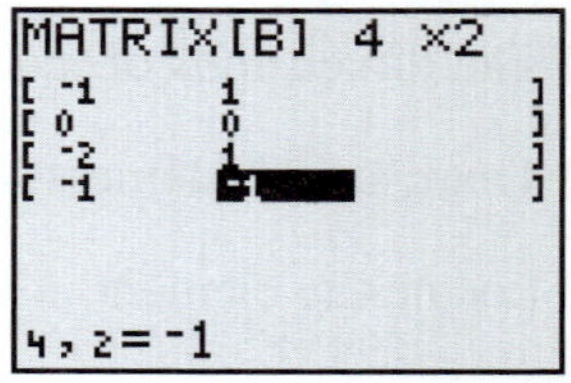

b) Add *A* and *B* to determine the new long-distance rates. The result is shown in the screen to the right.

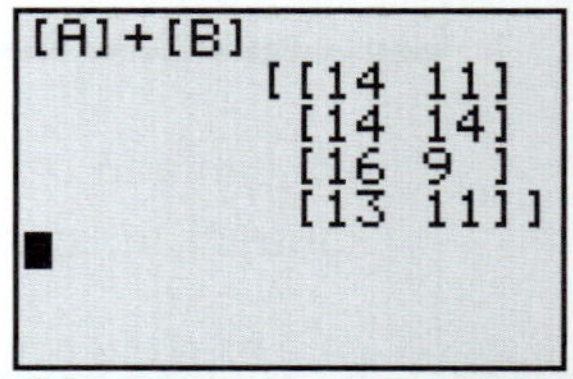

The adjusted long-distance rates are summarized in the following table.

	Adjusted Daytime Rates	Adjusted Evening Rates
Company I	14 cents	11 cents
Company II	14 cents	14 cents
Company III	16 cents	9 cents
Company IV	13 cents	11 cents

A *scalar* is a number. When we multiply a matrix by a scalar, each element in the matrix is multiplied by that scalar.

Example 3 **Use technology to multiply a matrix by a scalar.**

An American travel company offers the following prices on cruises.

	Caribbean (US$)	Alaska (US$)
5 days	579	689
7 days	689	809
10 days	869	999

a) Represent these prices in a matrix, *C*.

b) Suppose the exchange rate between the U.S. dollar and the Canadian dollar is US$1 = Can$1.46, determine the prices in Canadian dollars.

SOLUTION

a) Define matrix C (below left).

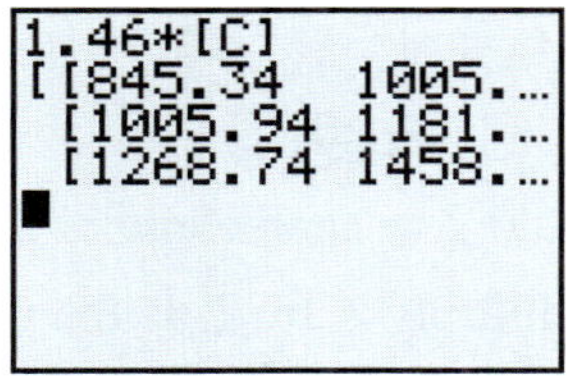

b) To find the prices in Canadian dollars, multiply each element in C by 1.46. The result is shown in the screen above right.

In Canadian dollars, the prices to the Caribbean are \$845.34, \$1005.94, and \$1268.74 for 5, 7, and 10 days, respectively. The prices to Alaska are \$1005.94, \$1181.14, and \$1458.54 for 5, 7, and 10 days, respectively.

DISCUSSING THE IDEAS

1. Why do you think the elements of a matrix are enclosed within brackets?

2. Why is the notation a_{ij} useful when you describe a particular element of matrix A? Does the order of the subscripts i and j matter? Explain.

3. How are the designation of an element of a matrix and the designation of a cell in a spreadsheet similar? How are they different?

4. When you add and subtract matrices, why must they have the same dimensions?

5. Explain why the dimensions of the matrix do not matter when you multiply by a scalar.

EXERCISES

Checking Your Skills

1. Given $A = \begin{bmatrix} 3 & -2 & 6 \\ 8 & 1 & 0 \end{bmatrix}$, $B = \begin{bmatrix} 3 \\ 2 \\ -5 \end{bmatrix}$, and $C = \begin{bmatrix} 1 & 10 & -3 \\ 4 & 6 & -2 \end{bmatrix}$, calculate (if possible):

a) $A + C$ **b)** $B + C$ **c)** $3B$ **d)** $2A - C$

2. In a magic square, all the rows, columns, and diagonals have the same sum. Matrix A is a magic square.

$$A = \begin{bmatrix} 3 & 8 & 1 \\ 2 & 4 & 6 \\ 7 & 0 & 5 \end{bmatrix}$$

a) Add 2 to each element in A. Is the result a magic square?

b) Multiply A by 2. Is the result a magic square?

c) Add A to your result from part a. Is the result a magic square?

3. Below are the CFL Western Division Standings for August 23.

Against Eastern Teams

	W	T	L	Points
BC	1	0	2	2
Calgary	2	0	0	4
Saskatchewan	1	0	1	2
Edmonton	1	0	2	2

Against Western Teams

	W	T	L	Points
BC	4	0	0	8
Calgary	3	0	2	6
Saskatchewan	1	0	4	2
Edmonton	1	0	3	2

Use matrix addition to combine these standings.

4. On January 9, five Manitoba cities recorded the temperature in degrees Celsius and the forecast changes in temperatures for the next 5 days.

	Temperature (°C)	Forecast Change in Temperature (°C)				
	Jan. 9	Jan. 10	Jan. 11	Jan. 12	Jan. 13	Jan. 14
Brandon	−1	−3	+1	−2	+1	−5
Dauphin	−1	−2	0	−1	+2	−3
Winnipeg	−1	+1	+2	0	+1	−2
The Pas	−7	−2	−1	0	+1	−1
Churchill	−18	−5	−2	+1	−1	−2

a) Construct a column matrix for each of these dates.

b) Use the matrices from part a to determine the forecast temperatures for January 14.

5. An arctic expedition outfitter has the following price list for tours of varying lengths and starting points. The prices are quoted in Canadian dollars.

Starting Point	Length of Tour and Cost			
	4 days	**7 days**	**10 days**	**15 days**
Yellowknife	$740	$1145	$1550	$2225
Edmonton	$1640	$2045	$2540	$3125
Toronto	$2240	$2645	$3050	$3725
Vancouver	$1940	$2345	$2750	$3425

a) Find the current exchange rate between Canadian dollars and
i) U.S. dollars **ii)** Japanese yen

b) Write a row matrix that represents the cost, in U.S. dollars, of the 4 tours that start in Toronto.

c) Write a matrix for the entire price list in Japanese yen.

6. A new Internet company, Hallogram, recorded the frequency with which clients used their various services for the months of October, November, and December.

OCTOBER	WorldExplore	InfoSearch	BudgetBuster
Client A	59	37	14
Client B	101	88	31
Client C	29	18	56
Client D	83	121	6

NOVEMBER	WorldExplore	InfoSearch	BudgetBuster
Client A	62	45	18
Client B	127	75	42
Client C	18	7	68
Client D	79	103	21

DECEMBER	WorldExplore	InfoSearch	BudgetBuster
Client A	75	59	28
Client B	122	77	39
Client C	6	9	49
Client D	62	118	15

a) Enter these three tables as matrices A, B, and C on your graphing calculator.

b) Determine the total number of hours each client used each service during the three-month period.

c) If Hallogram charges $2.79/h for its Internet services, determine the total charges for each client for each service.

7. General labourers at the pulp and paper mill are paid \$17.50/h for the first 8 h in a shift. After that, they are paid time and a half. The hours for one week for three employees are given in the table below.

	Sun.	Mon.	Tues.	Wed.	Thurs.	Fri.	Sat.
M. Cairns	3.75	9.5	8.0	8.0	9.0	9.75	10.0
D. Gieselman	0.0	8.0	9.0	9.0	8.75	9.0	6.5
S. Sahota	3.0	8.0	8.0	8.0	8.5	8.5	8.5

a) Enter two matrices in your graphing calculator, one to represent regular hours and the other to represent overtime hours for these employees.

b) Determine the daily pay for each worker.

8. Western World Farming Equipment has three outlets in Saskatchewan. At the beginning of February, the inventory at the North Battleford outlet was 12 cultivators, 4 swathers, 4 combines, and 10 tractors. The inventory at the Saskatoon outlet was 18 cultivators, 6 swathers, 8 combines, and 16 tractors. The inventory at the Regina outlet was 16 cultivators, 8 swathers, 9 combines, and 15 tractors.

a) Create a matrix to represent these inventories.

b) This table shows February's sales for the three outlets. Create a matrix to represent February's sales.

	Cultivators	Swathers	Combines	Tractors
North Battleford	4	2	1	0
Saskatoon	6	1	4	3
Regina	5	3	3	6

c) Calculate the inventory in each outlet at the end of February.

d) The company expects sales to increase dramatically in March as farmers prepare for spring. It will make a shipment of stock to each outlet approximately equal to 200% of its February sales. Determine the inventory at the beginning of March for each outlet.

9. A store sells items that are tax free, items with 7% GST added to the base price, and items with both 7% GST and 7% PST added to the base price. Last weekend's sales (before tax) are summarized in this table.

Weekend Sales Summary		
Sales by Tax Category	**Saturday Sales (\$)**	**Sunday Sales (\$)**
Tax free items	501.34	781.03
Items charged GST only	1250.56	417.79
Items charged GST and PST	822.60	707.28

a) Determine the amount of each tax (GST and PST) collected by the store last weekend.

b) Generate a matrix model to determine the total sales (including taxes) on each day.

c) How would your matrix model from part b change if GST was 5% and PST was 12%?

Extending Your Thinking

10. Create your own 4×4 magic square matrix.

a) Repeat parts a, b, and c of exercise 2 on this matrix. Were the resulting matrices magic squares?

b) Would this be the case with a magic square matrix with any dimensions? Why?

c) Are there any operations, or combinations of operations, you could perform on the magic square matrix that would result in a matrix that is not a magic square?

11. The Insurance Corporation of British Columbia (ICBC) uses a claim-rated scale to determine eligibility discounts on insurance premiums. All drivers start at a base rate that depends on the vehicle driven and other factors. A driver moves down one level on the claim-rated scale for each year without an at-fault crash. Drivers start at the base rate, and discounts increase by 10% for each year without an at-fault crash, until the discount reaches 40% after 4 years. Three new drivers want to determine the amount of money they will save on insurance if they avoid having an at-fault crash for 4 years. Currently, Katie pays \$1175 per year, Marina pays \$1437, and Sascha pays \$1782. Calculate their total discounts over a four-year accident-free period. Use scalars and matrices.

PROJECT

In this Tutorial you represented data in matrices and performed matrix operations. The skills and concepts you used relate to the Chapter Project and the following project in the Project Book.

- Creating and Selling Rustic Twig Furniture
- Organizing a Meeting

COMMUNICATING THE IDEAS

Write a paragraph to explain what matrices are, how to use them, and why they are useful.

TUTORIAL 2.2

Multiplying Matrices

The addition and subtraction of two matrices is straightforward. However, the multiplication of matrices is not equally so. Historically, the multiplication of matrices arose from a study of linear systems by the mathematician Arthur Cayley in the middle of the 19th century. In this Tutorial you will learn how to multiply matrices.

Investigation 1

Multiplying matrices

The table below shows the standings for four local hockey teams.

Team	Wins	Ties	Losses
Blazers	7	2	4
Hawks	5	1	7
Tigers	8	0	5
Eagles	4	1	8

Teams are awarded 2 points for a win, 1 point for a tie, and 0 points for a loss.

1. For each team, calculate the points for wins, ties, losses, and the total points. Show all your calculations.
2. Write to explain how you calculated the total points for the Blazers. Did you use the same process for the other three teams?
3. Create a 4×3 matrix, S, for the standings.
4. Create a 3×1 matrix, P, for the points per win, tie, or loss.
5. Create a 4×1 matrix, T, for the teams' total points.
6. Multiply $S \times P$. To find the product matrix, $S \times P$, multiply each of the elements in **row 1** of S by the corresponding elements in P and take their sum. Put this result in **row 1**, **column 1** of $S \times P$. Next, multiply each of the elements in **row 2** of S by the corresponding elements in P and take their sum. Put this result in **row 2**, **column 1** of $S \times P$. Repeat this process for **row 3** and **row 4** of S.
7. How does $S \times P$ compare to T? How is the process of matrix multiplication you used in exercise 6 similar to the process you described in exercise 2?
8. What are the dimensions of $S \times P$? How are the dimensions of $S \times P$ related to the dimensions of S and P?

We define multiplication of suitable matrices by multiplying "row on column." To multiply a row by a column, multiply the first element in the row by the first element in the column, then multiply the second element in the row by the second element in the column, and so on. Then, add the products.

For example:

If $A = \begin{bmatrix} 0 & 5 & 7 \\ 3 & -1 & 8 \end{bmatrix}$ and $B = \begin{bmatrix} 2 & 6 \\ 4 & 1 \\ 6 & 0 \end{bmatrix}$, then

(**Row 1** from A) × (**Column 1** from B) goes in **Row 1**, **Column 1** of $A \times B$

(**Row 1** from A) × (**Column 2** from B) goes in **Row 1**, **Column 2** of $A \times B$

$$A \times B = \begin{bmatrix} (0 \times 2) + (5 \times 4) + (7 \times 6) & (0 \times 6) + (5 \times 1) + (7 \times 0) \\ (3 \times 2) + (-1 \times 4) + (8 \times 6) & (3 \times 6) + (-1 \times 1) + (8 \times 0) \end{bmatrix}$$

(**Row 2** from A) × (**Column 1** from B) goes in **Row 2**, **Column 1** of $A \times B$

(**Row 2** from A) × (**Column 2** from B) goes in **Row 2**, **Column 2** of $A \times B$

$$= \begin{bmatrix} (0 + 20 + 42) & (0 + 5 + 0) \\ (6 + (-4) + 48) & (18 + (-1) + 0 \end{bmatrix}$$

$$= \begin{bmatrix} 62 & 5 \\ 50 & 17 \end{bmatrix}$$

When we multiply matrices A and B, we multiply each of the elements in **row 1** of A by the corresponding elements in **column 1** of B and take their sum. This result goes in **row 1**, **column 1** of $A \times B$. Then we multiply each of the elements in **row 1** of A by the corresponding elements in **column 2** of B and take their sum. This result goes in **row 1**, **column 2** of $A \times B$. This process is repeated until all the rows in A have been multiplied by all the columns in B.

Investigation 2 Dimensions of the product matrix

A movie theatre records these sales on a Saturday.

	Children	Students	Adults	Seniors
Sixth Sense	0	14	103	32
Pokémon	67	29	45	18
The Messenger	0	0	86	54

1. Create a 3×4 matrix, S, to represent the theatre's sales.

2. This theatre charges \$5 for children, \$6 for students, \$8 for adults, and \$6 for seniors. Create a column matrix, A, to represent admission prices.

3. Multiply $S \times A$. What are the dimensions of the resulting matrix? How are these dimensions related to the dimensions of S and A?

4. Interpret the meaning of the elements of the product matrix, $S \times A$.

5. Multiply $A \times S$. What do you notice?

6. What conclusions can you draw about the dimensions of the matrices that are being multiplied?

7. To test your conclusions, define two new matrices, M and N, and try to multiply $M \times N$ and $N \times M$.

Two matrices, A and B, can be multiplied to form $A \times B$ only if:

The number of columns in A	=	The number of rows in B

If this condition is satisfied, then the product matrix, AB, will have the same number of rows as A and the same number of columns as B.

$$A_{m \times n} \quad \times \quad B_{n \times p} \quad = \quad AB_{m \times p}$$

n and n: must be equal

m and p: dimensions of product matrix

In general, when you multiply matrices, $A \times B \neq B \times A$

UTILITY

If you need help to multiply matrices with a TI-83 graphing calculator, see Utility 26. If you are using a different graphing calculator, consult the user's manual that came with it.

Example

Multiplying matrices

This is the matrix of test marks from *Example 1* of *Tutorial 2.1*.

$$A = \begin{array}{c} \begin{array}{ccc} \textbf{T1} & \textbf{T2} & \textbf{T3} \end{array} \\ \left[\begin{array}{ccc} 86 & 59 & 45 \\ 73 & 65 & 50 \\ 76 & 72 & 56 \\ 71 & 68 & 52 \end{array}\right] \end{array} \begin{array}{c} \\ \textbf{A} \\ \textbf{B} \\ \textbf{C} \\ \textbf{D} \end{array}$$

The teacher feels that the scores on the first unit test were too high and the scores on the third unit test were quite low. Determine the adjusted scores if the teacher decreases the first test scores by 5% and increases the third test scores by 10%.

SOLUTION

Matrix B indicates how the test scores should be adjusted. To decrease only the first unit test results by 5%, the element in row 1, column 1 of B must be 95% or 0.95. Similarly, to ensure that the second unit test results remain the same, the element in row 2, column 2 must be 100% or 1. The adjusting numbers are along the diagonal and there are zeros everywhere else.

$$B = \begin{bmatrix} 0.95 & 0 & 0 \\ 0 & 1 & 0 \\ 0 & 0 & 1.1 \end{bmatrix}$$

To determine the adjusted scores, define matrices A and B on the TI-83 calculator then multiply $A \times B$. The screen shows the result.

The adjusted averages for the first test are 81.7, 69.35, 72.2, and 67.45 for classes A, B, C, and D, respectively. The adjusted class averages for the third test are 49.5, 55, 61.6, and 57.2 for classes A, B, C, and D, respectively. The test scores for the second test remain unchanged.

DISCUSSING THE IDEAS

1. What assumptions can you make about the number of rows and columns in two matrices if their product exists? If their product does not exist?

2. In *Investigation 2*, how else could you define matrices S and A so that the product matrix $A \times S$ is the same as your result in exercise 3?

3. In the *Example*, explain why the element in row 3, column 3 must be 1.1.

4. In the *Example*, why were there zeros everywhere except along the diagonal?

5. In the *Example*, describe the product $A \times B$ if B had ones along the diagonal and zeros everywhere else.

EXERCISES

Checking Your Skills

1. Calculate each product.

a) $\begin{bmatrix} -9 & 4 & -1 \\ 12 & 5 & -3 \end{bmatrix}\begin{bmatrix} 17 \\ -6 \\ -2 \end{bmatrix}$

b) $\begin{bmatrix} 5 & 6 \\ 2 & 1 \end{bmatrix}\begin{bmatrix} 1 & 0 \\ 1 & 1 \end{bmatrix}$

2. Given matrices A and B, calculate (if possible).

$$A = \begin{bmatrix} 0.5 & 2.5 & 1.75 \\ 2.9 & 3.0 & 1.8 \\ 4.1 & 3.3 & 0.65 \end{bmatrix} \qquad B = \begin{bmatrix} 0.8 & -4.2 \\ 5.0 & 3.25 \\ 2.2 & -1.6 \end{bmatrix}$$

a) $A \times B$

b) A^2

c) $B \times A$

3. This table shows sales for a ski shop during the winter months.

	Downhill Skis	Cross-Country Skis	Snowboards
November	43	14	32
December	76	26	59
January	52	19	51

a) Use a matrix to represent this store's sales for the winter months.

b) Assume that, on average, the profits on downhill skis are \$78, \$47 on cross-country skis, and \$65 on snowboards. Create a column matrix to represent the store's profit on each item.

c) Calculate the store's total monthly profit in the winter.

4. Singh's Grocery sells several different kinds of breakfast cereal, each at a different price. Cereal A is \$2.65/box, Cereal B is \$3.73/box, Cereal C is \$3.15/box, and Cereal D is \$2.99/box. On Wednesday, 5 boxes of Cereal A, 8 boxes of Cereal B, 7 boxes of Cereal C, and 10 boxes of Cereal D were sold.

a) Use an appropriate row matrix and an appropriate column matrix to model this information.

b) Use matrix multiplication to determine Wednesday's revenue from sales of cereal.

5. Soccer's administrators wish to discourage tie games, especially those in which no goals are scored. Traditionally, a win earned 2 points and a tie earned 1 point. Three different point systems have been proposed as alternatives:

i) 3 points for a win and 1 point for a tie

ii) 3 points for a win, 1 point for a tie with goals scored, and 0 points for a tie with no goals scored

iii) 5 points for a win, 3 points for a tie with goals scored, and 0 points for a tie with no goals scored

In a local soccer league, the records of the top 4 teams after 42 games are:

	Wins	Ties with Goals	Ties with no Goals	Losses
Tigers	30	2	8	2
Irish	24	9	2	7
Colts	25	7	0	10
Jets	26	1	10	5

a) Determine the points for each team. Use the traditional system.

b) Use matrix multiplication to model the solution in part a.

c) Use matrices to determine the points for each team, using the new system. Repeat for the other two proposed systems.

d) Which of the alternative scoring systems can make the Irish second in the standings?

e) Which of the alternative scoring systems can make the Colts second in the standings?

f) Which of the alternative scoring systems can make the Jets second in the standings?

g) Design a system that would drop the Tigers out of first place.

6. Three major grocery stores compete for customers. Each store advertises these prices.

	Saveway	I.M.A.	Supershop
Milk (1 L)	$1.13	$1.19	$1.21
Bread (1 loaf)	$1.09	$1.12	$0.99
Coffee (1 kg)	$7.89	$7.25	$6.17
Eggs (1 doz)	$1.22	$1.35	$1.33
Soup (can)	$0.55	$0.45	$0.60

A customer wants to buy 4 L of milk, 2 loaves of bread, 1 kg of coffee, 1 doz eggs, and 5 cans of soup. Use matrix multiplication to determine which store has the lowest total price for this order.

7. The Cuddly Toy Company manufactures 3 types of stuffed animals: pandas, kangaroos, and rabbits. Each toy must be cut, sewn, and finished. The table below gives the numbers of labour hours for each type of toy.

	Panda	Kangaroo	Rabbit
Cutting	0.5	0.8	0.4
Sewing	0.8	1.0	0.5
Finishing	0.6	0.4	0.5

a) Represent this information in a matrix, A.

b) In October, the company received orders for 1000 pandas, 600 kangaroos, and 800 rabbits. In November, the company received orders for 1100 pandas, 850 kangaroos, and 725 rabbits. Create a matrix, B, to represent these orders.

c) Multiply $A \times B$. Add all the elements of the product matrix to determine the total labour hours required to fill the orders for both October and November.

d) The company has 3 plants, one in each of Alberta, Saskatchewan, and Manitoba. The table below shows the hourly wage paid for each task in each plant. Determine the cost of labour to produce each type of toy at each plant.

	Cutting	Sewing	Finishing
Alberta	\$8.50	\$9.00	\$8.40
Saskatchewan	\$8.00	\$8.50	\$8.20
Manitoba	\$8.90	\$10.50	\$10.15

8. WeatherNet, an Internet weather service, provided this report on the Air Quality Index for various cities in British Columbia.

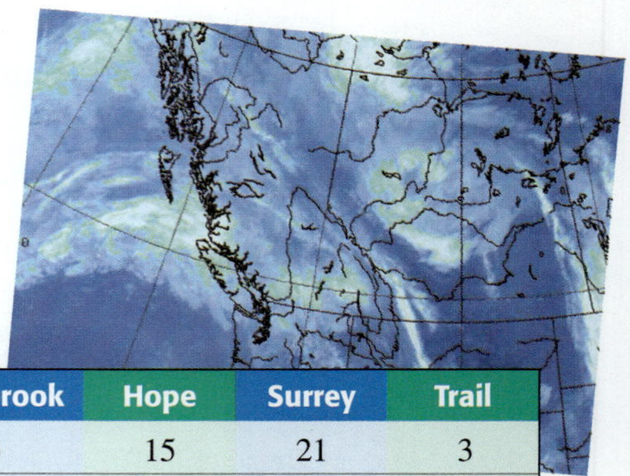

	Abbotsford	Cranbrook	Hope	Surrey	Trail
Monday	20	3	15	21	3
Tuesday	19	1	13	19	4
Wednesday	18	2	10	18	5
Thursday	19	2	11	19	4
Friday	18	2	10	21	5
Saturday	16	1	11	20	4
Sunday	17	0	12	18	3

a) Enter these data as matrix A on your graphing calculator.

b) WeatherNet predicts that, for the next week, Abbotsford's Air Quality Index will increase by 15% overall, Cranbrook's will remain the same, Hope's will increase by 5% overall, Surrey's will decrease by 10% overall, and Trail's will decrease by 10% overall. Enter an appropriate 5×5 matrix into your graphing calculator, to reflect these adjustments.

c) Use matrix multiplication to determine WeatherNet's predicted Air Quality Index for the following week.

9. Sales of economy cars were 200 in 1998 and fell by 4% in 1999. Sales of midsize cars were 300 in 1998 and rose by 10% in 1999. Sales of luxury cars were 40 in 1998 and rose by 30% in 1999.

a) Represent the 1998 sales with an appropriate row matrix.

b) Represent the change in sales from 1998 to 1999 with an appropriate matrix.

c) Use matrix multiplication to determine the 1999 sales.

Extending Your Thinking

10. A manufacturer makes two types of calculators, a business model and a scientific model. Each business model takes 3 min to assemble and 6 min to package and label. Each scientific model takes 6 min to assemble and 4 min to package and label. There are 8 h available for assembly and 8 h for packaging. All calculators will be sold. The profit on each business model is \$8 and the profit on each scientific model is \$10.

a) Use your knowledge of optimization and matrix operations to determine how many of each type of calculator should be made to maximize profit.

b) What is the maximum profit to be made?

PROJECT

In this Tutorial you multiplied matrices. The skills and concepts you used relate to the Chapter Project and the following projects in the Project Book.

- Creating and Selling Rustic Twig Furniture
- Organizing a Meeting

COMMUNICATING THE IDEAS

Use a clear diagram or flow-chart to describe the process of multiplying two 2×2 matrices.

Comparing the Competition

In the first stage of this project, you researched various long-distance telephone providers. In this stage, you will examine existing long-distance telephone plans, and then design your own plan.

1. Telephone at least 3 different companies that offer long-distance telephone service in your area. Ask them for the details of their plan, including rates and incentives offered to new long-distance clients. You could also visit their website. If you are unable to do this, you can use the information provided for these three fictitious companies.

- Dash Canada offers a plan called "The Ultimate." It charges users 10 cents per minute for any long-distance calls in Canada and 20 cents per minute for any long-distance calls to the United States. Calling card rates for calls within Canada or the United States are the same price as regular calls.
- For a monthly fee of $22, TellMe's "North American Plan" offers 250 min of long-distance calls to any place within Canada and the United States on weekday evenings (6 p.m. to 8 a.m.) and weekends (6 p.m. Friday to 8 a.m. Monday). Beyond 250 min, the long-distance rate for evenings and weekends is 10 cents per minute. TellMe charges 22 cents per minute for daytime, weekday calls anywhere in Canada and the United States and 27 cents per minute for calling card calls anywhere in Canada and the United States.

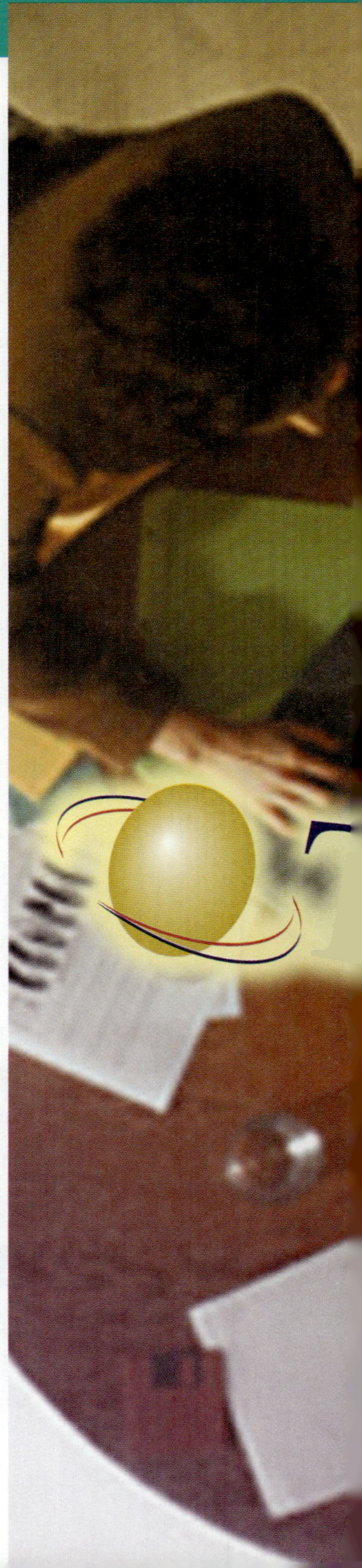

CHAPTER PROJECT

To complete this Chapter Project, you need to use transition matrices. Complete *Tutorials* 2.3 and 2.4 to develop your skills. You will return to the Chapter Project on page 86.

- With BellWest, calls less than 10 s are free. They also offer "by-the-second" billing and have no monthly service fee. The "Infinite Talk" plan charges 7 cents per minute for any long-distance calls within Canada on weekday evenings (6 p.m. to 8 a.m.) and weekends (6 p.m. Friday to 8 a.m. Monday), and 15 cents per minute for long-distance calls to the United States on weekday evenings and weekends. BellWest also charges 20 cents per minute for daytime, weekday calls anywhere in Canada and 25 cents for daytime, weekday calls anywhere in the United States. Calling card rates are the same price as regular calls anywhere in Canada or the United States.

CHAPTER PROJECT

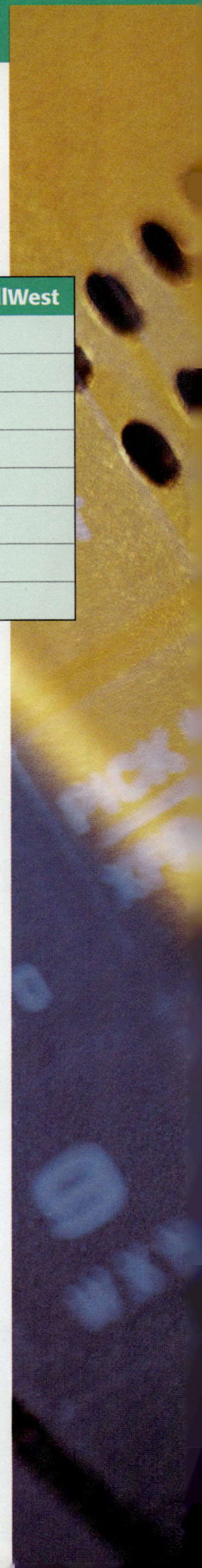

2. Each company offers advantages depending on your long-distance calling patterns. Create a table like the one below. Fill in the appropriate information for the companies you investigated.

	Dash	TellMe	BellWest
Canada: daytime, weekday			
Canada: night-time, weekend			
U.S.: daytime, weekday			
U.S.: night-time, weekend			
Calling Card Canada: daytime, weekday			
Calling Card Canada: night-time, weekend			
Calling Card U.S.: daytime, weekday			
Calling Card U.S.: night-time, weekend			

3. On your graphing calculator, enter a matrix to represent the above rates.

4. Use your family's last long-distance bill (or create one of your own) to determine the number of minutes in each of the 8 calling times. Enter an appropriate row matrix in your graphing calculator.

5. Use matrix multiplication to determine the amount of your family's telephone bill with each of the plans.

6. Ask your family if they would consider a transfer to the cheapest plan. What factors might you consider before you switch?

Develop your own long-distance plan. Provide a detailed outline of your plan with an explanation of each feature. Calculate the amount of your family's telephone bill under your long-distance plan. Save your answers for the final project.

Career Skills

Companies in a competitive industry must be aware of the incentives that competing companies offer to consumers. Market researchers and financial planners use this information to ensure profits for their company.

TUTORIAL 2.3

Solving Network Problems with Matrices

Matrices can be used to describe and analyze networks.

An airline flies between cities as shown in the network diagram below.

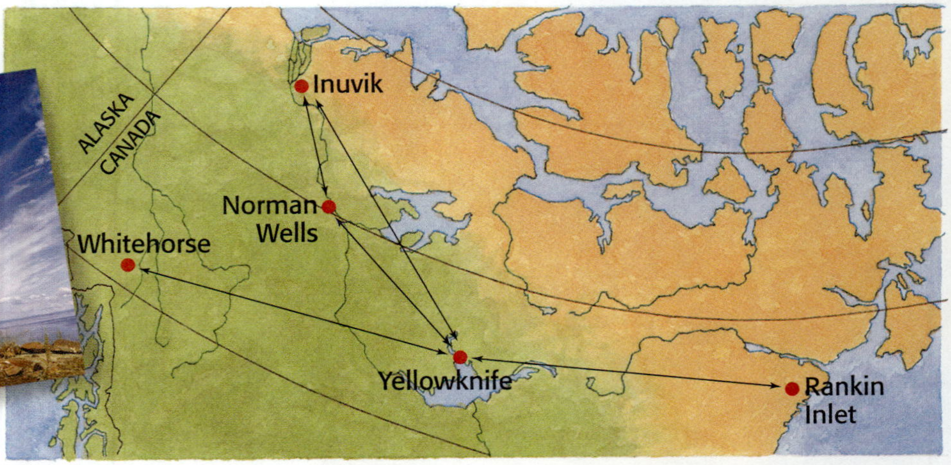

You can travel directly from Inuvik to Norman Wells, but you have to go through Yellowknife to get from Norman Wells to Rankin Inlet.

This network can be summarized in a table or a matrix. An entry of 1 indicates that direct travel from a given city to another is possible. The entry 0 indicates that direct travel from a given city to another is not possible. The order in which the cities are listed in the matrix does not matter as long as the order is the same vertically and horizontally.

		TO City				
		Y	**I**	**NW**	**RI**	**W**
FROM City	**Y**	0	1	1	1	1
	I	1	0	1	0	0
	NW	1	1	0	0	0
	RI	1	0	0	0	0
	W	1	0	0	0	0

$$
\begin{array}{c} \\ A = \end{array}
\begin{array}{c}
\begin{array}{ccccc} \mathbf{Y} & \mathbf{I} & \mathbf{NW} & \mathbf{RI} & \mathbf{W} \end{array} \\
\begin{bmatrix} 0 & 1 & 1 & 1 & 1 \\ 1 & 0 & 1 & 0 & 0 \\ 1 & 1 & 0 & 0 & 0 \\ 1 & 0 & 0 & 0 & 0 \\ 1 & 0 & 0 & 0 & 0 \end{bmatrix}
\end{array}
\begin{array}{l} \\ \mathbf{Y} \\ \mathbf{I} \\ \mathbf{NW} \\ \mathbf{RI} \\ \mathbf{W} \end{array}
$$

Investigation

Determining the Number of Stopovers

Refer to the network diagram of airline flights and matrix A from the previous page.

1. Copy and complete the table to show the number of flights between cities with exactly 1 stopover. The first row has been completed for you. For example, there is 1 route from Yellowknife to Inuvik with 1 stopover at Norman Wells.

Number of Routes Between Cities with Exactly One Stopover						
		TO City				
		Yellowknife	Inuvik	Norman Wells	Rankin Inlet	Whitehorse
FROM City	Yellowknife	4	1	1	0	0
	Inuvik					
	Norman Wells					
	Rankin Inlet					
	Whitehorse					

Write a matrix, B, to represent the number of flights between cities with exactly one stopover. (Note that the flights on the main diagonal of B are return flights from the city of origin to the destination and back to the origin. The flights off the main diagonal refer to flights from one city to another with one stop over in between.)

2. Enter matrix A in your graphing calculator and calculate A^2. How does A^2 compare to matrix B from exercise 1?

UTILITY

If you need help to calculate a matrix to a power using a TI-83 calculator, Utility 26 shows you how. If you are using a different graphing calculator, consult the user's manual.

3. Write a matrix, C, to represent the number of routes between cities with exactly 2 stopovers.

4. Calculate A^3. How does A^3 compare to matrix C from exercise 3?

5. How could you find the matrix that represents the number of routes between cities with exactly 4 stopovers?

6. Add $A + A^2$. What do the entries in this matrix represent?

7. How could you find the matrix that represents the number of routes between cities with **at most** 2 stopovers?

Example

Create a spy network.

A network of spies has been created according to these specifications.

- Alison is able to contact Greg and Kari directly.
- Greg is able to contact both Steve and Kari directly.
- Kari can contact Alison, Greg, and Toni directly.
- Steve can contact Greg directly.
- Toni can contact everyone directly.
- Each spy can contact himself/herself.

If a spy cannot make direct contact with another spy, an intermediate spy can be used according to the above specifications.

a) Draw the network diagram.

b) Complete a table like the one below. Use 1 to represent direct contact and 0 to represent no direct contact.

Message FROM \ Message TO	Alison	Greg	Kari	Steve	Toni
Alison					
Greg					
Kari					
Steve					
Toni					

c) Create a matrix, S, to represent this spy network.

d) How many ways can one spy contact another and use **exactly** 1 intermediate spy?

e) How many ways can one spy contact another using **at most** 2 intermediate spies?

SOLUTION

a) Draw double arrows between Greg and Steve, Greg and Kari, Toni and Kari, and Alison and Kari to show they can contact each other.

Since Alison can contact Greg directly, but Greg cannot contact Alison directly, draw a single arrow from Alison to Greg. Similarly, draw single arrows from Toni to Steve, Greg, and Alison.

Since each spy can contact himself/herself, draw an arrow from each spy to himself/herself.

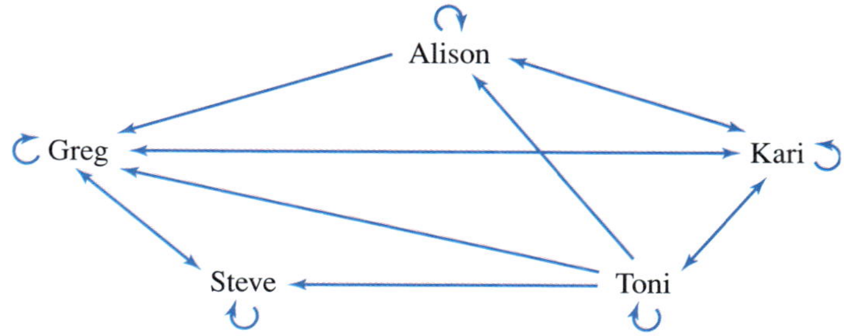

b)

Message FROM	Message TO				
	Alison	**Greg**	**Kari**	**Steve**	**Toni**
Alison	1	1	1	0	0
Greg	0	1	1	1	0
Kari	1	1	1	0	1
Steve	0	1	0	1	0
Toni	1	1	1	1	1

c)

$$\begin{array}{c} \\ S = \end{array} \begin{array}{c} \begin{array}{ccccc} \mathbf{A} & \mathbf{G} & \mathbf{K} & \mathbf{S} & \mathbf{T} \end{array} \\ \begin{bmatrix} 1 & 1 & 1 & 0 & 0 \\ 0 & 1 & 1 & 1 & 0 \\ 1 & 1 & 1 & 0 & 1 \\ 0 & 1 & 0 & 1 & 0 \\ 1 & 1 & 1 & 1 & 1 \end{bmatrix} \end{array} \begin{array}{c} \\ \mathbf{A} \\ \mathbf{G} \\ \mathbf{K} \\ \mathbf{S} \\ \mathbf{T} \end{array}$$

d) The matrix S^2 shows the number of ways one spy can contact another and use exactly 1 intermediate spy.

$$\begin{array}{c} \\ S^2 = \end{array} \begin{array}{c} \begin{array}{ccccc} \mathbf{A} & \mathbf{G} & \mathbf{K} & \mathbf{S} & \mathbf{T} \end{array} \\ \begin{bmatrix} 2 & 3 & 3 & 1 & 1 \\ 1 & 3 & 2 & 2 & 1 \\ 3 & 4 & 4 & 2 & 2 \\ 0 & 2 & 1 & 2 & 0 \\ 3 & 5 & 4 & 3 & 2 \end{bmatrix} \end{array} \begin{array}{c} \\ \mathbf{A} \\ \mathbf{G} \\ \mathbf{K} \\ \mathbf{S} \\ \mathbf{T} \end{array}$$

e) To determine the number of ways one spy can contact another and use at most 2 intermediate spies, add S (the number of ways to contact with **no** intermediate spies), S^2 (the number of ways to contact with **one** intermediate spy), and S^3 (the number of ways to contact with **two** intermediate spies).

$$\begin{array}{c} \\ S + S^2 + S^3 = \end{array} \begin{array}{c} \begin{array}{ccccc} \mathbf{A} & \mathbf{G} & \mathbf{K} & \mathbf{S} & \mathbf{T} \end{array} \\ \begin{bmatrix} 9 & 14 & 13 & 6 & 5 \\ 5 & 13 & 10 & 9 & 4 \\ 13 & 20 & 18 & 10 & 9 \\ 1 & 8 & 4 & 7 & 1 \\ 13 & 23 & 19 & 14 & 9 \end{bmatrix} \end{array} \begin{array}{c} \\ \mathbf{A} \\ \mathbf{G} \\ \mathbf{K} \\ \mathbf{S} \\ \mathbf{T} \end{array}$$

DISCUSSING THE IDEAS

1. Describe the difference between "exactly" and "at most."

2. Explain why the entries on the diagonal of matrix A from the introduction to this Tutorial are all zero, but the entries on the diagonal of matrix S from the *Example* are all one.

3. In the *Example*, what are some advantages and disadvantages if you have one or more spy intermediaries?

4. Explain why negative entries are not possible in a network matrix.

EXERCISES

Checking Your Skills

1. Write a matrix to summarize each network.

a)

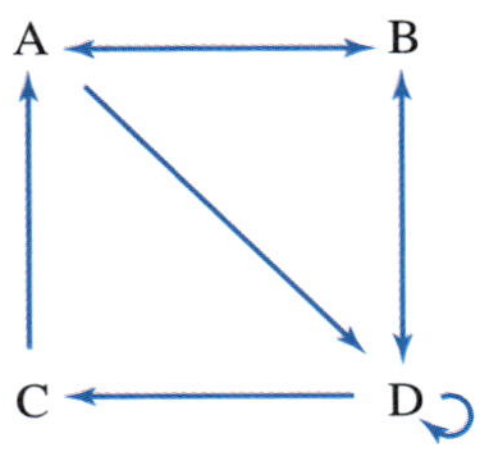

b)

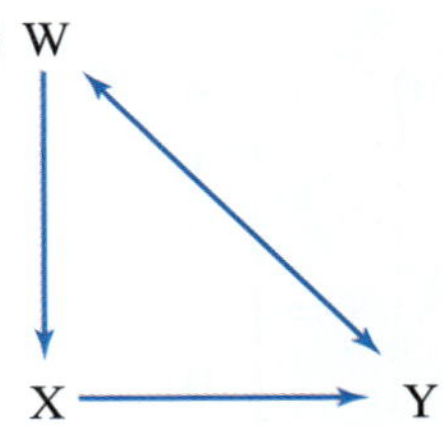

2. Draw each network.

a)

$$\begin{array}{cc} & \textbf{To} \\ & \begin{array}{ccc} \textbf{A} & \textbf{B} & \textbf{C} \end{array} \\ \textbf{From} \begin{array}{c} \textbf{A} \\ \textbf{B} \\ \textbf{C} \end{array} & \begin{bmatrix} 0 & 1 & 1 \\ 1 & 0 & 1 \\ 0 & 1 & 0 \end{bmatrix} \end{array}$$

b)

$$\begin{array}{cc} & \textbf{To} \\ & \begin{array}{cccc} \textbf{P} & \textbf{Q} & \textbf{R} & \textbf{S} \end{array} \\ \textbf{From} \begin{array}{c} \textbf{P} \\ \textbf{Q} \\ \textbf{R} \\ \textbf{S} \end{array} & \begin{bmatrix} 0 & 1 & 1 & 1 \\ 1 & 1 & 1 & 1 \\ 1 & 1 & 0 & 0 \\ 1 & 1 & 0 & 0 \end{bmatrix} \end{array}$$

3. Air freight is transported among several cities in Western Canada, with Calgary as the hub. This means that all air freight either originates in Calgary or is transported through Calgary.

Cities in the Network

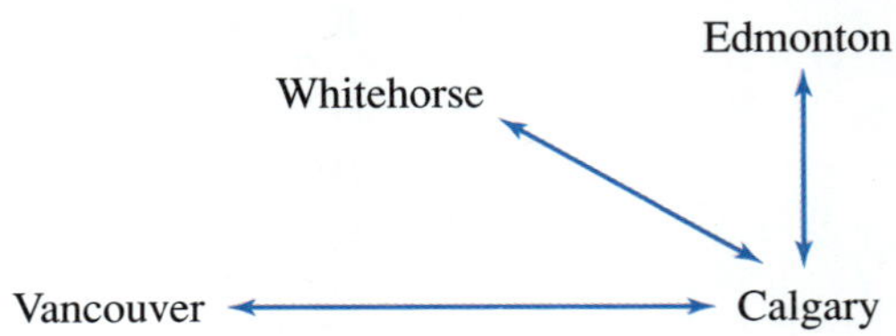

a) Write a matrix, A, to represent this network.

b) Calculate the matrix A^2. What do the elements in this matrix represent?

c) Explain why there is one element equal to 3, and why there are nine elements equal to 1 and six elements equal to 0 in the matrix A^2.

4. The diagram shows a network that connects computer terminals in 5 offices. The arrows indicate the directions in which messages can be sent.

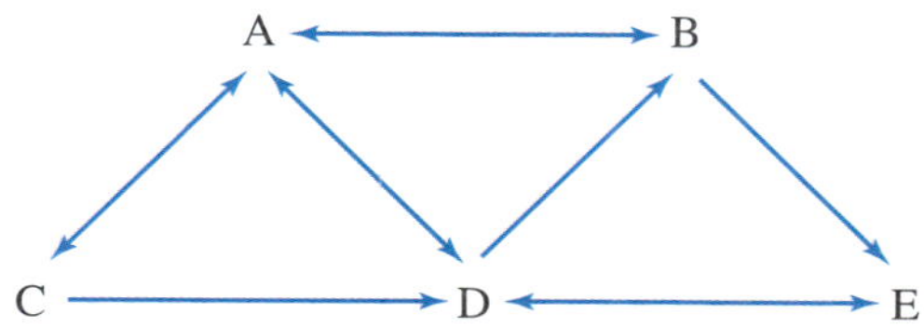

a) Write the matrix, C, for the communications network.

b) Find the matrix that represents the number of ways a message can be sent from one office to another, using exactly one relay.

c) Find the matrix that represents the number of ways a message can be sent from one office to another, using at most one relay.

5. An Alberta oil company, OilWell Drilling, is re-evaluating its system of communications. The present system is described below.

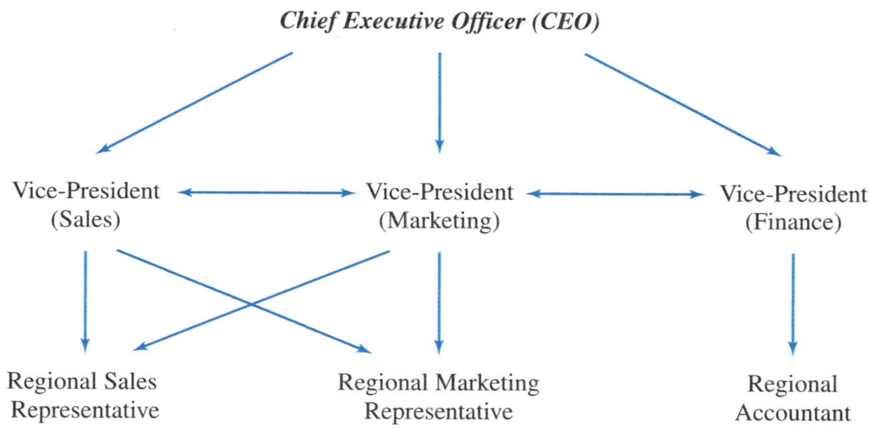

a) Construct a network matrix to represent this communications system.

b) How many ways can a message from the CEO reach the regional sales representative with exactly 1 intermediary?

c) How many ways can a message from the CEO reach the regional marketing representative with at most two intermediaries?

6. To reduce traffic congestion, a city planner proposed the conversion of some downtown streets to one-way streets. The new plan is shown below.

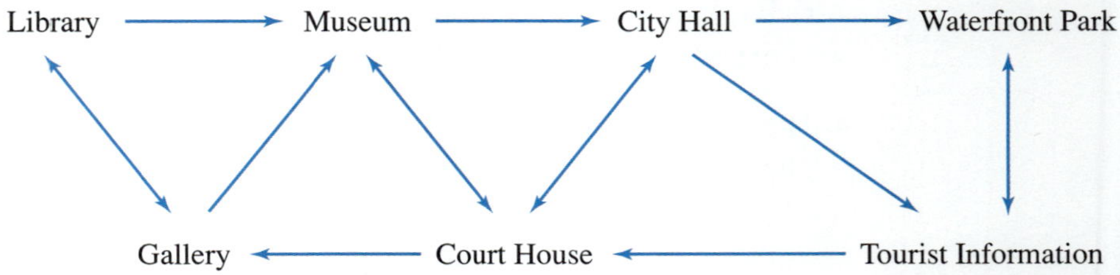

a) Construct a network matrix to represent this plan.

b) Determine the number of ways a tourist could get from the Gallery to the Tourist Information centre and see exactly 3 sites along the way.

7. WestJet Airlines offers these direct flights.

a) How many ways can you travel from Thunder Bay to Calgary with exactly 1 stop?

b) How many ways can you travel from Thunder Bay to Prince George with at most 2 stops?

Flight FROM \ Flight TO	Prince George	Vancouver	Edmonton	Regina	Winnipeg	Thunder Bay	Calgary
Prince George	0	1	0	0	0	0	1
Vancouver	1	0	1	0	0	0	1
Edmonton	0	1	0	1	0	0	1
Regina	0	0	1	0	1	1	1
Winnipeg	0	0	0	1	0	1	1
Thunder Bay	0	0	0	1	1	0	1
Calgary	1	1	1	1	1	1	0

c) If WestJet Airlines began a direct flight between Edmonton and Prince George, how would your answer in part b change?

d) There is a severe snowstorm that has temporarily closed the Calgary airport. How many ways can you get from Vancouver to Winnipeg with at most 2 stops?

Extending Your Thinking

8. Five warehouses are used to store dangerous goods and other goods. A perimeter road and diagonal roads link the warehouses as shown to the right. The perimeter road is used to transport both types of goods, while the diagonal roads are only used to transport goods that are not dangerous.

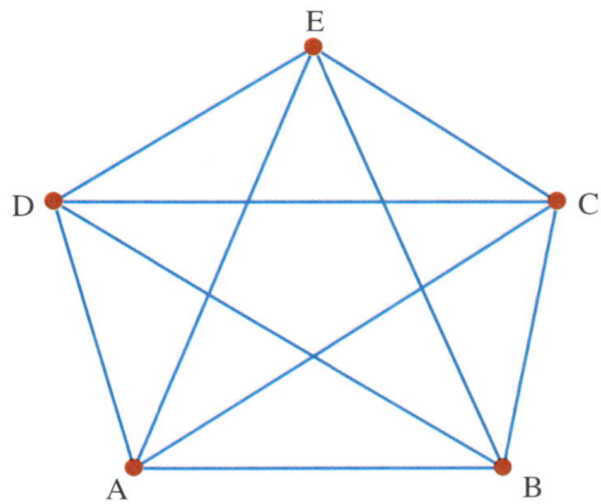

a) Create two matrices, one to represent the transportation network for dangerous goods and the other to represent the transportation network for all other goods.

b) Determine the number of ways dangerous goods can be transported from warehouse A to warehouse D if at most 2 relay stations are used.

c) Determine the number of ways all other goods can be transported from warehouse A to warehouse D if at most 2 relay stations are used.

d) The network is changed so that all roads to and from warehouse A are used to transport both dangerous goods and other goods. All other roads are used for non-dangerous goods only. Modify your matrices from part a to reflect these changes.

PROJECT

In this Tutorial, you used matrices to solve network problems. The skills and concepts you used relate to the Chapter Project and the following project in the Project Book.

- Creating and Selling Rustic Twig Furniture
- Organizing a Meeting

COMMUNICATING THE IDEAS

Create a spy network with at least 4 spies. Write to explain how the corresponding communication matrix can be used to determine the number of ways a message can be sent. Use at most 1 intermediate spy.

TUTORIAL 2.4

Solving Transition Problems with Matrices

Business analysts rely on the assumption that a person's decision can be predicted based on prior behaviour and statistical analysis. In *Tutorial 1.4*, you calculated the probabilities of independent events. Matrices can be used to calculate the probabilities of repeated experiments where the outcomes are dependent.

Consider this situation. Dana and Joe are playing squash. They start with an even chance of winning. If Dana wins, her confidence increases and she has a 65% chance to win the next game. If she loses, she has a 30% chance to win the next game. There cannot be a tie game.

The *initial probability matrix*, P_0, is a row matrix that contains the probabilities at the outset of the experiment. In this case, Dana and Joe are equally likely to win the first game so the probabilities are both 0.5.

$$\begin{array}{cc} & \begin{array}{cc} \textbf{D} & \textbf{J} \end{array} \\ P_0 = & \begin{bmatrix} 0.5 & 0.5 \end{bmatrix} \end{array}$$

To calculate the probabilities for subsequent games, we need to create the *transition matrix*, T. This is a square matrix that shows how the probability of one event is dependent on another event occurring.

If Dana wins the first game, the probability that she wins the next one is 0.65 and the probability that Joe wins the next one is 0.35. These are the entries for the first row of the transition matrix.

If Joe wins the first game, the probability that Dana wins the next one is 0.3 and the probability that Joe wins the next one is 0.7. These are the entries for the second row of the transition matrix.

$$\begin{array}{ccc} & \begin{array}{cc} \text{Dana wins } 2^{\text{nd}} \text{ game} & \text{Dana loses } 2^{\text{nd}} \text{ game} \end{array} & \\ T = & \begin{bmatrix} 0.65 & 0.35 \\ 0.3 & 0.7 \end{bmatrix} & \begin{array}{l} \text{Dana wins 1st game} \\ \text{Dana loses 1st game} \end{array} \end{array}$$

Investigation

Finding *n*-step probabilities

Use the initial probability matrix and the transition matrix for Dana and Joe.

1. Calculate $P_0 \times T$. Label the resulting matrix P_1. What do the elements in P_1 represent?

2. Calculate $P_1 \times T$. Label the resulting matrix P_2. What do the elements in P_2 represent?

3. Calculate $P_2 \times T$. Label the resulting matrix P_3. What do the elements in P_3 represent?

4. Calculate $P_3 \times T$. Label the resulting matrix P_4. What do the elements in P_4 represent?

5. Calculate $P_0 \times T^2$. How does the result compare to your answer to exercise 2?

6. Calculate $P_0 \times T^3$ and $P_0 \times T^4$. How do the results compare to your answers to exercises 3 and 4?

7. Describe two methods of finding the probabilities after 20 games. Which method would you choose to use? Explain your choice.

To find the probabilities, P_n, after n repetitions of an experiment, multiply the initial probability matrix by the transition matrix raised to the power n.

$$P_n = P_0 \times T^n$$

Example

Predict population shifts

As of January 1999, Alberta's population was approximately 2 968 992. Assume that 60% of the population lives in cities, 19% in suburbs, and 21% in rural areas.

a) Write the initial population matrix.

b) In January 1999, how many people lived in each area?

i) cities **ii)** suburbs **iii)** rural areas

c) A company that specializes in population statistics has predicted that each year the population would shift as shown in the table.

FROM	TO		
	City	Suburb	Rural Area
City	74%	21%	5%
Suburb	24%	61%	15%
Rural Area	8%	10%	82%

Write the transition matrix that represents these shifts.

d) Assuming that the total population remains constant and based on these predictions, find the population in Alberta cities, suburbs, and rural areas in

i) January 2000 **ii)** January 2001

SOLUTION

a)

$$P_0 = \begin{matrix} \mathbf{C} & \mathbf{S} & \mathbf{R} \\ [0.6 & 0.19 & 0.21] \end{matrix}$$

b) i) The population in cities in January 1999 is $0.6 \times 2\,968\,992$ or about 1 781 395.

ii) The population in the suburbs in January 1999 is $0.19 \times 2\,968\,992$ or about 564 109.

iii) The population in rural areas in January 1999 is $0.21 \times 2\,968\,992$ or about 623 488.

c)

$$T = \begin{matrix} & \mathbf{C} & \mathbf{S} & \mathbf{R} & \\ \end{matrix}$$
$$T = \begin{bmatrix} 0.74 & 0.21 & 0.05 \\ 0.24 & 0.61 & 0.15 \\ 0.08 & 0.10 & 0.82 \end{bmatrix} \begin{matrix} \mathbf{C} \\ \mathbf{S} \\ \mathbf{R} \end{matrix}$$

d) i) $P_1 = P_0 \times T$

$$= [0.6 \quad 0.19 \quad 0.21] \times \begin{bmatrix} 0.74 & 0.21 & 0.05 \\ 0.24 & 0.61 & 0.15 \\ 0.08 & 0.10 & 0.82 \end{bmatrix}$$

$$= \begin{matrix} \mathbf{C} & \mathbf{S} & \mathbf{R} \\ [0.5064 & 0.2629 & 0.2307] \end{matrix}$$

In January 2000, the population in cities will be $0.5064 \times 2\,968\,992$, or about 1 503 498; the population in the suburbs will be $0.2629 \times 2\,968\,99$, or about 780 548; and the population in rural areas will be $0.2307 \times 2\,968\,992$ or about 684 946.

ii) $P_2 = P_0 \times T^2$

$$= [0.6 \quad 0.19 \quad 0.21] \times \begin{bmatrix} 0.74 & 0.21 & 0.05 \\ 0.24 & 0.61 & 0.15 \\ 0.08 & 0.10 & 0.82 \end{bmatrix}^2$$

$$= [0.6 \quad 0.19 \quad 0.21] \times \begin{bmatrix} 0.6020 & 0.2885 & 0.1095 \\ 0.3360 & 0.4375 & 0.2265 \\ 0.1488 & 0.1598 & 0.6914 \end{bmatrix}$$

$$= \begin{matrix} \mathbf{C} & \mathbf{S} & \mathbf{R} \\ [0.456\,288 & 0.289\,783 & 0.253\,929] \end{matrix}$$

In January 2001, the population in cities will be $0.456\,288 \times 2\,968\,992$, or about 1 354 715; the population in the suburbs will be $0.289\,783 \times 2\,968\,992$, or about 860 363; and the population in rural areas will be $0.253\,929 \times 2\,968\,992$, or about 753 913.

DISCUSSING THE IDEAS

1. In the *Example*, what will happen to the reliability of your predictions as you project farther into the future?

2. How reliable are predictions that are made when you use transition matrices? What factors affect the reliability of a prediction?

3. Why is the sum of the entries in any probability matrix or any row of a transition matrix always 1?

EXERCISES

Checking Your Skills

1. Given $P_0 = [\,0.6 \quad 0.25 \quad 0.15\,]$ and $T = \begin{bmatrix} 0.50 & 0.18 & 0.32 \\ 0.92 & 0.01 & 0.07 \\ 0.12 & 0.43 & 0.45 \end{bmatrix}$, calculate

   a) P_1 **b)** P_3 **c)** P_7

2. Market research shows that if a person buys Brand X, the probability that they will buy Brand X the next time is 0.75. If a person buys Brand Y, the probability that they will buy Brand Y the next time is 0.55. Keith bought Brand X today.

   a) Write the initial probability matrix and the transition matrix.

   b) What is the probability that Keith will switch to Brand Y on his next purchase?

3. Wendy drives an old car. If it starts easily today, the probability that it will start easily tomorrow is 0.9. If it does not start easily today, the probability that it will start easily tomorrow is only 0.85. Write the transition matrix for this situation.

4. If Regan passes his math test, the probability that he will pass the next one is 97%. If he fails his math test, the probability that he will pass the next one is only 85%. Regan has worked hard on this unit so the probability that he will pass the first test is 95%.

   a) Write the initial probability matrix and the transition matrix.

   b) What is the probability that Regan will pass the fifth test?

5. A car manufacturer makes three models: full-size, midsize, and compact. Of the buyers of full-size cars, 13% will switch to midsize on a subsequent purchase, 2% will switch to compact, and 85% will purchase another full-size car. Of the buyers of midsize cars, 5% will switch to full-size, 4% will switch to compact, and 91% will purchase another midsize. Of the buyers of compact cars, 21% will switch to midsize, 3% will switch to full-size, and the remainder will purchase another compact.

a) If the initial market share is 30% full-size, 20% midsize, and 50% compact, write the initial probability matrix.

b) Write the transition matrix.

c) Determine the market share for second-time buyers.

6. The Westgate Tigers Senior Girls volleyball team is playing in a tournament. The probability that they will win their first match is 60%. The coach has noticed that when the team wins a match, the probability that they will win the next match rises to 80%. If they lose a match, the probability that they will win the next match falls to 25%.

a) Write the initial probability matrix and the transition matrix.

b) Determine the probability that the team will win its second match.

c) What is the probability that the team will win its third match?

7. Kenyon Insurance has completed a study on its insured drivers. Based on the results of this study, they made these predictions:

- 25% of their insured drivers who had been involved in an accident in the last year, will be in another accident in the next year. The remaining 75% will not.
- 10% of their insured drivers who had not been involved in an accident in the last year will be in an accident in the next year. The remaining 90% will not.

Kenyon Insurance currently insures 70 drivers who have been in an accident in the last year and 249 drivers who have not been in accident in the last year.

a) Write the initial probability matrix and the transition matrix for this situation.

b) Predict the number of drivers who will be involved in accidents next year.

8. A washing powder is sold in 6-L and 10-L packages. Market research shows that 40% of the buyers of the 6-L size switch to the 10-L size for their next purchase, and 20% of the buyers of the 10-L size switch to the 6-L size for their next purchase.

a) If the original market share was 60% for 6 L and 40% for 10 L, what is the market share for each size in the next round of purchases?

b) What is the market share for each size for the third round of purchases?

9. In 1905, the Saskatchewan government held an election in which 13 constituencies elected a Liberal candidate and 9 constituencies elected a Conservative candidate. A government analyst determined that if a constituency elected a Liberal candidate in this election, the probability that they would elect a Liberal candidate in the next election was 80%. If a constituency elected a Conservative candidate in this election, there was a 65% probability that they would elect a Conservative candidate in the next election. Predict the outcome of the next election based on the analyst's forecasts.

Extending Your Thinking

10. While rivalry between cola companies is nothing new, competition in the diet cola industry has recently become fierce. Market research shows that each month, 20% of Brand A drinkers will switch to Brand B, while 80% will continue to drink Brand A. Of Brand B drinkers, 40% will switch to Brand A, while 60% will continue to drink Brand B.

a) A survey of 437 Brand A drinkers and 332 Brand B drinkers was conducted. Write the initial probability matrix and the transition matrix for this situation.

b) The *steady state* is reached when P_n approaches fixed values. In other words, $P_{n+1} = P_n$. Find the steady state probability matrix. This represents the final market share for each product.

PROJECT

In this Tutorial you used transition matrices to make predictions. The skills and concepts you used relate to the Chapter Project and the following project in the Project Book.

- Organizing a Meeting

COMMUNICATING THE IDEAS

Explain to a friend how transition matrices are used and why they are useful.

CHAPTER PROJECT

Determining Your Competitiveness

In your previous work on the Chapter Project, you compared different long-distance telephone packages and developed your own long-distance plan.

You will now complete your Chapter Project. Complete the following tasks for your final project.

- Write an unbiased summary of each of the existing long-distance telephone plans you investigated plus your own. Be sure to include information about rates, special incentives, bonuses, etc.
- Survey at least 30 adults. Ask which of these 4 plans is closest to their current long-distance plan and whether they would change to a different plan based on the information you presented.
- Use the statistical information from this survey to create a transition matrix. Use the transition matrix to predict the number of people in your city that would convert to your long-distance plan.

Career Skills

Companies conduct market research to predict market trends and consumer demand for various products and services. Production decisions are made based on the results of the market research.

TONE
PULSE
ABC
DEF
MNO
WXY
JKL
GHI
TUV
PRS
OPER
REDIAL
FLASH
MUTE
RESET
GHI

CONSOLIDATING YOUR SKILLS

What Do I Need To Know?

Tutorial	Skills and Concepts	Important Results and Formulas
2.1	• Model and solve problems using technology to perform addition, subtraction, and scalar multiplication	• When adding or subtracting matrices, the dimensions must be the same. • A scalar is a constant.
2.2	• Model and solve problems that involve matrix multiplication	• To multiply two matrices, A and B, the number of columns in A must equal the number of rows in B. • $A_{m \times n} \times B_{n \times p} = AB_{m \times p}$ • In general, $A \times B \neq B \times A$
2.3	• Model and solve network problems; apply the use of the network matrix A^2	• In a network matrix, A, the element 1 represents direct contact or travel; the element 0 indicates direct contact or travel is impossible. • The network matrix A^2 represents the number of ways to contact or travel using exactly one relay.
2.4	• Model and solve transition problems	• $P_n = P_0 \times T^n$

What Should I Be Able To Do?

When you have completed the work in Chapter 2, you should be able to solve the problems that follow. Part A is investigative. As you work through the exercises, refer to the *Student Reference*, the *Utilities*, and the *Tutorials* as necessary.

Part A

1. This graph compares provincial retail sales taxes.

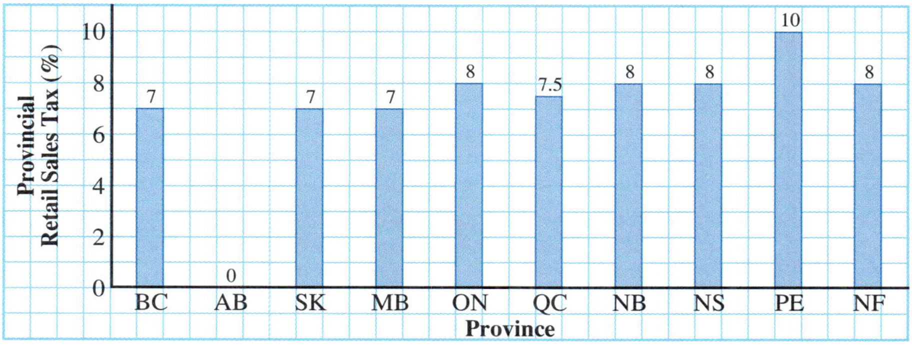

a) Combine each province's sales tax with the goods and services tax (7%). Enter these as row matrix A into your graphing calculator.

b) Research provincial spending habits on the Internet, or use the following information. Enter this as column matrix B into your graphing calculator.

AVERAGE YEARLY CLOTHING EXPENDITURES PER HOUSEHOLD, 1998	
British Columbia	$2082
Alberta	$2479
Saskatchewan	$1873
Manitoba	$1905
Ontario	$2439
Québec	$1986
New Brunswick	$1873
Nova Scotia	$1874
Prince Edward Island	$1870
Newfoundland	$2219

c) Assume all clothing items were taxable. What was the average tax revenue per household on clothing expenditures in each province?

d) In 1998, Alberta had the highest average clothing expenditures per household and Prince Edward Island had the lowest. Did Alberta and Prince Edward Island pay the highest and lowest amounts in taxes, respectively?

Part B

2. Given matrices M and N, calculate (if possible).

$$M = \begin{bmatrix} -5 & 2 & -9 \\ 6 & -3 & 3 \\ 0 & 1 & 8 \end{bmatrix} \qquad N = \begin{bmatrix} 4 & 7 \\ 6 & 0 \\ 3 & 9 \end{bmatrix}$$

a) $2.5M$ **b)** $M \times N$ **c)** M^2 **d)** $0.05N^3$

3. The final marks for a mathematics class will be calculated according to this formula:

- 30% for assignments
- 30% for projects
- 40% for quizzes and tests

Use matrices to determine the final marks for these students.

	Assignments	Projects	Quizzes/Tests
Teya, C.	82%	79%	73%
Lauren, F.	68%	75%	56%
Tyler, O.	53%	45%	88%
Cassidy, P.	95%	90%	84%
Jayson, Z.	48%	52%	55%

4. For the 1996 Census, Statistics Canada gathered this information about the value of private dwellings in the Western provinces.

Value of Dwelling	Manitoba	Saskatchewan	Alberta	British Columbia
less than \$20 000	4%	9%	4%	2%
\$20 000 – \$49 999	13%	20%	5%	3%
\$50 000 – \$99 999	52%	49%	26%	8%
\$100 000 – \$149 999	21%	16%	38%	17%
\$150 000 – \$199 999	7%	4%	16%	19%
more than \$200 000	3%	2%	11%	51%

a) Create a matrix, A, to represent this information.

b) In a randomly selected sample of 1000 homeowners, 130 are from Manitoba, 110 are from Saskatchewan, 310 are from Alberta, and 450 are from British Columbia. Display this information in a column matrix.

c) Determine how many of the people surveyed in the western provinces have dwellings in each of the above categories.

5. A young entrepreneur decides he will sell tank tops, short-sleeved T-shirts, long-sleeved T-shirts, and sweatshirts. The table below shows the cost to produce each item.

	Tank Tops	Short-Sleeved T-Shirts	Long-Sleeved T-Shirts	Sweatshirts
Materials	$8.50	$11.75	$15.00	$24.75
Labour	$1.75	$2.00	$2.50	$3.50

a) Create a matrix to represent these costs.

b) The young entrepreneur plans to make 350 tank tops, 300 short-sleeved T-shirts, 150 long-sleeved T-shirts, and 75 sweatshirts. Create a column matrix to represent this information.

c) Determine the total projected costs.

6. Statistics Canada has documented this census information for the western provinces.

Year	Manitoba	Saskatchewan	Alberta	British Columbia
1998	1 138 100	1 025 700	2 910 000	4 002 800
1999	1 143 400	1 028 100	2 969 000	4 029 300

a) Enter the 1999 population into your graphing calculator as a row matrix.

b) Determine the percent increase for each province from 1998 to 1999. Enter this as a column matrix into your graphing calculator.

c) Use these matrices to predict the total population of the western provinces in 2000.

d) What assumptions are you making when you use this method to forecast population growth? Is this a valid process by which to predict population?

7. Write the matrix that corresponds to this network.

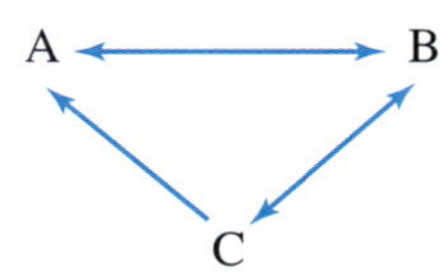

8. This is a section of Route #5 of the Regina Transit System.

a) Construct a matrix, A, to represent this network.

b) Determine the number of ways one can travel from one station to another with exactly 1 relay stop.

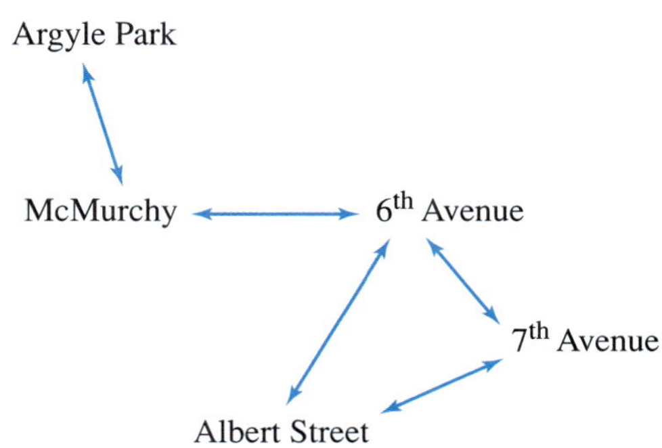

c) Determine the number of ways one can travel from one station to another with at most 1 relay stop.

d) What does the matrix $A + A^2 + A^3 + A^4$ represent?

9. When a complaint at a local telephone company is received, it is dealt with in the following manner.

- The customer service representative receives the complaint from a customer.
- The customer service representative relays the complaint to the manager.
- The manager either discusses the complaint with the customer or refers it to the customer relations officer.
- The customer relations officer either discusses the complaint with the customer or refers it to the public relations vice-president.
- The public relations vice-president provides a response to the customer relations officer.

a) Construct a matrix to represent this network.

b) Determine the number of ways one person can contact another and use exactly 2 intermediaries.

10. The Kelowna Regional Library has 5 branches. Initially, a customer is equally likely to visit any one branch. A study determined that patrons do not always return to the same branch. The table summarizes the results of the study.

FROM Branch	TO Branch				
	A	B	C	D	E
A	0.75	0.10	0.10	0.00	0.05
B	0.30	0.70	0.00	0.00	0.00
C	0.25	0.00	0.65	0.10	0.00
D	0.10	0.00	0.10	0.80	0.00
E	0.10	0.00	0.00	0.00	0.90

For example, if a patron visits Branch A, there is a 75% chance that he will return to Branch A the next time, a 10% chance he will go to Branch B, a 10% chance he will go to Branch C, and a 5% chance he will go to Branch E.

a) Construct the transition matrix for this problem.

b) Use the transition matrix to predict the probabilities for a customer's second visit to the library.

c) Predict the probabilities for a customer's third visit.

Part C

The Chapter project combines many skills and concepts in a single problem. Here is a sample of student work from different stages of the Project. Review these samples to gain insights into your own project work.

11. Here is a student's work that compares long-distance plans and calculates her family's long-distance telephone bill. Is her work complete and accurate? Explain.

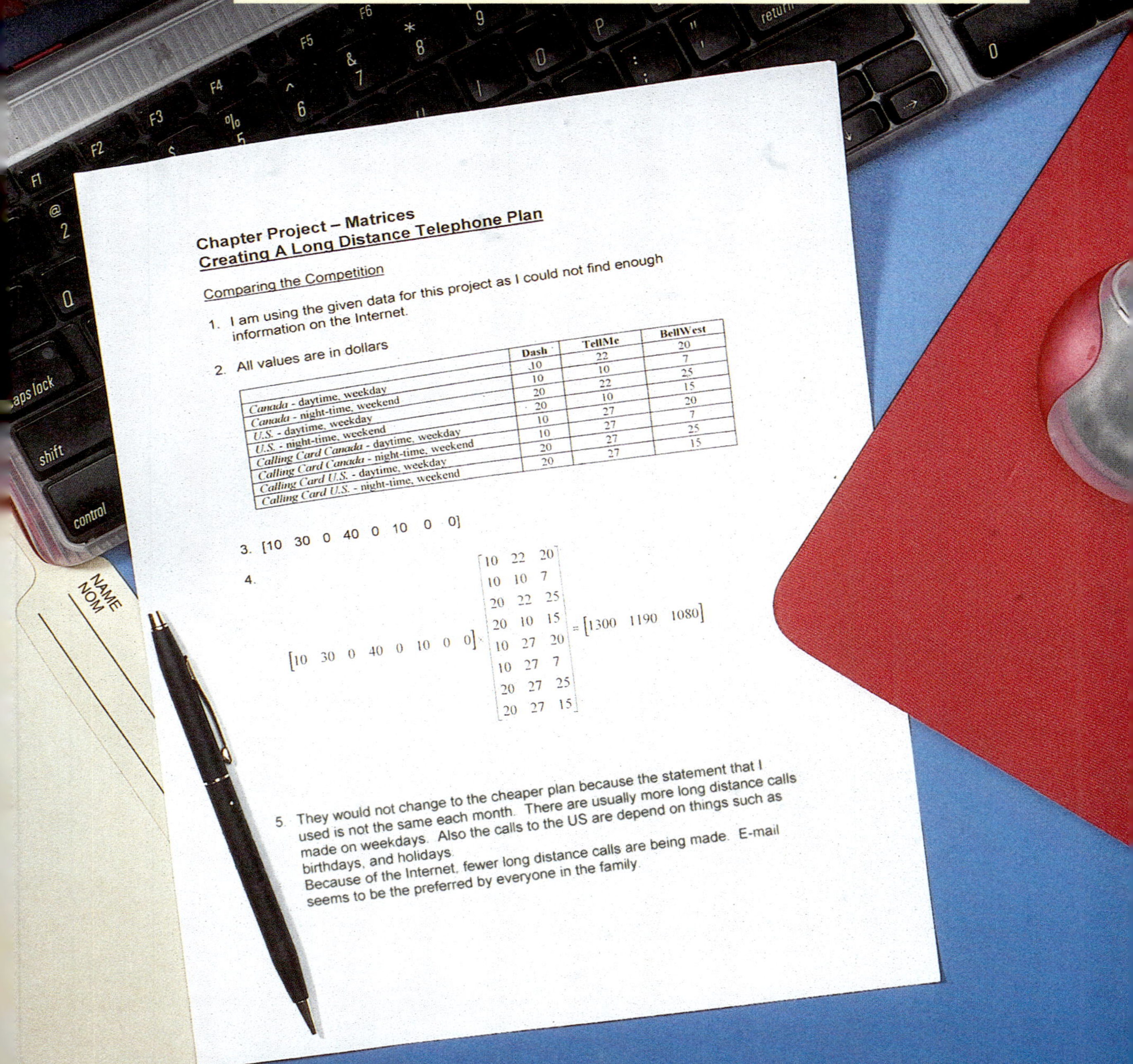

Chapter Project – Matrices
Creating A Long Distance Telephone Plan

Comparing the Competition

1. I am using the given data for this project as I could not find enough information on the Internet.

2. All values are in dollars

	Dash	TellMe	BellWest
Canada - daytime, weekday	10	22	20
Canada - night-time, weekend	10	10	7
U.S. - daytime, weekday	20	22	25
U.S. - night-time, weekend	20	10	15
Calling Card Canada - daytime, weekday	10	27	20
Calling Card Canada - night-time, weekend	10	27	7
Calling Card U.S. - daytime, weekday	20	27	25
Calling Card U.S. - night-time, weekend	20	27	15

3. [10 30 0 40 0 10 0 0]

4.

$$\begin{bmatrix}10 & 30 & 0 & 40 & 0 & 10 & 0 & 0\end{bmatrix} \cdot \begin{bmatrix}10 & 22 & 20\\ 10 & 10 & 7\\ 20 & 22 & 25\\ 20 & 10 & 15\\ 10 & 27 & 20\\ 10 & 27 & 7\\ 20 & 27 & 25\\ 20 & 27 & 15\end{bmatrix} = \begin{bmatrix}1300 & 1190 & 1080\end{bmatrix}$$

5. They would not change to the cheaper plan because the statement that I used is not the same each month. There are usually more long distance calls made on weekdays. Also the calls to the US are depend on things such as birthdays, and holidays.
Because of the Internet, fewer long distance calls are being made. E-mail seems to be the preferred by everyone in the family.

12. This is the long-distance plan developed by the same student. Did she explain her plan clearly?

3 Statistics

Overview

Chapter Section	Skills and Concepts
Chapter Project: Opinion Polls, page 96	Research methods of reporting survey results
Tutorial 3.1: Distributions, page 98	Construct probability distributions; use a binomial distribution to solve problems
Tutorial 3.2: Mean and Standard Deviation, page 105	Calculate the mean and standard deviation of a data set and a binomial distribution
Tutorial 3.3: The Normal Distribution, page 111	Use the normal distribution to solve problems
Tutorial 3.4: Standard Normal Distribution, page 119	Use *z*-scores and the standard normal distribution to calculate probabilities
Tutorial 3.5: The Normal Approximation to a Binomial Distribution, page 126	Use the normal distribution to approximate the binomial distribution when sample sizes are large
Chapter Project: Conducting a Survey, page 132	Design a survey and calculate statistics from the data gathered
Tutorial 3.6: Confidence Intervals, page 134	Construct confidence intervals; calculate the margin of error
Chapter Project: Reporting the Results, page 142	Determine the margin of error and the correct statistical inferences from your sample statistics
Consolidating Your Skills, page 144	Review the work of the chapter

PROJECT

You should do at least one project that uses the skills you learn in this chapter. Recommended projects in the Project Book are:

- Reaction Time
- Quality Control
- Un-Designer Jeans

CHAPTER PROJECT

Opinion Polls

Public opinion researchers use surveys of carefully selected samples of the population to estimate public opinion about various issues. Businesses, the media, and politicians rely on the results of surveys to plan marketing and campaign strategies.

In this Chapter Project, you will examine reported survey results, conduct your own survey, and then report the results. Here are some initial questions to consider.

- What are some factors that affect the accuracy of the results of a survey?
- What kind of information would a politician be interested in gathering? What kind of information would a business be interested in gathering?
- How large a sample is necessary to ensure good results?

To begin this project, collect as many articles as you can from newspapers, magazines, and the Internet that report the results of surveys and opinion polls. Look for articles that describe the average or percent of a sample, and that give an indication of the accuracy of the results. Keep all your articles for use in the next stage of this project.

CHAPTER PROJECT

In the next stage of the Chapter Project, you will design a survey, analyze the data, and interpret the results. Complete *Tutorials 3.1* to *3.5* to develop an understanding of distributions, their graphs, and their statistical measures. You will return to this Chapter Project on page 132.

Observers view focus group through one-way mirror.

FYI Visit: www.awl.ca/school/connections

You can start your research from our Internet site: Click on MATHLINKS, followed by the *Applied Mathematics 12* logo. Then select a topic under Opinion Polls.

TUTORIAL 3.1

Distributions

In previous grades, you used graphs to analyze data and calculated the *mean*, *median*, *mode*, and *range*. In Chapter 1 you studied probability. In this chapter, you will combine these concepts to work with *probability distributions*. A probability distribution gives the theoretical probability for each outcome of an experiment.

STUDENT REFERENCE

If you need to review "mean," "median," "mode," and "range," look in the Student Reference.

To construct a probability distribution, list each outcome in the sample space and calculate the associated probability. In the Investigation you will calculate and estimate a probability distribution.

Investigation

Constructing Probability Distributions

When 4 coins are tossed, what is the probability distribution for the number of heads?

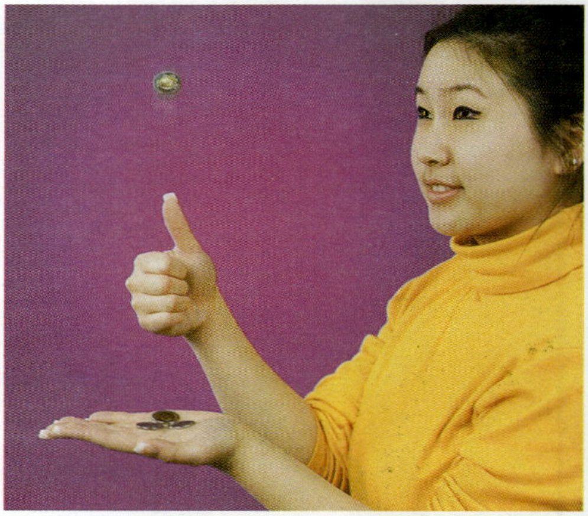

Part A: Calculating the Probability Distribution

1. Use a tree diagram to list the sample space for the experiment.

2. To construct the probability distribution, fill in a table like this one. List all outcomes in the sample space in the first column. In the second column, enter the theoretical probability of each outcome as a fraction. Do not reduce the fraction.

Outcome	Probability

3. What is the value of the lowest probability in the distribution from exercise 2? What is the highest? Between what two integers are these values?

4. What is the sum of all the probabilities from exercise 2?

5. Draw a histogram of the probability distribution. Put the outcomes on the horizontal axis and the probabilities on the vertical axis. Describe the histogram.

Part B: Estimating the Probability Distribution

6. Work with a partner. Have one person toss 4 coins and the other person record the number of heads. Repeat this experiment 50 times.

7. Use your results from exercise 1 to calculate the experimental probability for each outcome as a fraction. Record your results in a table. Do not reduce the fractions. How do the probabilities compare to those calculated in exercise 2?

8. Repeat exercises 3 and 4 using the estimated probability distribution from exercise 7.

9. Draw a histogram. Put the outcomes on the horizontal axis and the probabilities on the vertical axis. How does this histogram compare to the histogram from exercise 5?

10. Combine your results with those of other students. Obtain data for as many repetitions of this experiment as possible. Calculate the new experimental probability for each outcome and graph these results on a histogram. How do these probabilities compare to the probabilities you calculated in exercise 2?

11. Repeat exercises 3 and 4 using the probability distribution from exercise 10.

12. What happens to the estimated probability distribution as the number of repetitions increases?

A *probability distribution* gives the theoretical probability for each outcome of an experiment. It can be displayed as a table or as a graph. Here are two important properties of probability distributions.

- The probabilities are between 0 and 1.
- The sum of the probabilities is always 1.

The estimate of a probability distribution gets closer to the true probability distribution as the number of repetitions increases.

A *uniform distribution* is a probability distribution in which the probabilities are all equal. For example, an experiment consists of tossing a single 6-sided die and noting the number on the face of the die. The possible outcomes of this experiment are 1, 2, 3, 4, 5, or 6. The probability of each of these outcomes is $\frac{1}{6}$. The probability distribution for this experiment is shown below as a table and as a histogram.

Number on Face of Die	Probability
1	$\frac{1}{6}$
2	$\frac{1}{6}$
3	$\frac{1}{6}$
4	$\frac{1}{6}$
5	$\frac{1}{6}$
6	$\frac{1}{6}$

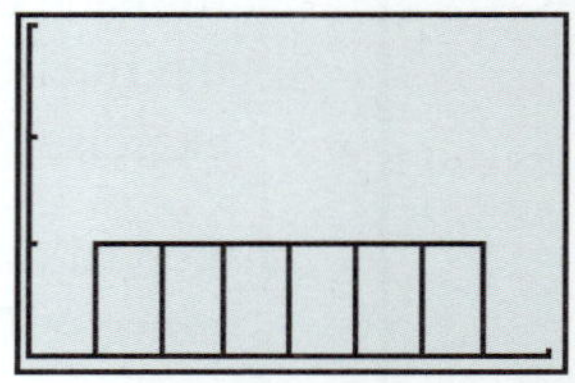

A *binomial experiment* has a fixed number of independent trials in which the outcomes can be classified as successes or failures. The experiment in the Investigation is an example of a binomial experiment.

A *binomial distribution* is the probability distribution for a binomial experiment. The probability distribution from exercises 2 and 5 of the Investigation is a binomial distribution.

Example

Classify probability distributions.

Experiment A: A 6-sided die is tossed 2 times and the number of 6s is recorded.

Experiment B: A card is drawn from a well-shuffled standard deck of cards and the suit is recorded.

1. Construct the probability distribution for each experiment.
2. Classify each distribution as a uniform distribution, a binomial distribution, or neither. Explain.

SOLUTION

1. Experiment A: The sample space is given in the table below.

First Roll

Second Roll	1	2	3	4	5	6
1	(1, 1)	(2, 1)	(3, 1)	(4, 1)	(5, 1)	(6, 1)
2	(1, 2)	(2, 2)	(3, 2)	(4, 2)	(5, 2)	(6, 2)
3	(1, 3)	(2, 3)	(3, 3)	(4, 3)	(5, 3)	(6, 3)
4	(1, 4)	(2, 4)	(3, 4)	(4, 4)	(5, 4)	(6, 4)
5	(1, 5)	(2, 5)	(3, 5)	(4, 5)	(5, 5)	(6, 5)
6	(1, 6)	(2, 6)	(3, 6)	(4, 6)	(5, 6)	(6, 6)

There are 36 different outcomes, 25 have no 6s, 10 have one 6, and 1 has two 6s. The probability distribution is given in the table.

Number of 6's	Probability
0	$\frac{25}{36}$
1	$\frac{10}{36}$
2	$\frac{1}{36}$

Experiment B: The sample space is: spades, clubs, hearts, diamonds. There are 13 cards in each suit and 52 cards in the deck, so the probability for each suit is $\frac{13}{52}$ or $\frac{1}{4}$. The probability distribution is given in the table to the right.

Suit	Probability
Spades	$\frac{13}{52}$
Clubs	$\frac{13}{52}$
Hearts	$\frac{13}{52}$
Diamonds	$\frac{13}{52}$

2. The probability distribution for Experiment A is a binomial distribution. Tossing a 6 is considered a success. The probability distribution shows the probabilities for 0 successes, 1 success, and 2 successes.

   The probability distribution for Experiment B is a uniform distribution because all the probabilities are equal.

We can use the **binompdf(** command on a graphing calculator to generate a binomial distribution. The syntax is **binompdf(**number of trials, probability of success on each trial).

UTILITY

If you need help constructing a binomial distribution with a TI-83, Utility 27 shows you how. If you are using a different graphing calculator, consult the user's manual.

To generate the binomial distribution for the experiment of tossing a coin 4 times and recording the number of heads, enter binompdf(4,0.5) and store the results in list L2. Enter the outcomes of tossing 0 heads, 1 head, 2 heads, 3 heads, and 4 heads in list L1. The screen shows the binomial distribution in table form.

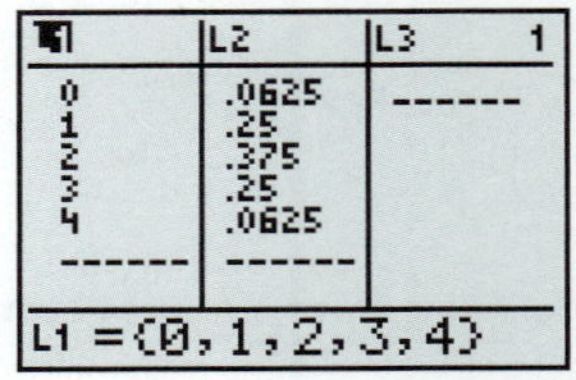

DISCUSSING THE IDEAS

1. In a probability distribution, explain why the probabilities are between 0 and 1.

2. In a probability distribution, explain why the sum of the probabilities is always 1.

3. In what situations might you estimate a probability distribution using experimental probabilities rather than calculating the theoretical probabilities?

EXERCISES

Checking Your Skills

1. Describe some possible binomial outcomes in a survey that involves each *population*. (A population is defined as the entire set of objects, people, or processes being studied.)

a) students in grade 12

b) quality testing

c) educational testing

d) vehicles

2. Explain your answer to each question for the graphs below.

   a) Which graph shows a uniform distribution?

   b) Which graphs show binomial distributions?

   c) Which graph shows neither a uniform distribution nor a binomial distribution?

Distribution 1

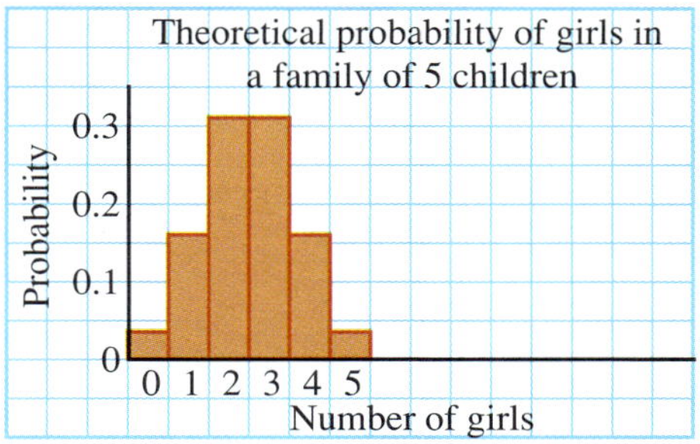

Distribution 2

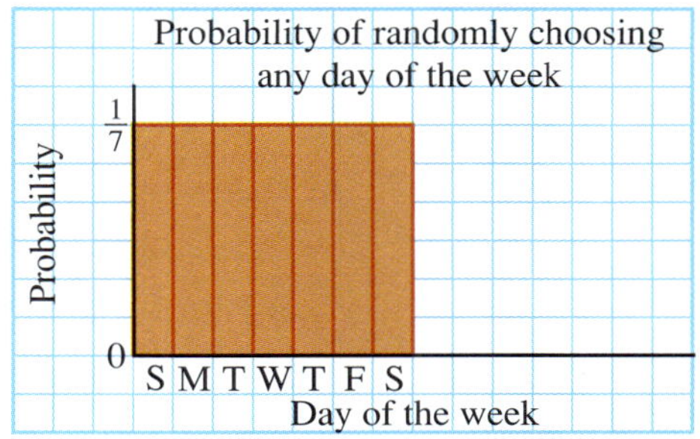

Distribution 3

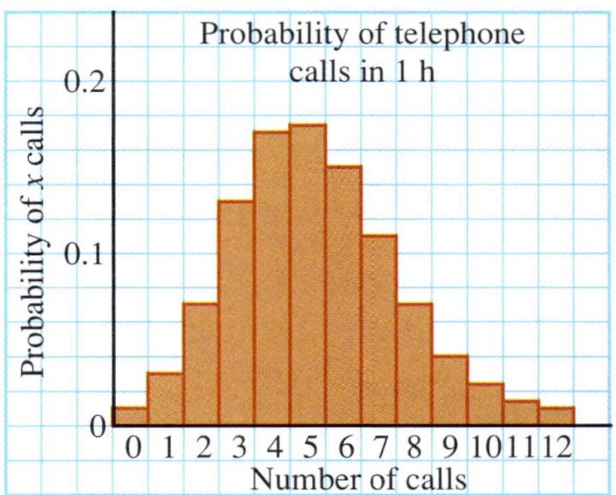

Distribution 4

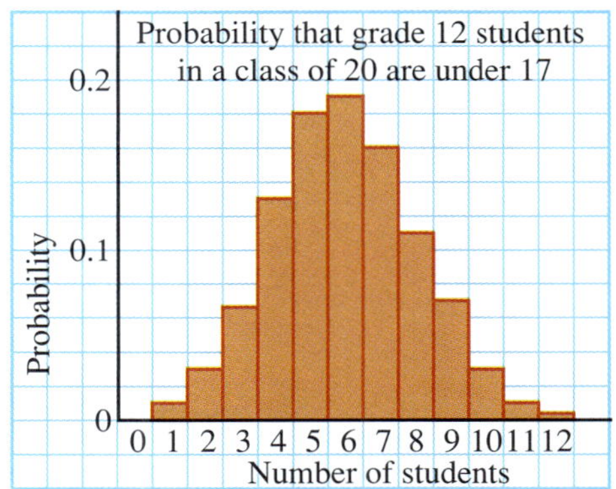

3. A charity sells 5000 raffle tickets for $1 each. There is 1 grand prize of $1000, 2 second prizes of $500 each, and 10 third prizes of $20 each. Each ticket is eligible to win only one prize.

   a) Show the probability distribution for one ticket.

   b) Is this a binomial distribution? Explain.

4. Consider the experiment of rolling two 6-sided dice and recording the sum of the numbers on the faces of the dice.

   a) List the sample space.

   b) Is this a binomial experiment? Explain.

   c) Construct a histogram to show the probability distribution for this experiment.

5. A department store is offering customers scratch and save coupons. Of the 15 000 coupons released, 1000 "win" a 50% discount, 4000 win a 25% discount, and the remainder win a 10% discount. Calculate the probability distribution for one coupon.

6. The traffic signal at an intersection is programmed so the light is green for 75 s, amber for 5 s, and red for 60 s.
   a) List the sample space.
   b) Is this a binomial experiment? Explain.
   c) Construct a histogram to show the probability distribution for this traffic signal.

7. A bag contains 7 red marbles, 10 green marbles, and 8 blue marbles. One marble is drawn from the bag, its colour is noted, then it is returned to the bag. This procedure is repeated 2 more times. Construct a histogram to show the probability distribution for the number of red marbles.

8. Design your own binomial experiment, list the sample space, and construct a histogram to show the probability distribution for the experiment.

Extending Your Thinking

9. The *expected value* is the number we would expect to be the average when we repeat an experiment many times. The mean of a probability distribution is the expected value. Find the expected value for the experiment in the Investigation.

PROJECT

In this Tutorial, you constructed a histogram to show probability distributions. The skills and concepts you used relate to the Chapter Project, and to the following project in the Project Book.

- Un-Designer Jeans

COMMUNICATING THE IDEAS

Explain the characteristics of the various types of probability distributions. How are they similar? How are they different?

TUTORIAL 3.2

Mean and Standard Deviation

In manufacturing, the actual measurements of the finished product are often slightly different from the specifications. Small deviations may be acceptable. People who work in quality control gather data on randomly selected samples of output. They test the data for acceptability. If necessary, adjustments are made to the manufacturing process.

A quantity used to describe a set of data is called a *statistic*. In this Tutorial, we will use two statistics—the *mean* and the *standard deviation* — to analyze data.

- The *mean* is a measure of central tendency. The mean of a set of values is the sum of the values divided by the number of values in the set.
- The *standard deviation* is a measure of the extent to which the data are spread from the mean.

In *Investigation 1*, you will examine two sets of data that have the same mean but different standard deviations.

UTILITY

If you need help to calculate the mean and standard deviation of a data set using the TI-83, Utility 28 shows you how. If you use a different graphing calculator, consult the user's manual.

Investigation 1 Compare Sets of Data.

Data Set 1

Value	Frequency
24	1
26	2
28	3
30	5
32	7
34	8
36	5
38	4
40	2
42	2
44	1

Data Set 2

Value	Frequency
28	2
30	5
32	9
34	12
36	7
38	4
40	1

1. Construct a histogram for each data set. Use the same horizontal and vertical scales for both histograms.

2. How are the histograms from exercise 1 similar? How are they different?

3. Use your calculator to calculate the mean and standard deviation for each data set.

4. Which histogram is more spread out? Which data set has the higher standard deviation? What does the standard deviation tell you about the spread of the data?

5. Test your conclusions from exercise 4 by sketching histograms for Data Sets 3 and 4 and calculating the mean and standard deviation. Did your conclusion hold true?

Data Set 3

Value	Frequency
0	4
1	12
2	30
3	36
4	28
5	10
6	6

Data Set 4

Value	Frequency
0	1
1	6
2	37
3	38
4	37
5	6
6	1

When comparing sets of data, if the standard deviation is low, the data are close to the mean. However, if the standard deviation is high, the data are spread out from the mean.

You have calculated the mean and standard deviation of a data set. In *Investigation 2* you will calculate the mean and standard deviation of a binomial distribution.

Investigation 2 Mean and Standard Deviation of a Binomial Distribution

A multiple-choice test has 30 questions. There are 4 choices for each question; only one is correct. A student answers each question by randomly guessing.

1. Use the **binompdf(** command to construct the binomial distribution for the number of correct answers.

If you need help, Utility 27 shows you how to construct a binomial distribution.

UTILITY

2. Calculate the mean and standard deviation of the distribution.

UTILITY

If you need help to calculate the mean and standard deviation of a binomial distribution using the TI-83, Utility 28 shows you how. If you use a different graphing calculator, consult the user's manual.

3. Multiply the number of trials by the probability of success on a given trial. How does your answer compare to the mean?

4. Multiply your answer to exercise 3 by the probability of failure on a given trial, then take the square root of the result. How does your answer compare to the standard deviation?

5. Develop a rule by which to estimate the mean and the standard deviation for a binomial experiment.

6. Test your rule. Use the experiment of tossing a 6-sided die 25 times and recording the number of 3s that appear.

For any binomial distribution,

- Mean $= n \times p$
- Standard deviation $= \sqrt{n \times p \times (1 - p)}$

where n represents the number of trials, p represents the probability of success on a given trial, and $(1 - p)$ represents the probability of failure on a given trial.

Example **Select a product.**

A company needs fasteners for general shop supplies. A buyer examines two boxes, each of which contains 2000 fasteners. A sample is tested from each box. The fasteners in crate A have a mean length of 6.5 cm with a standard deviation of 0.9 cm. The fasteners in crate B also have a mean length of 6.5 cm, but the standard deviation is 0.01 cm. Which crate should the buyer choose?

SOLUTION

The fasteners in both crates have the same mean. However, the very small standard deviation of the fasteners in crate B suggests they have uniform size. The buyer should choose crate B.

DISCUSSING THE IDEAS

1. Explain why it is reasonable to use $n \times p$ to represent the mean of a binomial distribution.

2. In the Example, explain why the small standard deviation suggests that the fasteners in crate B are uniform in size.

3. Describe a situation in which it would be good for data to have the following.

a) a high standard deviation

b) a low standard deviation

EXERCISES

Checking Your Skills

1. Without calculation, determine which set of data is likely to have the greatest standard deviation and which set is likely to have the least. Explain.

a)

Length	Frequency
14	2
15	5
16	12
17	5
18	1

b)

Length	Frequency
14	5
15	5
16	5
17	5
18	5

c)

Length	Frequency
14	1
15	3
16	18
17	3
18	0

2. Calculate the mean and standard deviation for each set of data in exercise 1. How accurate were your predictions?

3. Examine each histogram below. Without calculation, determine which set of data is likely to have the greatest standard deviation and which is likely to have the least. Explain.

a)

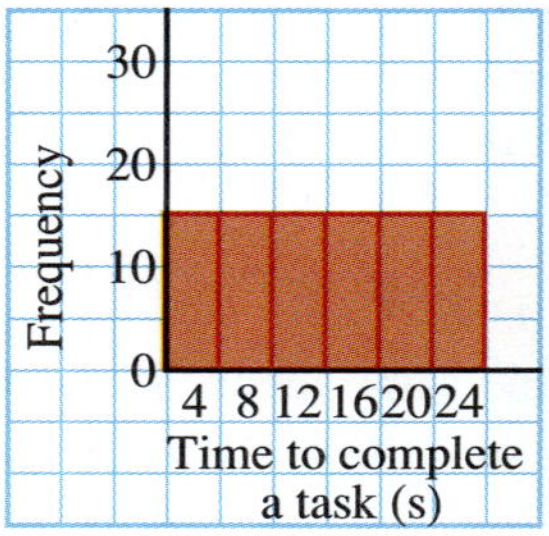

b)

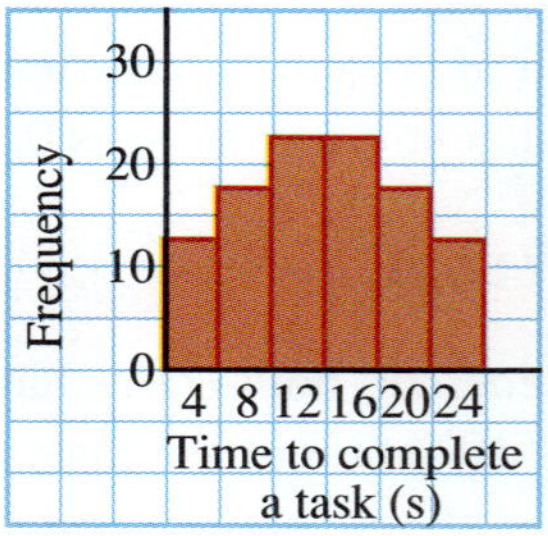

c)

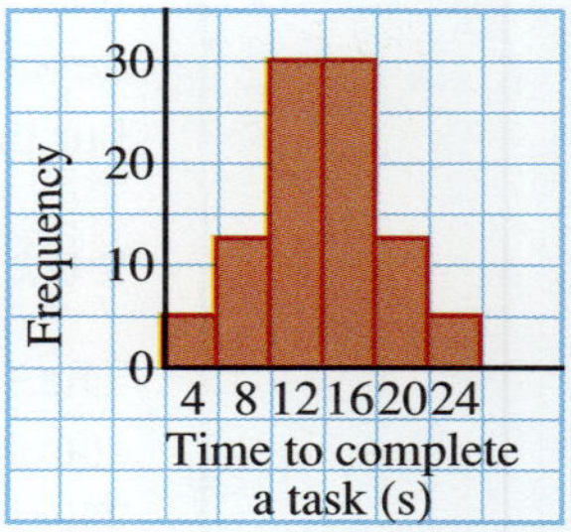

4. Thirty-five randomly selected members of a high school basketball league were asked to make 10 free throws. The rate of success for each player was recorded. The standard deviation was calculated for the group. This process was repeated with 35 randomly selected people at a local shopping mall. Which group do you expect would have the greater standard deviation? Explain.

5. Forty randomly selected people at a hockey game were asked to play a certain video game. Each score was recorded and the standard deviation was computed for the group. This process was repeated with 40 randomly selected people at a video-game arcade. Which group do you expect would have the lesser standard deviation? Explain.

6. A soft drink company uses an automated device to fill 355-mL cans of soda pop. The machine must be checked frequently to ensure each can contains 355 mL. In a sample of 30 cans of pop, these volumes (in mL) were found.

352	354	353	351	357	358
352	356	354	352	353	351
353	355	351	353	353	350
356	358	358	358	354	356
355	355	352	357	354	356

a) Calculate the mean and standard deviation of these data.

b) What problems will be encountered if the standard deviation gets too high?

7. Measure the height, to the nearest millimetre, of each student in the class. Calculate the mean and standard deviation.

8. The lifetimes of 30 batteries from each of two brands of batteries are given.

Measured Lifetimes of 30 Brand X Batteries (in years)

5.0	7.2	6.8	4.6	4.5	6.1	6.3	5.4	4.8	6.8
5.9	4.7	4.0	5.2	8.0	6.2	7.4	4.9	5.6	9.2
3.2	3.0	4.2	5.8	6.5	5.7	4.9	6.0	4.5	5.6

Measured Lifetimes of 30 Brand Y Batteries (in years)

5.3	6.2	4.9	5.8	5.5	4.6	5.9	3.2	6.5	5.9
4.9	6.4	5.7	5.3	4.8	5.6	6.7	5.5	4.8	5.9
4.8	5.6	6.1	7.4	5.7	6.7	5.8	5.2	5.5	5.8

a) Calculate the mean lifetime and the standard deviation of the lifetimes for each brand.

b) Which is the better battery? Explain.

9. A manufacturer of switches has found that about 2.8% of the switches it produces are defective.

a) Find the mean and standard deviation for the number of defective switches in a batch of 300.

b) A consultant is hired. She claims that she can improve the production process and reduce the rate of defective switches to 1.5%. Find the mean and standard deviation for the number of defective switches in a batch of 300.

Extending Your Thinking

10. Refer to exercise 9. Suppose 300 switches are produced using the consultant's methods and 6 are defective. Use your results from part a to determine whether this result supports the consultant's claim.

PROJECT

In this Tutorial, you calculated the mean and the standard deviation of a data set. The skills and concepts you used relate to the Chapter Project, and to the following projects in the Project Book.

- Un-Designer Jeans
- Reaction Time
- Quality Control

COMMUNICATING THE IDEAS

Explain how the standard deviation and the mean help to describe a set of data. Explain the difficulties that could arise if you know one value and not the other.

TUTORIAL 3.3

The Normal Distribution

In *Tutorial 3.1*, you constructed probability distributions for *discrete data*. In this Tutorial, you will explore distributions for *continuous data*.

STUDENT REFERENCE

If you need to review "discrete data" and "continuous data," look in the Student Reference.

An inspector for a potato chip manufacturer measured the masses of 500 bags of potato chips to the nearest gram. The results are displayed in the two graphs below. The frequency graph shows the number of bags with each mass. The probability graph shows the probability that a randomly selected bag from the 500 bags has each mass. The probabilities for this graph were calculated by dividing each frequency in the frequency graph by 500.

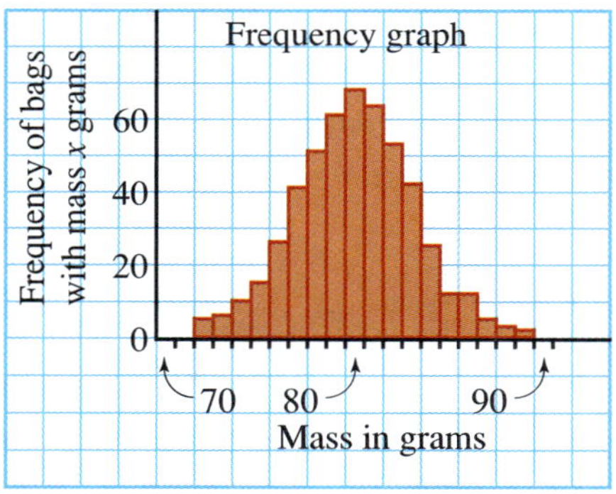

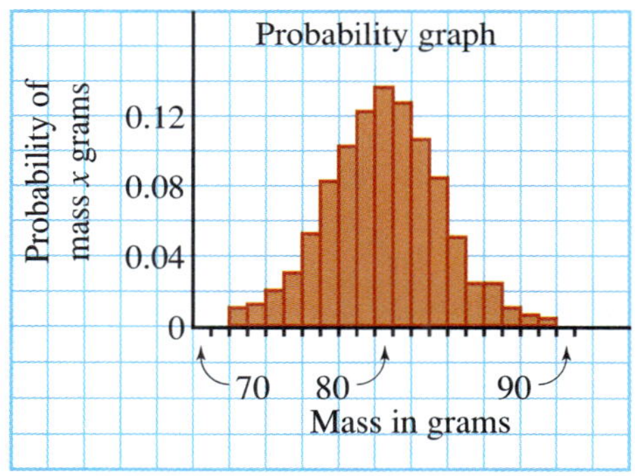

The company makes thousands of bags of potato chips. If we measure the mass of each bag to the nearest milligram and create a probability graph like the one on page 111, the midpoints of the tops of the bars would form a curve similar to the one in the graph below.

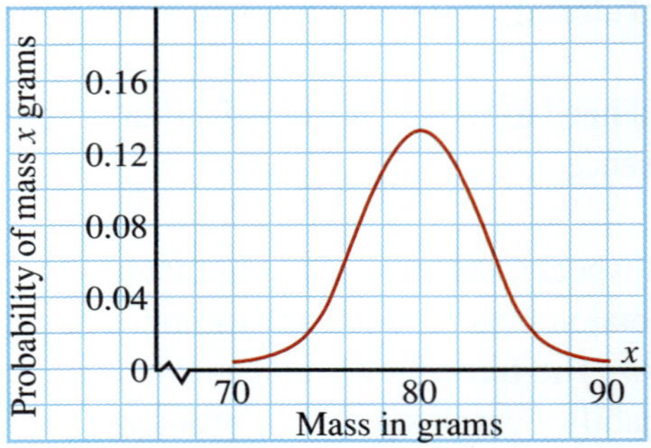

This bell-shaped curve is the graph of a *normal distribution*. Many characteristics of manufactured products (such as volume, mass, and length) have probability graphs that closely approximate a normal distribution. Human characteristics such as height, mass, and intelligence have probability graphs that closely approximate a normal distribution. Every normal distribution is identified by its mean, μ, and its standard deviation, σ. Examine the normal distributions in the diagram below.

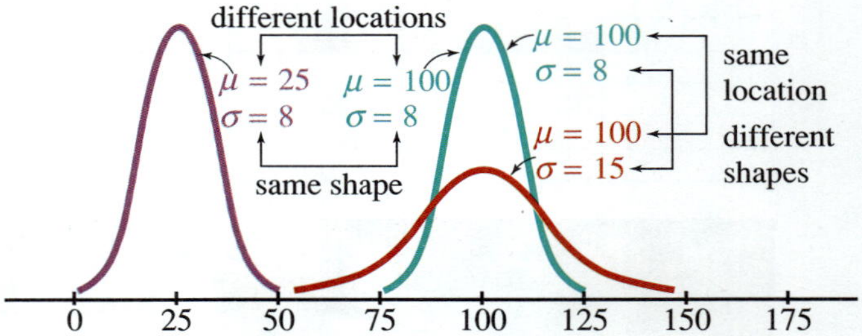

A change in the mean causes the curve to shift left or right. A change in the standard deviation causes the curve to become broader or narrower.

Properties of a Normal Distribution

- Every normal distribution has a mean, μ, and a standard deviation, σ.
- The graph is symmetrical about the mean.
- Almost all the data lie within 3 standard deviations of the mean.
- The curve approaches the x-axis, but does not touch it.
- The area under the curve represents probability so the total area under the curve is 1.

The normal distribution is useful because certain percents of the data values fall within specific distances from the mean. Some examples of these relationships form the *68-95-99 Rule*.

68-95-99 Rule

- About 68% of the data are within 1 standard deviation of the mean.
- About 95% of the data are within 2 standard deviations of the mean.
- About 99.7% of the data are within 3 standard deviations of the mean.

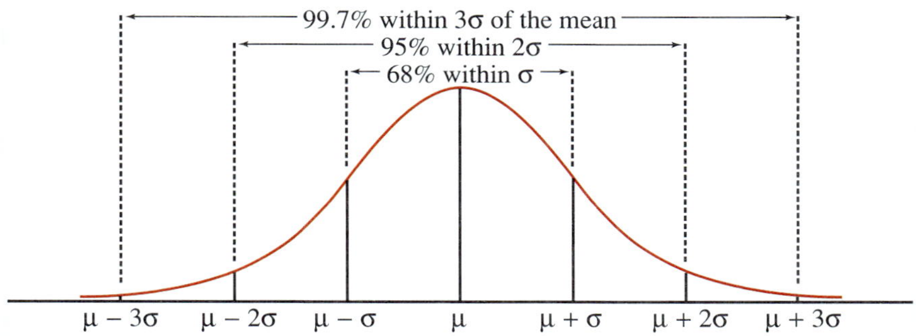

Example

Determine proportions.

A bag of Cadbury's Mini Eggs has a label that reads 43 g. Every hour, 10 bags are taken from the production line and weighed. Here are some of the results.

Time	Mass of bag in grams									
7:00	44.8	44.9	44.8	44.7	43.5	44.9	43.1	44.8	43.3	45.0
8:00	43.6	43.7	43.2	43.6	43.1	43.1	43.2	43.2	43.6	44.8
9:00	43.3	43.4	43.1	43.3	44.0	45.0	43.2	43.2	43.2	45.2

1. Enter the data in list L1 of your calculator and calculate the mean and standard deviation. Round your answers to two decimal places.

2. If the distribution is normal, find the proportion of bags of Mini Eggs with mass that is:

a) less than 42.34 g

b) greater than 44.62 g

c) between 43.10 g and 45.38 g

3. Estimate the range of masses of 43-g bags of Cadbury Mini Eggs.

SOLUTION

1. Use **1-Var Stats** to determine the mean and standard deviation.

The mean is 43.86 and the standard deviation is approximately 0.76.

2. Draw a sketch to help visualize the problem.

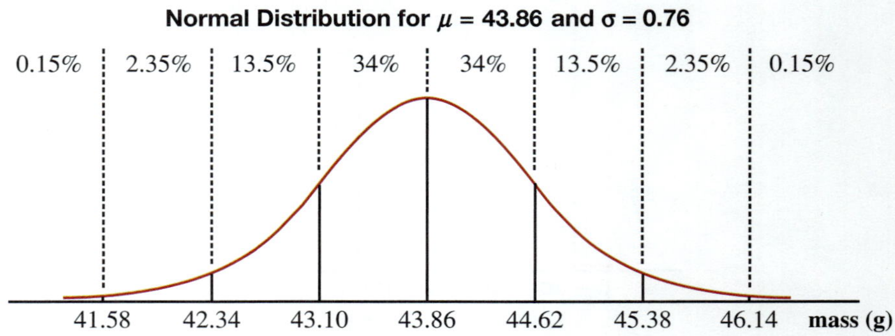

The 68-95-99 Rule tells us that about 68% of the data lie within 1 standard deviation from the mean. The graph is symmetrical, so 34% of the data lie between μ or 43.86 g and $(\mu + \sigma)$ or 44.62 g and 34% lie between μ and $(\mu - \sigma)$ or 43.10 g.

The percent of the data that lies between $(\mu + \sigma)$ or 44.62 g and $(\mu + 2\sigma)$ or 45.38 g is:

$$\frac{95 - 68}{2}\% = 13.5\%$$

This is also the percent of the data that lies between $(\mu - \sigma)$ or 43.10 g and $(\mu - 2\sigma)$ or 42.34 g.

A similar calculation gives us the percent that lies between $(\mu + 2\sigma)$ or 45.38 g and $(\mu + 3\sigma)$ or 46.14 g, and $(\mu - 2\sigma)$ or 42.34 g and $(\mu - 3\sigma)$ or 41.58 g:

$$\frac{99.7 - 95}{2}\% = 2.35\%$$

The percent of data at each end of the distribution beyond 3 standard deviations of the mean is:

$$\frac{100 - 99.7}{2}\% = 0.15\%$$

a)

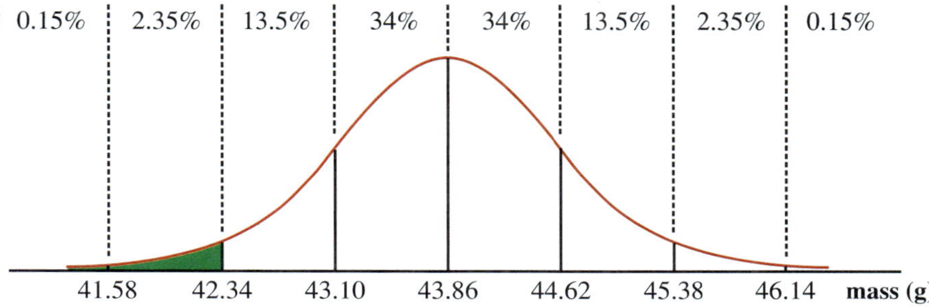

About 2.35% + 0.15% = 2.5% of bags of Mini Eggs have a mass less than 42.34 g.

b)

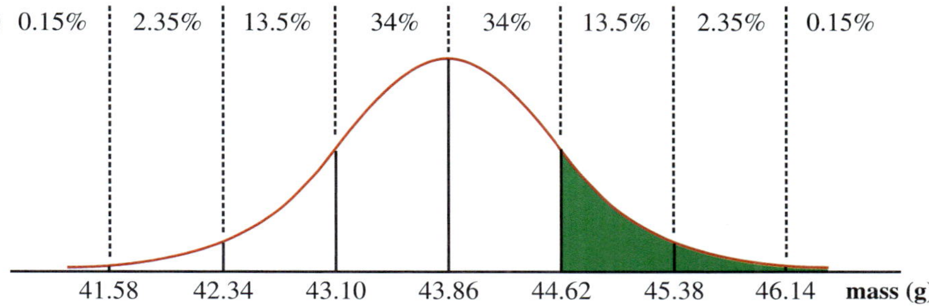

About 13.5% + 2.35% + 0.15% = 16% of bags of Mini Eggs have a mass greater than 44.62 g.

c)

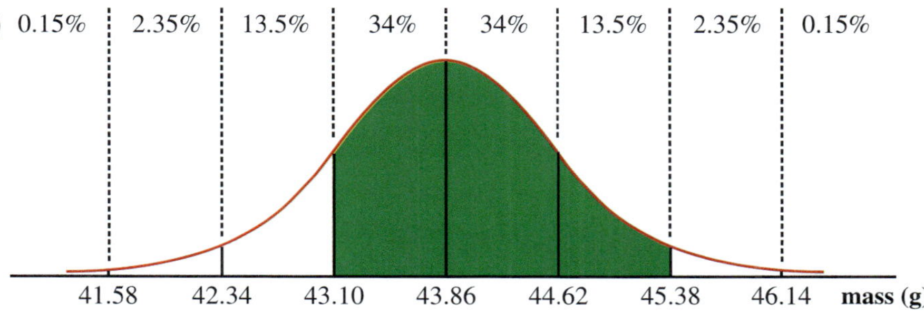

About 34% + 34% + 13.5% = 81.5% of bags of Mini Eggs have a mass between 43.10 g and 45.38 g.

3. We know that about 99.7% of the data lie within 3 standard deviations of the mean. We can use this to estimate the range of masses of bags of Mini Eggs.

46.14 – 41.58 = 4.56 g

A reasonable estimate of the range of masses is 4.56 g.

Since the total area under the normal curve is 1, the areas of regions under the curve can be considered as probabilities. In the Example, the probability that the mass of a bag of Cadbury Mini Eggs will be less than 42.34 g is 0.025, and the probability that the mass of a bag of Mini Eggs will be greater than 44.62 g is 0.16.

Discussing the Ideas

1. How can you tell if data are discrete or continuous? Give at least one example of each type of data.
2. Look at the normal curves on page 112. Explain how the first curve would change in each case.
   a) The mean is the same and the standard deviation is 16.
   b) The mean is 75 and the standard deviation is the same.
   c) The mean is 50 and the standard deviation is 11.
3. In exercise 2 of the Example, an estimate of the range, 4.56 g, was found by subtracting 41.58 g from 46.14 g.
   a) Describe a different way to estimate this range.
   b) Explain why the range cannot be determined exactly.
4. The total area under the normal curve is 1. Explain why this allows us to consider the areas of regions under the normal curve as probabilities.
5. Describe the kinds of problems that can be solved with the information about a normal distribution that is presented in this Tutorial.

Exercises

Checking Your Skills

1. Three sets of data are described. Match each data set with the graph most likely to represent its distribution. Use each graph only once. Explain your pairings.
   a) The masses of the posts in a bundle of 200 wooden fence posts
   b) The ages of people in a municipal park on a Sunday afternoon
   c) The marks received by students on a standardized achievement test

i) 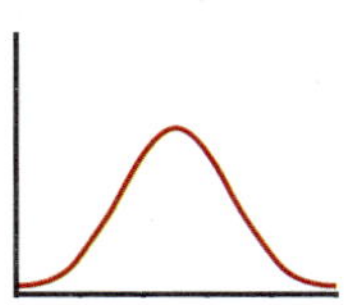

ii) 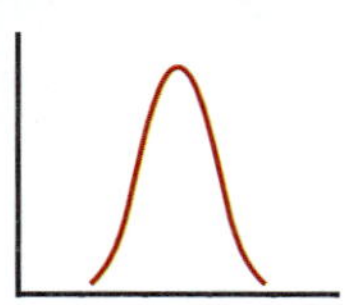

iii) 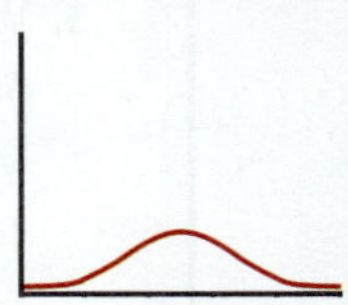

2. A barge contains a load of gravel. Assume the distribution of the masses of stones is normal. The mean mass is 5.0 g and the standard deviation is 0.5 g.
   a) Sketch a normal curve to show the distribution of the masses of the stones. Mark the points that are 1, 2, and 3 standard deviations from the mean.

b) What percent of the shipment has the masses indicated?

i) greater than 3.5 g
ii) less than 3.5 g
iii) less than 4.0 g
iv) between 4.5 g and 5.5 g
v) between 4.0 g and 6.0 g
vi) between 5.0 g and 5.5 g

3. The Stanford-Binet IQ scores of Canadians are normally distributed with a mean of 100 and a standard deviation of 16. According to this test, what is the probability that a randomly selected Canadian will have the following IQs?

a) an IQ less than 100
b) an IQ less than 116
c) an IQ greater than 116
d) an IQ between 84 and 116

4. A shipment of logs has a mean diameter of 15 cm and a standard deviation of 2 cm. Assume the diameters of the logs are normally distributed.

a) By how many standard deviations does a log with a diameter of 11 cm differ from the mean?

b) What is the probability that a randomly selected log will have a diameter less than 11 cm?

c) Estimate the range of diameters of the logs.

5. A soft-drink bottler has a machine that fills 375-mL cans. The machine is designed to place 377.0 mL in the can with a standard deviation of 0.5 mL.

a) What percent of the cans will have less than 376.0 mL?

b) What percent of the cans will have less than 375.5 mL?

c) Cans that contain less than 375.5 mL are rejected. In a production run of 50 000, how many cans will be rejected?

d) What is the probability that a randomly selected can contains a volume greater than 376.0 mL?

6. A brand of electric light bulbs has an average life of 1 year and a standard deviation of 3 months. Assume the distribution is normal. What is the probability that a randomly selected light bulb will have a life between 9 months and 18 months?

7. A hundred squash balls are dropped from a height of 250 cm and the height of the bounce is recorded. The height of bounce was normally distributed, with mean 75 cm and standard deviation 1.9 cm.

a) A ball is considered "fast" if it bounces above 78.8 cm. What is the probability that a ball selected at random will be fast?

b) Estimate the range of heights of bounce.

8. Sacks of grain packed by an automatic loader have an average mass of 61.0 kg. Ninety-five percent of the sacks have masses between 56.4 kg and 65.6 kg. Assume that the masses are normally distributed. What is the standard deviation?

9. A manufacturer of electric pencil sharpeners has found that the average life of their product is 3.5 years, with a standard deviation of 18 months. Pencil sharpeners that last less than 6 months are replaced free. Assume that the distribution of life spans is normal. For every 200 pencil sharpeners sold, how many can be expected to be replaced free?

Extending Your Thinking

10. John got 75% on a statistics test where the mean mark was 70% and the standard deviation was 6%. Janet got 70% on a different statistics test where the mean mark was 60% and the standard deviation was 8%.

a) How many standard deviations from the mean was John's mark? Janet's mark?

b) John and Janet both claimed to have the better performance on the test.
i) What argument might John make to justify his claim?
ii) What argument might Janet make to justify her claim?

PROJECT

In this Tutorial, you worked with normal distributions. The skills and concepts you used relate to the Chapter Project, and to the following projects in the Project Book.

- Un-Designer Jeans
- Quality Control

COMMUNICATING THE IDEAS

Write to explain what is meant by a normal distribution and why normal distributions are useful. Include some examples to illustrate your explanation.

TUTORIAL 3.4 Standard Normal Distribution

In *Tutorial 3.3*, you used the normal distribution to calculate probabilities. In each problem, the normal distribution had a different mean and standard deviation. We could only calculate probabilities for data values that were exactly 1, 2, or 3 standard deviations away from the mean. In this Tutorial, we introduce the standard normal distribution that can be used in every problem for any data values.

The *standard normal distribution* is a normal distribution that has a mean of 0 and a standard deviation of 1. It has the same properties as any other normal distribution and it obeys the 68-95-99 Rule.

Properties of a Standard Normal Distribution

- The mean is 0 and the standard deviation is 1.
- The graph is symmetrical about the mean.
- Almost all the data lie within 3 standard deviations of the mean.
- The curve approaches the x-axis, but does not touch it.
- The total area under the curve is 1.
- The distribution of data follows the 68-95-99 Rule.

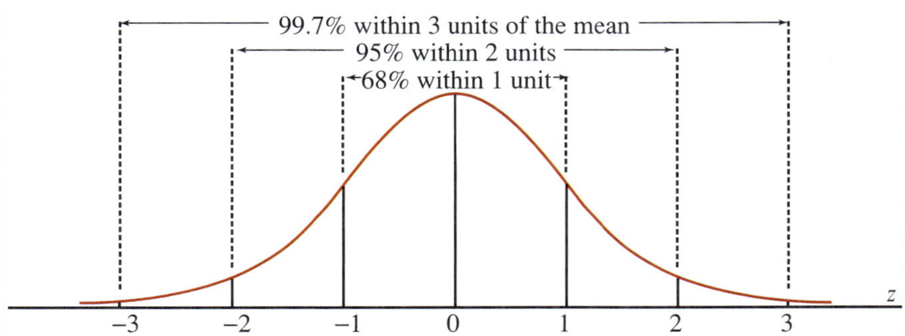

Standard deviation scale

When we use the standard normal distribution, we use the letter z to represent the numbers along the horizontal axis. The values of z, called *z-scores*, represent the number of standard deviations from the mean.

To calculate the z-score, subtract the mean from the given value, x, then divide by the standard deviation.

$$z = \frac{x - \mu}{\sigma}$$

Converting data to z-scores allows us to calculate probabilities for data values that are not exactly 1, 2, or 3 standard deviations from the mean and to compare data from different distributions.

In the Example in *Tutorial 3.3*, we assumed that data about the actual masses of 43-g bags of Cadbury Mini Eggs were distributed normally. The mean mass was 43.86 g with a standard deviation of 0.76.

Normal Distribution for μ = 43.86 and σ = 0.76

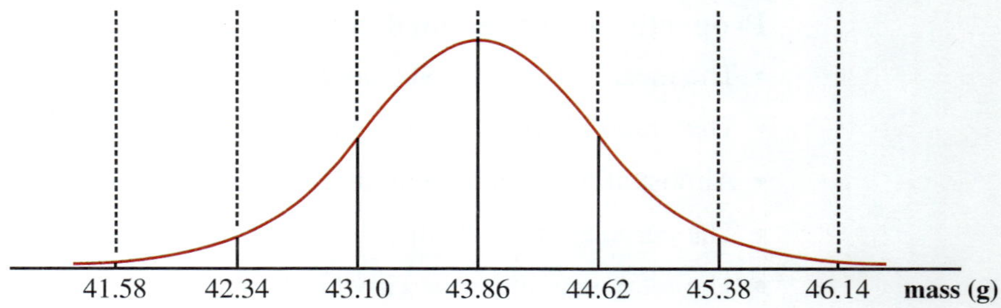

We can use these data to estimate the probability that a randomly selected bag will have a mass less than 43 g.

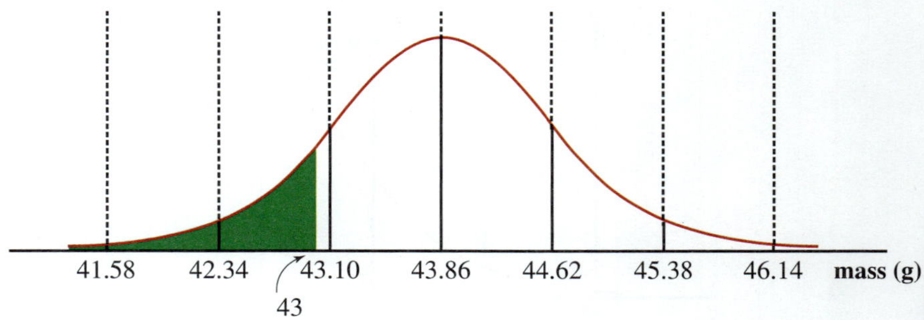

A mass of 43 g is between 1 and 2 standard deviations below the mean. Calculate the z-score.

$$\begin{aligned} z &= \frac{x - \mu}{\sigma} \\ &= \frac{43 - 43.86}{0.76} \\ &\doteq -1.13 \end{aligned}$$

This tells us that a mass of 43 g is 1.13 standard deviations below the mean.

We can sketch this on a standard normal distribution.

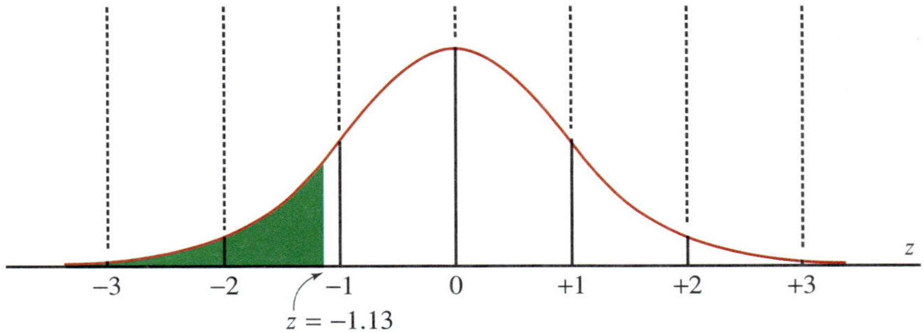

The area of the shaded region to the left of $z = -1.13$ on the standard normal distribution is the same as the area to the left of 43 g in the normal distribution on the previous page. This area represents the probability that a randomly selected bag will have a mass less than 43 g.

UTILITY

If you need help using the TI-83 to calculate probabilities with a standard normal distribution, see Utility 30. If you are using a different graphing calculator, consult the user's manual.

If you do not have a graphing calculator, Utility 29 shows you how to use *z*-tables to find probabilities with a standard normal curve.

Use the **ShadeNorm(** command with lower bound –5 and upper bound –1.13 to find the probability.

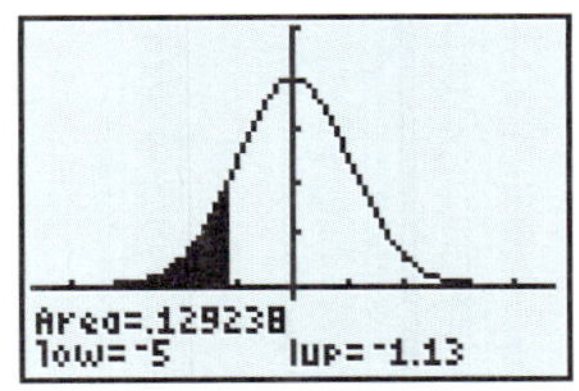

The area is approximately 0.1292.

The probability that a randomly selected bag of Mini Eggs will have a mass less than 43 g is 0.1292.

Example 1

Determine the number of ball bearings that are expected to fail.

A manufacturer produces ball bearings with a mean diameter of 3.00 mm and a standard deviation of 0.20 mm. Assume the distribution is normal. Ball bearings are rejected if their diameter is less than 2.75 mm or greater than 3.25 mm.

a) In a production run of 50 000 ball bearings, how many are expected to be rejected?

b) What can the manufacturer do to reduce the number of rejected ball bearings?

SOLUTION

a) Draw the normal distribution.

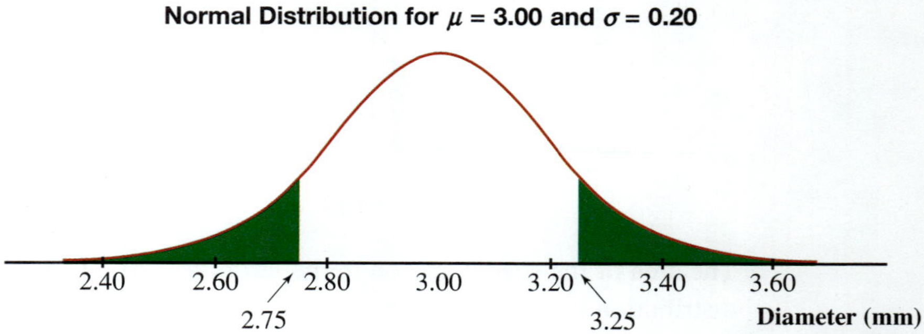

The values 2.75 and 3.25 are between 1 and 2 standard deviations away from the mean. Calculate the z-scores.

$$z = \frac{x - \mu}{\sigma} = \frac{2.75 - 3}{0.2} \doteq -1.25 \qquad z = \frac{x - \mu}{\sigma} = \frac{3.25 - 3}{0.2} \doteq 1.25$$

To find the area to the left of $z = -1.25$, use **ShadeNorm(** with lower bound –5 and upper bound –1.25 (shown below left).

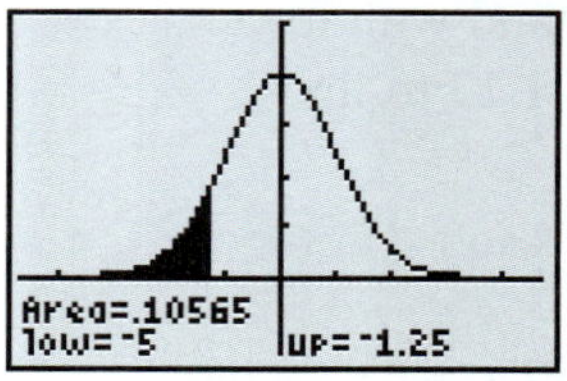

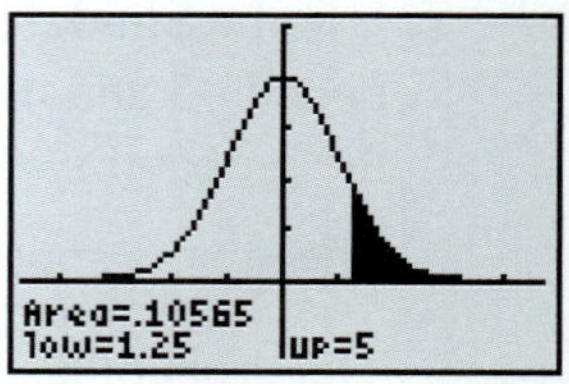

To find the area to the right of $z = 1.25$, use **ShadeNorm(** with lower bound 1.25 and upper bound 5 (shown above right).

The probability that a randomly selected ball bearing has a diameter less than 2.75 mm is 0.105 65. The probability that a randomly selected ball bearing has a diameter greater than 3.25 mm is also 0.105 65.

The probability that a ball bearing will be rejected is 0.105 65 + 0.105 65, or 0.2113.

In a run of 50 000 ball bearings, it is expected that 0.2113 × 50 000, or 10 565 bell bearings will be rejected.

b) The manufacturer needs to adjust the production process to lower the standard deviation. The mean should remain the same.

The standard normal distribution is useful when we compare data from different distributions.

Example 2 **Compare data.**

Every student who applies to an MBA program must write the Graduate Management Admissions Test (GMAT). Last year, Marilyn scored 746 on her GMAT. The mean was 504 and the standard deviation was 110. This year, Colin scored 730 on his GMAT. The mean of this year's exam was 515, with a standard deviation of 100. Who had the better score on the GMAT?

SOLUTION

Convert the results to z-scores:

Marilyn

$$z = \frac{x - \mu}{\sigma} = \frac{746 - 504}{110} = +2.2$$

Colin

$$z = \frac{x - \mu}{\sigma} = \frac{730 - 515}{100} = +2.15$$

Marilyn's score was 2.2 units above the mean and Colin's score was 2.15 units above the mean. Marilyn performed better on the GMAT than Colin did.

DISCUSSING THE IDEAS

1. In the calculation of a z-score, explain why it is important that we find $x - \mu$, not $\mu - x$.

2. Explain why the formula $z = \frac{x - \mu}{\sigma}$ expresses a data value, x, in terms of the number of standard deviations it is from the mean.

3. When we use the graphing calculator to find the probability that a particular value will be less than a given z-value, why do we use –5 as the lower bound?

EXERCISES

Checking Your Skills

1. Convert each measurement to a z-score. All data are normally distributed.

   a) In a shipment of lumber, the pieces have a mean thickness of 2.4 cm, with a standard deviation of 0.1 cm. One particular piece has a thickness of 2.5 cm.

b) The results of a mathematics test had a mean of 78 and a standard deviation of 12. Jill's score was 85.

c) In a load of cod, the mean mass was 2.7 kg and the standard deviation was 0.3 kg. A particular fish had a mass of 1.8 kg.

d) A certain brand of CD player has an MTBF (mean time between failures) of 8500 h and a standard deviation of 250 h. A test item failed after 6375 h.

2. Bean plants in a field have heights that are normally distributed, with a mean of 20.0 cm and a standard deviation of 3.5 cm. Find the proportion of plants that are:

a) between 20.0 cm and 24.9 cm tall

b) between 15.8 cm and 24.2 cm tall

c) taller than 29.1 cm

3. The mean score on a math test is 115 and the standard deviation is 20. The teacher decides to give an A to all students who score 135 or more. Assume the scores are normally distributed. What is the probability that a randomly selected student will receive an A?

4. For entry into the Canadian Armed Forces, the standards used for height are set at 158 cm to 194 cm for males, and 152 cm to 184 cm for females. Assume that heights for males and females are normally distributed, with means of 176 cm and 163 cm and standard deviations of 8 cm and 7 cm, respectively. Are these two height standards equivalent? Explain.

5. The mass of the contents of a box of animal crackers is normally distributed about a mean of 210 g, with a standard deviation of 2.5 g.

a) Calculate the z-score for a box with mass 214 g.

b) What is the probability that the mass of a randomly selected box will be between 210 g and 214 g?

c) What is the probability that the mass of a randomly selected box will be less than 214 g?

d) A box must be rejected if the mass is less than 204 g. What is the probability that a box will be rejected?

6. In a sample of 122 people, the mean body temperature is 36.8°C, with a standard deviation of 0.35°C. Assume a normal distribution to find the expected number of people in the sample with temperatures:

a) above 37.0°C

b) below 36.0°C

7. The results on a district-wide grade 11 math achievement test were distributed normally around a mean of 220 and a standard deviation of 40. The district decided to provide remediation for all students who scored less than 116. What is the probability that a randomly selected student will be eligible for remediation?

8. The table shows the mean and standard deviation for the final course marks for three grade 12 courses.

Course	Mean	Standard Deviation
English	60	14
Mathematics	65	10
Physics	59	8

A student scored 60 in Physics, 67 in Mathematics, and 70 in English. Compare his standing in the three subjects.

9. In a shipment of eggs, the eggs have a normal distribution, with a mean height of 5.20 cm and a standard deviation of 0.12 cm. The eggs are to be placed in containers that can accommodate heights between 4.92 cm and 5.48 cm. What percent of the eggs must be stored in another container?

Extending Your Thinking

10. A machine is programmed to fill sacks with grain. The masses are normally distributed about a mean of 114 kg. Tests show that 10% of the bags have masses greater than 116 kg. Find the standard deviation.

PROJECT

In this Tutorial, you calculated probabilities using the standard normal distribution. The skills and concepts you used relate to the Chapter Project, and to the following projects in the Project Book.

- Un-Designer Jeans
- Reaction Time
- Quality Control

COMMUNICATING THE IDEAS

Use an example to explain how the standard normal distribution is used to solve problems that involve data that are distributed normally.

TUTORIAL 3.5

The Normal Approximation to a Binomial Distribution

In *Tutorial 3.1*, you used the binomial distribution to calculate probabilities. This may not always be practical. In this Tutorial, you will use the normal distribution to approximate the binomial distribution.

Investigation

Calculating Probabilities

A basketball player is successful on her free throws 75% of the time.

1. Determine the probability that she makes at least 7 out of her next 10 free throws in these two ways:

a) Calculate the Probability Using the Binomial Distribution.
Use the **binompdf(** command on your graphing calculator to generate the binomial distribution for this experiment. To find the probability that she makes at least 7 of her next 10 free throws, add the probabilities that she makes exactly 7, exactly 8, exactly 9, and exactly 10. Record your answer to 5 decimal places.

UTILITY

Utility 27 shows you how to calculate and graph a binomial distribution using a TI-83.

b) Estimate the Probability Using the Normal Approximation to the Binomial Distribution.
Recall from *Tutorial 3.2* that the mean and standard deviation of a binomial distribution are:

$$\mu = n \times p \qquad \sigma = \sqrt{n \times p \times (1 - p)}$$

i) Determine the mean and standard deviation for this experiment, then sketch the normal distribution.

ii) Calculate the z-score.

iii) Use the **ShadeNorm(** command to estimate the probability to 5 decimal places. What is the difference between this probability and the probability you calculated in part a?

UTILITY

Utility 30 shows you how to calculate probabilities with a standard normal distribution using a TI-83.

Utility 29 shows you how to find areas under the standard normal curve using tables.

c) What is the value of $n \times p \times (1 - p)$?

2. Determine the probability that she makes at least 21 out of her next 30 free throws in these two ways:

a) Calculate the Probability Using the Binomial Distribution.

Generate the binomial distribution for this experiment. Add the probabilities that she makes exactly 21, exactly 22, exactly 23, and so on up to exactly 30. Record your answer to 5 decimal places.

b) Estimate the Probability Using the Normal Approximation to the Binomial Distribution.

i) Determine the mean and standard deviation for this experiment then sketch the normal distribution.

ii) Calculate the z-score.

iii) Use the **ShadeNorm(** command to estimate the probability to 5 decimal places. What is the difference between this probability and the probability you calculated in exercise 2, part a?

c) What is the value of $n \times p \times (1 - p)$?

3. Determine the probability that she makes at least 42 out of her next 60 free throws in these two ways:

a) Calculate the Probability Using the Binomial Distribution.
Generate the binomial distribution for this experiment. Record your answer to 5 decimal places.

b) Estimate the Probability Using the Normal Approximation to the Binomial Distribution.

i) Determine the mean and standard deviation for this experiment then sketch the normal distribution.

ii) Calculate the z-score.

iii) Use the **ShadeNorm(** command to estimate the probability to 5 decimal places. What is the difference between this probability and the probability you calculated in exercise 3, part a?

c) What is the value of $n \times p \times (1 - p)$?

4. Copy and complete the table.

Exercise	Value of $n \times p \times (1 - p)$	Difference between Actual and Estimated Probability
1		
2		
3		

5. a) In which exercise was the difference between the actual probability and the estimate the smallest? the largest?

b) In which exercise was the value of $n \times p \times (1 - p)$ the smallest? the largest?

c) What inference can you make about the relationship between the value of $n \times p \times (1 - p)$ and the difference between the actual probability and the estimated probability.

The binomial distribution is a distribution of discrete data. The graph of the binomial distribution consists of individual points or bars. The normal distribution is a distribution of continuous data. The graph of the normal distribution is a smooth curve.

The graphs below show the binomial distribution along with the normal distribution that approximates it for each exercise from the Investigation.

Number of Successful Free Throws When $n = 10$

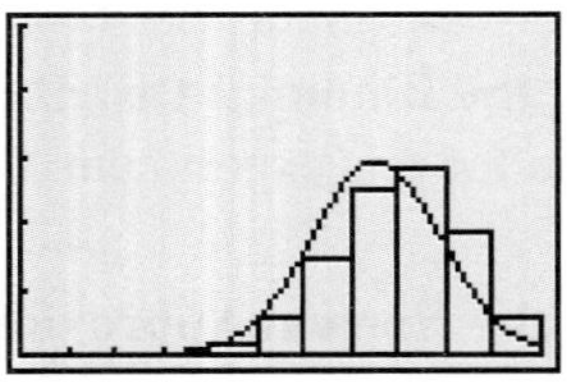

Number of Successful Free Throws When $n = 30$

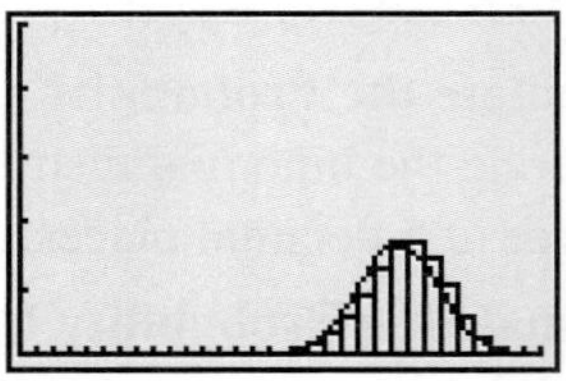

Number of Successful Free Throws When $n = 60$

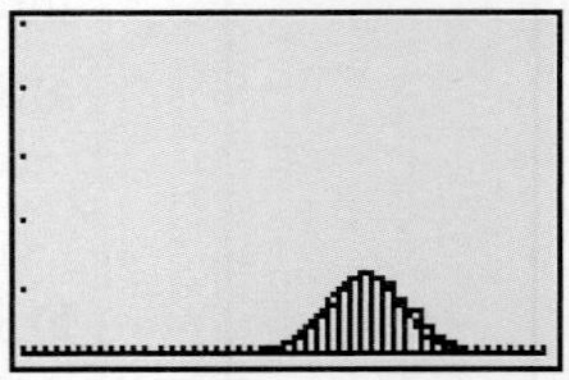

As the sample size increases, the value of $n \times p \times (1 - p)$ increases and the normal distribution "fits" the binomial distribution more closely.

In a binomial distribution, if $n \times p \times (1 - p) \geq 10$, the number of successes is approximately normally distributed and we can use the normal distribution to approximate the binomial distribution.

The following procedure can be used to estimate binomial probabilities with the normal distribution.

i) Determine the values of n and p.

ii) Calculate $n \times p \times (1 - p)$ to determine if the normal approximation to the binomial distribution can be used.

iii) To calculate the mean and standard deviation, use $\mu = n \times p$ and $\sigma = \sqrt{n \times p \times (1 - p)}$.

iv) Convert the endpoints of the intervals to z-scores.

v) To calculate the probability, find the area of the interval under the standard normal curve.

Example **Use the normal approximation to a binomial distribution.**

A clothing manufacturer estimates from past quality control inspections that 4.8% of sweatshirts produced are irregular. If an inspector randomly selects 650 sweatshirts from the warehouse, estimate the probability that at least 30 of them are irregular.

SOLUTION

i) $n = 650$ and $p = 0.048$

ii) $n \times p \times (1 - p) = 650 \times 0.048 \times (1 - 0.048)$
$\doteq 29.7$

The normal approximation to the binomial distribution is appropriate.

iii) $\mu = n \times p$
$= 31.2$

$\sigma = \sqrt{n \times p \times (1 - p)}$
$= \sqrt{650 \times 0.048 \times (1 - 0.048)}$
$\doteq 5.45$

iv) $z = \frac{x - \mu}{\sigma}$
$= \frac{30 - 31.2}{5.45}$
$\doteq -0.22$

v) Use the **ShadeNorm(** command with lower bound –0.22 and upper bound 5.

The probability that at least 30 of the 650 sweatshirts are irregular is approximately 0.5871.

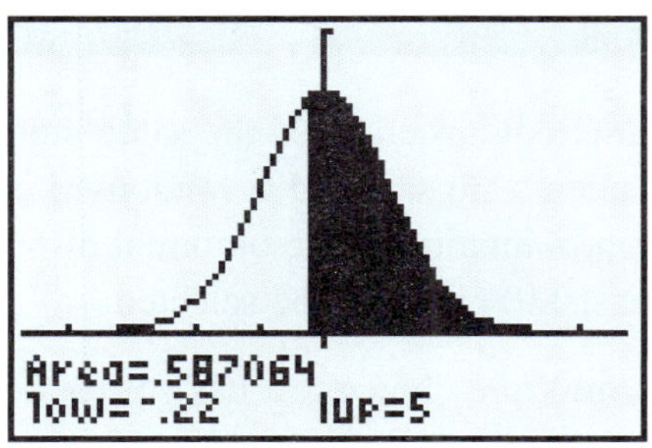

DISCUSSING THE IDEAS

1. Suggest some reasons why a normal approximation to the binomial distribution should not be used if $n \times p \times (1 - p)$ is less than 10.

2. Describe some factors that affect how close an estimate of the probability is to the actual probability.

EXERCISES

Checking Your Skills

1. In each case, state whether or not it is appropriate to use the normal approximation to the binomial distribution.
   a) In a survey of 750 students, 30% like anchovies on their pizza.
   b) A film manufacturer estimates that 1.6% of all rolls of film produced are defective.
      i) Fifty rolls of film are randomly selected and tested.
      ii) Six hundred and fifty rolls of film are randomly selected and tested.

2. A school has enrolled the same number of boys and girls. Two hundred students are selected at random to participate in a survey. Use a normal approximation to the binomial distribution to estimate the probability that at least 110 girls will be selected.

3. A multiple-choice test has 75 questions. Each question has 5 choices, only one of which is correct. If each question is answered by a guess, estimate the probability that more than 30 answers are correct.

4. Estimate the probability that, in 1000 tosses of a fair 6-sided die, there would be at most 600 heads.

5. Market research shows that 12% of consumers choose a particular brand of pain reliever. In a survey, a random sample of 325 consumers is chosen. Estimate the probability that at least 30 of the consumers surveyed purchase this brand.

6. According to Canadian Blood Services, 46% of Canadians have type O blood. If 500 volunteers donate blood, estimate the probability that more than one-half of the volunteers have type O blood.

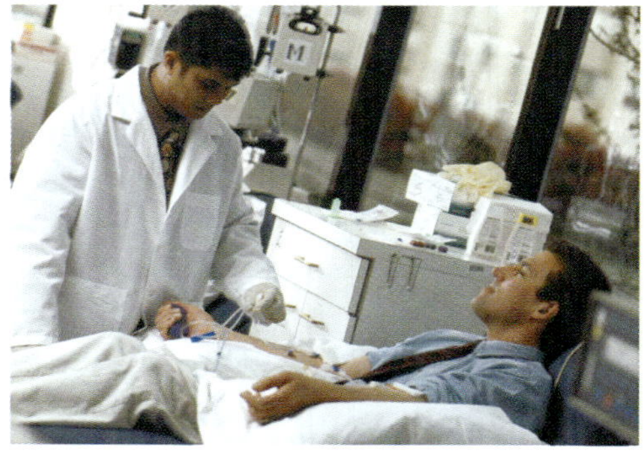

7. The Canadian Automobile Association conducted a survey of its members and found that 38.1% of car buyers bought a used car. If 280 car buyers are selected at random, determine the probability that fewer than 90 of them purchased a used car.

8. A mechanical failure caused one-quarter of a machine shop's production of pipe connectors to be defective. In a random sample of 75 connectors, estimate the probability that no more than 8 were defective.

Extending Your Thinking

9. Determine the number of times you would have to roll two 6-sided dice so that the probability you would get a sum of 5 at least three times is 0.5.

PROJECT

In this Tutorial, you used the normal distribution to approximate the binomial distribution. The skills and concepts you used relate to the Chapter Project, and to the following projects in the Project Book.

- Un-Designer Jeans
- Quality Control

COMMUNICATING THE IDEAS

Compare and contrast the normal distribution and the binomial distribution. Describe as many similarities and differences as you can and explain how they affect the normal approximation to the binomial distribution.

Conducting a Survey

In this stage of the Chapter Project you will conduct a survey to estimate public opinion about an issue of interest to you, then you will analyze the results of your survey.

1. Select an issue that interests you. Here are some possibilities.
   - The number of hours of television students watch per week
   - Whether students in your school are in favour of, or against, an issue such as school uniforms
   - How much allowance students get per week

   Construct a suitably worded questionnaire. Select an appropriately sized random sample of the population.

2. Conduct your survey and record the responses.

3. Calculate the mean and standard deviation for the data that you collected.

4. What type of distribution do your data fit? If the distribution is binomial, is it appropriate to use the normal approximation to the binomial distribution to estimate probabilities?

5. Use the distribution of your data to calculate some probabilities relevant to the issue you selected.

CHAPTER PROJECT

In the next stage of the Chapter Project, you will report the statistics and margin of error for the data you collected in your survey. Complete *Tutorial 3.6* to learn about confidence intervals and margin of error. You will return to the Chapter Project on page 142.

Career Skills

Businesses, organizations, and individuals hire polling firms to conduct surveys to estimate public opinion about many issues. Many newspaper articles now contain excellent reports of the results of such surveys. Employees of such firms use many of the strategies that you have considered so far.

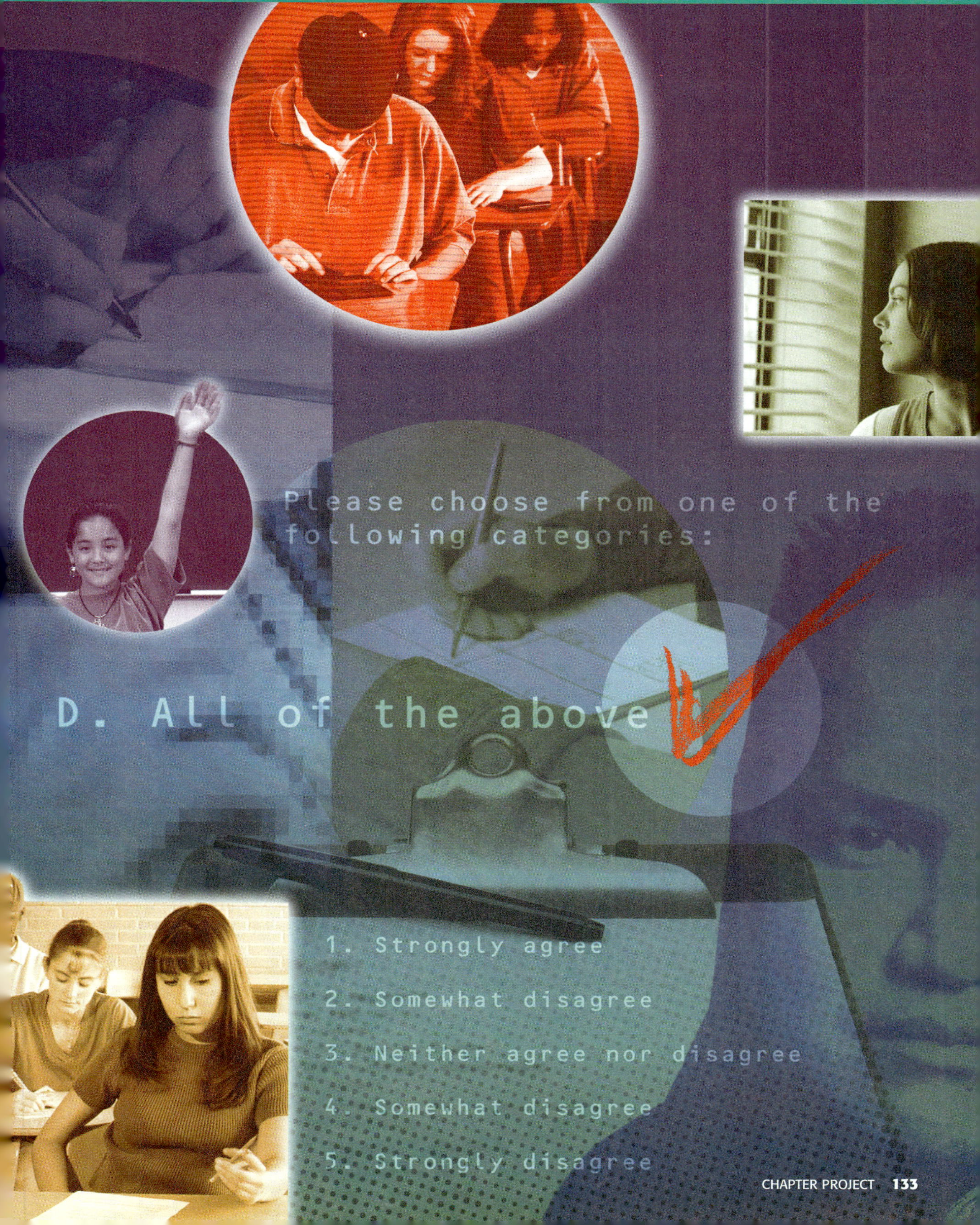
Please choose from one of the following categories:
D. All of the above
1. Strongly agree
2. Somewhat disagree
3. Neither agree nor disagree
4. Somewhat disagree
5. Strongly disagree

TUTORIAL 3.6 Confidence Intervals

Many surveys and polls are conducted to estimate the opinions and preferences of the general public. A good report on a survey gives the results along with an indication of their accuracy.

This excerpt is from a report that appeared in *The Vancouver Sun* on June 3, 2000.

> While Canadians are gravitating toward the Internet, they don't trust it as a source for information... Internet chat rooms scored the lowest of all in credibility... Independent Web sites fared better, but still lagged as a credible source, with only 14 percent of those surveyed viewing them as very credible. Another 37 percent said they were somewhat credible... The survey of 1613 Canadians was conducted over 10 days in late April and is considered accurate to within 3.1 percent, 19 times out of 20.

This survey showed that only 14% of the respondents in a sample of 1613 people surveyed viewed Internet web sites as a very credible source of information. The last sentence tells us that, if the survey were conducted many times, with different samples of 1613 people, on average 19 times out of 20 (or 95% of the time), the percent would be between 10.9% and 17.1%. Notice that 19 out of 20 is equivalent to 95%.

The interval 14% ± 3.1% is the 95% *confidence interval* for this survey, and ±3.1% is the *margin of error*.

In *Tutorials 3.3* and *3.4*, you worked with the 68-95-99 Rule. This rule stated that, in a normal distribution, *approximately* 95% of the data lie within 2 standard deviations of the mean. Complete this Investigation to find the range of z-scores that contains 95% of the data.

Investigation **Find the Symmetric Interval under the Standard Normal Curve That Contains 95% of the Data.**

1. Use your graphing calculator to determine the area under the standard normal curve between $z = -2$ and 2. What percent of the data is in this range?

2. Use your graphing calculator to find the area between $z = -1.99$ and 1.99.

3. Repeat exercise 2 several times. Each time subtract 0.01 from the upper and lower bound. Continue until you find a range of z-values for which the area under the standard normal curve is less than 0.95.
   a) Which calculation resulted in an area that was extremely close to 0.95? What were the z-values of the upper and lower bound?
   b) Why is the area not exactly 0.95? Describe how we could find an area that is exactly 0.95.
   c) Within how many standard deviations of the mean are 95% of the data located?

For data that have a normal distribution with mean μ and standard deviation σ, a 95% *confidence interval* is:

$$\mu \pm 1.96\sigma$$

This is the range of values that lie within 1.96 standard deviations of the mean. The probability that a particular data value lies in that range is 0.95.

The graph shows a 95% confidence interval for a normal distribution. The probability that a data value falls within 1.96 standard deviations of the mean is 0.95. The probability that it falls in one of the tails is 0.05.

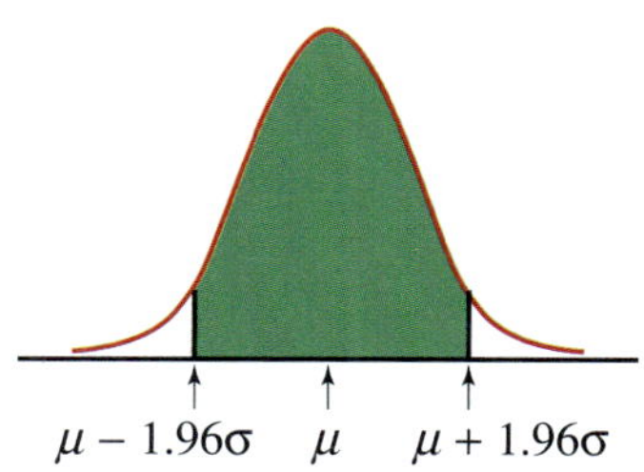

Example 1 **Construct a confidence interval.**

The Students' Council is planning a spring dance. They need to sell 110 tickets in order to cover their costs. Records show that, on average, 10% of the students at the school attend the dances. This year there are 1350 students enrolled at the school.

a) Construct the 95% confidence interval for the number of students who will attend the spring dance.

b) Will the Students' Council sell enough tickets to cover their costs?

SOLUTION

a) This is a binomial distribution with $n = 1350$ and $p = 0.1$. It is appropriate to use the normal approximation to the binomial distribution because $n \times p \times (1 - p)$ is greater than 10.

$$\mu = n \times p = 1350 \times 0.1 = 135 \qquad \sigma = \sqrt{n \times p \times (1-p)} = \sqrt{1350 \times 0.1 \times (1 - 0.1)} \doteq 11.02$$

$$95\% \text{ confidence interval} = \mu \pm 1.96\sigma = 135 \pm 1.96(11.02) = 135 \pm 21.6$$

$$= 135 - 21.6 \doteq 113 \quad \text{or} \quad = 135 + 21.6 \doteq 157$$

With 95% confidence we know that between 113 and 157 out of 1350 students will attend the spring dance.

b) The lower bound of the 95% confidence interval is 113. The Students' Council can be 95% certain that they will sell enough tickets to cover their costs.

To construct the confidence interval in *Example 1*, we found the values that were 1.96 standard deviations above and below the mean. The value 1.96×11.02 or 21.6 represents the half-width of the interval. This is the margin of error. To express this as a percent, we divide by the sample size, in this case 1350, and multiply by 100.

$$= \pm\frac{21.6}{1350} \times 100 = \pm 1.6\%$$

This is the percent margin of error for the problem in *Example 1*.

The *margin of error* is the proportion that we add to, and subtract from, the mean to construct the confidence interval.

For a 95% confidence interval:

$$\text{Margin of error} = \pm 1.96\sigma$$

When you conduct a survey, the smaller the margin of error, the better the results of the survey.

Example 2 **Construct the confidence interval and calculate the margin of error.**

Jette surveyed 200 randomly selected people at the shopping mall and found that 55% needed glasses or contact lenses.

a) Construct the 95% confidence interval for the percent of people that needed corrective lenses.

b) Calculate the percent margin of error.

c) If Jette surveys 1000 people and finds that 55% of them need corrective lenses, what is the percent margin of error?

d) What happens to the margin of error as the sample size increases?

SOLUTION

a) This is a binomial distribution, with $n = 200$ and $p = 0.55$. We can use the normal approximation to the binomial distribution because $n \times p \times (1 - p)$ is greater than 10.

$$\begin{aligned} \mu &= n \times p \\ &= 200 \times 0.55 \\ &= 110 \end{aligned} \qquad \begin{aligned} \sigma &= \sqrt{n \times p \times (1 - p)} \\ &= \sqrt{200 \times 0.55 \times (1 - 0.55)} \\ &\doteq 7.04 \end{aligned}$$

$$\begin{aligned} 95\% \text{ confidence interval} &= \mu \pm 1.96\sigma \\ &= 110 \pm 1.96(7.04) \\ &= 110 \pm 13.8 \end{aligned}$$

$$\begin{aligned} &= 110 - 13.8 \\ &\doteq 96 \\ &\doteq \frac{96}{200} \times 100\% \\ &\doteq 48\% \end{aligned} \qquad \text{or} \qquad \begin{aligned} &= 110 + 13.8 \\ &\doteq 124 \\ &\doteq \frac{124}{200} \times 100\% \\ &\doteq 62\% \end{aligned}$$

With a 95% confidence, we know that between 48% and 62% of people require corrective lenses.

b) Percent margin of error $= \pm\frac{13.8}{200} \times 100\%$
$= \pm 6.9\%$

The percent margin of error is ±6.9%.

c) $\sigma = \sqrt{n \times p \times (1 - p)}$
$= \sqrt{1000 \times 0.55 \times (1 - 0.55)}$
$\doteq 15.73$

Margin of error $= \pm 1.96\sigma$
$= \pm 1.96(15.73)$
$= \pm 30.8$

Percent margin of error $= \pm\frac{30.8}{1000} \times 100\%$
$= \pm 3.1\%$

The percent margin of error is ±3.1%.

d) The margin of error decreases as the sample size increases.

DISCUSSING THE IDEAS

1. Confidence intervals can be constructed for any level of confidence. In this Tutorial, we used only 95% confidence intervals because they are the most common.

a) Explain how 90% and 99% confidence intervals would be different from a 95% confidence interval.

b) Why do you think the 95% confidence interval is the most common?

2. When you report survey results, why is it important to give the margin of error?

3. Explain why the margin of error decreases as the sample size increases.

EXERCISES

Checking Your Skills

1. For each set of information, construct a 95% confidence interval and calculate the margin of error. Give answers to 3 decimal places.

a) $\mu = 68.08$, $\sigma = 4.2$, $n = 92$

b) $\mu = 102.446$, $\sigma = 6.67$, $n = 181$

c) $\mu = 68.404$, $\sigma = 6.67$, $n = 196$

d) $\mu = 102.384$, $\sigma = 8.368$, $n = 324$

2. Suppose a random sample of 150 teenagers were polled. Sixty-six said they watched *The National* at least once a week.

a) What is the 95% confidence interval for the percent of teenagers who watch the show?

b) What is the margin of error?

3. Kara surveyed 200 people, chosen at random on a busy street, to try to determine what percent of British Columbians are skiers. Eighty-six people in her sample said they were skiers.

a) State the 95% confidence interval for the percent of British Columbians who are skiers.

b) Calculate the margin of error.

4. Saul surveyed 150 randomly selected households in his city to see if any member of the household had watched an NHL game within the past week. Ninety-six respondents said yes. State the 95% confidence interval for the percent of households in Saul's city who watched a hockey game within the past week.

5. Based on past experience, a car salesperson will complete sales to 10% of the customers that enter the showroom. If the salesperson has 200 customers in the next month, establish a 95% confidence interval for the number of completed sales for the month and calculate the margin of error.

6. Ava surveyed 50 randomly selected students at her school to see what percent of the student body was in favour of extending the lunch break from 40 min to 1 h. She found 23 students in favour of the change.

a) Calculate the 95% confidence interval for the percent of the student population in favour of lengthening the lunch break.

b) Calculate the margin of error.

c) Suppose Ava increases her sample size to 100 and the same proportion of students favour the change, what is the margin of error?

7. Modern Pharmaceuticals Inc. surveyed 3000 users of its new extra strength painkiller. It was judged to be effective by 2752 respondents.

a) Calculate the 95% confidence interval for the percent of people who are likely to judge the new improved painkiller to be effective.

b) Suppose the regular strength painkiller had been judged effective by 90% of the respondents. Should Modern Pharmaceuticals Inc. replace it with its new improved version? Explain.

8. Jari surveyed a random sample of 60 people who had attended a local folk festival. Thirty-five respondents indicated they had enjoyed the concert and would attend another, even if the admission charge were tripled.

a) Calculate the 95% confidence interval for the percent of previous attendees who would likely attend the folk festival if the admission charge were tripled.

b) Would tripling the ticket price be a wise move by the festival promoters? Explain.

9. A total of 85 randomly selected patients were surveyed after they had undergone a new surgical procedure. Sixty-three of the respondents were found fit to return to work after two weeks of recuperation.

a) Calculate the 95% confidence interval for the percent of patients who undergo the procedure and will be fit to return to work after two weeks of recuperation.

b) Suppose 58% of the patients who had undergone the surgical procedure had been fit to return to work after two weeks. Would it be worthwhile to make all candidates for the procedure aware of it?

Extending Your Thinking

10. In a well-known British survey on the heights of 200 couples, it was found that in only 12 couples was the wife taller than her husband. This indicates that, for all British couples, between 3% and 9% of wives are taller than their husbands 95% of the time.

a) What is the margin of error for this survey?

b) How would you change the sample size to reduce the percent margin of error in this survey to ±1.5%?

11. Confidence intervals can be constructed for any level of confidence. A 95% confidence interval is the most common. Two other confidence intervals that are often used are 99% and 90%.

a) Follow the process from the Investigation to construct a 99% confidence interval. Start by finding the area under the standard normal curve between $z = -2.60$ and 2.60.

b) Follow the process from the Investigation to construct a 90% confidence interval. Start by finding the area under the standard normal curve between $z = -1.70$ and 1.70.

c) Describe the similarities and differences among 95%, 99%, and 90% confidence intervals.

d) Determine the formula for margin of error for each level of confidence.

PROJECT

In this Tutorial, you determined confidence intervals and margins of error. The skills and concepts you used relate to the Chapter Project, and to the following projects in the Project Book.

- Un-Designer Jeans
- Quality Control

COMMUNICATING THE IDEAS

In your journal, explain what is meant by the terms "confidence interval" and "margin of error." Why are these concepts useful when you interpret survey results?

Reporting the Results

At this stage of the Chapter Project, you will use all the statistics you gathered in your survey. Produce a report to summarize your results. First, examine the articles you collected and complete these exercises.

1. For each article, if the margin of error and the level of confidence are stated, construct the appropriate confidence interval for the result.

2. Examine each article. Does it accurately report the results of the survey, given the margin of error? If the article is misleading, rewrite it to accurately reflect the results.

For your survey, calculate the 95% confidence interval for the result, then write a complete report that accurately states the results. Be sure to include the margin of error and the level of confidence in your report.

Your completed project will consist of:

- The questionnaire
- A description of the sample population you surveyed and the method you used to select it
- The statistics and probabilities you calculated for your survey results
- Your collection of articles and your answers to exercises 1 and 2 above
- Your written report of the survey results

Career Skills

Employees of polling firms must calculate the margin of error and state the level of confidence when they tabulate the results of surveys. Journalists who report the results need to interpret this information in order to write a clear and unbiased report.

Professional tele-operators

What Do I Need To Know?

Tutorial	Skills and Concepts	Important Results and Formulas
3.1	• Construct probability distributions	• In a probability distribution, the probabilities are between 0 and 1 and the sum of the probabilities is always 1
3.2	• Calculate the mean and standard deviation of a data set and a binomial distribution	• For a binomial distribution, $\mu = n \times p$ and $\sigma = \sqrt{n \times p \times (1 - p)}$
3.3	• Use the normal distribution to solve problems	• About 68% of the data are within 1 standard deviation of the mean • About 95% of the data are within 2 standard deviations of the mean • About 99.7% of the data are within 3 standard deviations of the mean
3.4	• Use *z*-scores and the standard normal distribution to calculate probabilities	• The mean of the standard normal distribution is 0 and the standard deviation is 1 • $z = \frac{x - \mu}{\sigma}$
3.5	• Use the normal distribution to approximate the binomial distribution when sample sizes are large	• If $n \times p \times (1 - p) \geq 10$, we can use the normal distribution to approximate the binomial distribution
3.6	• Construct confidence intervals; calculate the margin of error	• The 95% confidence interval is $\mu \pm 1.96\sigma$ • The margin of error for a 95% confidence interval is $\pm 1.96\sigma$

What Should I Be Able To Do?

When you have completed the work in Chapter 3, you should be able to solve the problems that follow. Part A is investigative. As you work through the exercises, refer to the *Student Reference*, the *Utilities*, and the *Tutorials* as necessary.

Part A

1. Design a survey to collect data of interest to you; for example, the fraction of students who practise martial arts, roller blade, or have siblings. Survey about 30 people.

a) Determine the 95% confidence interval for the percent of the population that has the characteristic for which you surveyed.

b) Calculate the margin of error for your survey.

c) Survey about 30 more people. Combine the data with the data from your original survey. Use the combined data to determine the 95% confidence interval for the percent of the population that has the characteristic for which you surveyed. How does this confidence interval compare to the one you determined in part a?

d) Calculate the new margin of error. What happened to the margin of error when you increased the sample size?

Part B

2. A spinner is divided into 5 equal sectors labelled 1, 2, 3, 4, and 5. Construct a histogram to show the probability distribution of the outcomes for each experiment. In each case, state the type of probability distribution.

a) One spin of the spinner

b) The number of 4s in five spins of the spinner

3. A sample of 20 bags of potato chips produced by Crispy Chips Company was found to have these masses, in grams.

80	85	88	92	76	85	94	76	78	80
84	88	75	78	77	68	70	72	80	74

A sample of 20 bags of potato chips produced by Special Spuds Company was found to have these masses, in grams.

85	96	90	69	78	77	79	82	80	81
78	81	77	80	75	79	76	81	80	76

a) Calculate the mean and standard deviation for each brand of chips.

b) Which brand has the smaller standard deviation?

c) What problems may the manufacturer encounter if the standard deviation gets too high? Explain.

4. Determine the standardized score for a raw score of 712, if the mean is 504 and the standard deviation is 112.

5. Find the probability that a randomly selected individual has an IQ between 108 and 130, if the population is distributed normally with mean 100 and standard deviation 16.

6. In an examination, the mean score is 52 and the standard deviation is 8. What percent of students have a score less than 63?

7. A coin is tossed 100 times. Use the normal approximation to a binomial distribution to find the probability you will obtain

a) 60 or more heads

b) exactly 50 heads

8. A manufacturer of light bulbs finds that on average 5% of light bulbs are defective. A quality control inspector selects 1000 light bulbs at random. What is the probability that 30 or more are defective?

9. "Mixed laterality" is the term given to left- or right-handed individuals whose opposite side is the dominant one. It can be tested very simply. Have a person extend the index finger of the dominant hand to cover a corner of the room where two walls meet the ceiling, with both eyes open. If the finger "moves" as the eye opposite to the dominant hand is closed, the individual has mixed laterality.

a) Determine the percent of students in your class that have mixed laterality.

b) From the data collected in class, determine the 95% confidence interval for the percent of students who have mixed laterality and calculate the margin of error.

c) How can the margin of error be reduced?

Part C

The Chapter Project combines many skills and concepts in a single problem. Here is a sample of possible students' work from different stages of the Project. Review these samples to gain insight into your own project work.

10. Here are the results of one student's survey.

Chapter Project – Statistics

Conducting A survey

The issue that I surveyed was regarding the health care system in Alberta.

The question I asked was:

"If you had to rate our health care system, how would you describe it"

1. Excellent   2. Very Good   3. Good   4. Fair   5. Poor

2. I surveyed 200 people at random at my local mall. The following table shows the results:

Response	Mark assigned	Score
Excellent	5	14
Very Good	4	46
Good	3	78
Fair	2	44
Poor	1	20

In order to use this data for calculations, I assigned a mark of 5 for Excellent, 4 for Very Good, 3 for Good, 2 for Fair, and 1 for Poor.

3. Using a graphing calculator it was determined that

   Mean = 2.95 and standard deviation = 1.06

4. This data fits a normal distribution.

5. Using the graphing calculator I was able to determine the following probabilities:

   a. The probability that a randomly selected person would not rate the health system as excellent is 83%

   b. The probability that a randomly selected person would rate the system higher than Good is 48%

   c. The probability that a person pick between Very Good and Fair is 65.

11. Here are the same student's comments on the articles collected as part of this project. What additional information could the student have included?

ANGUS REID
WORLD WIDE

These are the findings of an Angus Reid Group / Calgary Herald poll conducted between March 17th and March 21st, 2000. The poll is based on a randomly selected sample of 1,071 adult Albertans. The results can be considered accurate to within +/- 3.0 percentage points, 19 times out of 20, of what they would have been had the entire adult Alberta population been polled. Within each of Alberta's major urban centres, Calgary and Edmonton, a total of 400 interviews were conducted, yielding a maximum margin of error of +/- 4.9 percentage points, 19 times out of 20, in those cities. The margin of error will be larger within various sub-groupings of the survey population. These data were statistically weighted to ensure the sample's regional, age and gender composition reflects the actual Alberta population according to Statistics Canada Census Data.

At 72% on the Alberta Issue Agenda, Top-Of-Mind Concerns About the Healthcare System Continue to be the Highest Ever Recorded in the Province

When asked about the most important issues facing the province today, a total of 72 percent of Albertans spontaneously mention healthcare (as either the number one or number two issue of concern, total mentions). This finding mirrors the February 2000 findings and represents the continued highest ranking of healthcare on the Alberta Issue Agenda since the Angus Reid Group began its tracking in 1989, and a 25 percentage point jump in mentions of this issue since October 1999. Continuing down the list, education (30%), the economy and taxes (12% each), government overspending and debt (5%), unemployment and government spending cutbacks (4% each), the environment (3%), inflation, crime, bad politicians and farming concerns (2% each) are also indicated. (SEE TABLE 1)

Reporting the results

1 Confidence interval = ±3.0%

$\frac{\pm 3.0}{100} \times 1701$

$= \pm 5.103$ this represents $1.96\,\sigma$

$\sigma = \frac{5.103}{1.96}$

$\sigma = 2.606$

Because this confidence interval was summarised for a wide range of questions and it is not known how values were assigned to the questions, a mean ($\bar{x}$) cannot be calculated.

2. The report is okay and did not need to be rewritten

4 Personal Finance

PROJECT

You should do at least one project that uses the skills that you have learned in this chapter. The recommended projects in the Project Book are:

- To Lease or Purchase?
- Investing in the Stock Market
- Landscape Gardening
- House Building Costs

Overview

Chapter Section	Skills and Concepts
Chapter Project: Major Decisions: Lease or Buy a Car? Rent or Buy a House? page 150	Research renting versus buying an expensive item
Tutorial 4.1: Investing Money, page 152	Use good spreadsheet design to create a template for investments
Tutorial 4.2: Investing for the Future, page 160	Investigate the effect of sheltering savings within an RRSP
Tutorial 4.3: Investment Portfolios, page 166	Analyze an investment portfolio; determine rate of return and total return on an investment
Tutorial 4.4: Leasing, page 174	Analyze the costs and benefits of leasing or buying a decreasing asset
Chapter Project: Lease or Buy a Car? page 182	Decide how to finance a vehicle
Tutorial 4.5: Mortgages, page 184	Investigate the cost of a mortgage
Tutorial 4.6: Renting versus Buying, page 191	Analyze the costs and benefits of renting or buying an increasing asset
Chapter Project: Major Decision: Rent or Buy a Home? page 198	Finance a home
Consolidating Your Skills, page 200	Review the work of the chapter

Major Decisions: Lease or Buy a Car? Rent or Buy a House?

One of the largest purchases you will likely make is a house. You will decide whether to purchase your own home or to rent one. In this project, you will compare the costs of renting and buying a house.

First, make a list of the features you would like in your future home. This list may include:

- Number of bedrooms and bathrooms
- Style of home (bungalow, 2-story, split-level, condominium)
- Garage (1- or 2-car)
- Preferred location of home
- Local amenities (schools, stores, public transportation)

Research the rental section of your local newspaper or rental guide and clip out advertisements of suitable homes. In particular, note the rent for each home. Look in your local newspaper, real estate guide, or on the Internet for houses that are for sale and clip or print out suitable advertisements.

Before you decide to buy a house there are several important questions that you must consider. How much can you afford? How much do you have in savings? What type of job are you likely to have in the future? Will you buy the house on your own, or with a partner? How much will your partner contribute? How much monthly income can you realistically expect to have?

Another expensive item you will consider purchasing is a car. What type of a vehicle do you want or need? Will it be a new vehicle or an older one? Is leasing this vehicle an option? Look through your local newspaper, and clip out advertisements for the type of vehicle that interests you, together with its purchase price. You may also find lease prices and terms listed. Keep the "fine print" of the leasing information.

CHAPTER PROJECT

In the next stage of this Chapter Project you will research the cost of leasing a car and compare it to the cost of financing. Complete *Tutorials 4.1* to *4.4* to develop the related skills. You will return to this project on page 182.

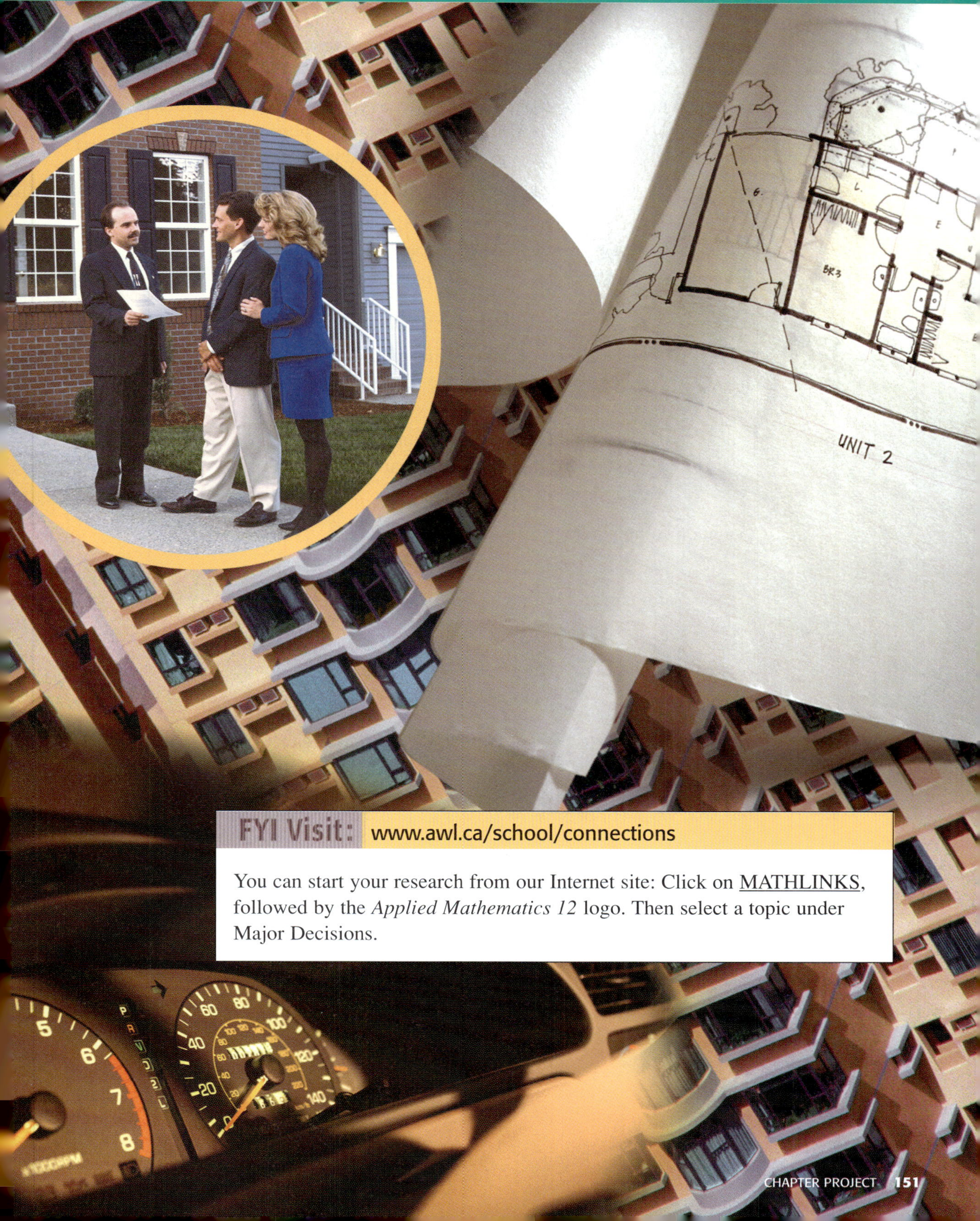

FYI Visit: www.awl.ca/school/connections

You can start your research from our Internet site: Click on MATHLINKS, followed by the *Applied Mathematics 12* logo. Then select a topic under Major Decisions.

TUTORIAL 4.1

Investing Money

Have you ever thought about all the different ways you can invest money? In this Tutorial, you will investigate some investment options. You will also practise spreadsheet design as well as create and use a template.

Most investments earn *compound interest*. This means that at the end of each interest period the earned interest is added to the principal. The formula used to calculate the future or accumulated amount, A, is:

$$A = P(1 + i)^n$$

where P is the present amount or principal;

i is the interest rate per compounding period expressed as a decimal; it is found by dividing the annual interest rate by the number of compounding periods per year; and

n is the total number of compounding periods; it is found by multiplying the number of years by the number of compounding periods per year.

Common compounding periods are:

Annual:	1 time a year
Semi-annual:	2 times a year, or every 6 months
Quarterly:	4 times a year, or every 3 months
Monthly:	12 times a year, or every month
Daily:	365 times a year, or every day

For an investment of \$500 at 8% per annum, compounded annually, for 3 years, $P = 500$, $i = 0.08$, and $n = 3$

For an investment of \$500 at 8% per annum, compounded quarterly, for 3 years, $P = 500$, $i = \frac{0.08}{4} = 0.02$, and $n = 3 \times 4 = 12$

Practise Your Prior Skills

You will work with compound interest formulas in this chapter. Try these exercises as preliminary review.

1. \$100 is invested. Determine the values of P, A, i, and n, for each situation.

a) An annual interest rate of 4%, compounded semi-annually, for 3 years

b) An annual interest rate of 6%, compounded monthly, for 2 years

2. Suppose \$1000 is invested for 10 years. Calculate the compound amount, A, under the following circumstances.

a) The annual interest rate is 6%, compounded annually.

b) The annual interest rate is 8.5%, compounded semi-annually.

c) The annual interest rate is 7.75%, compounded monthly.

3. You can use the Rule of 72 to estimate the time, t, in years, needed for an investment to double in value at a given annual interest rate.

$$t = \frac{72}{\text{Annual interest rate}}$$

a) Use the Rule of 72 to estimate the doubling time for these interest rates:

i) 4% per annum, compounded annually

ii) 8% per annum, compounded annually

iii) 24% per annum, compounded annually

b) Use the compound interest formula to calculate the compound amount of a \$100 investment for the doubling times estimated in part a.

c) What general conclusion can be drawn about the accuracy of the Rule of 72 calculations?

UTILITY

If you need help to use a graphing calculator to make financial calculations, Utility 22 shows you how to use the TVM Solver of a TI-83 calculator. You can access this feature on a TI-83 Plus calculator under the APPS menu.

4. Use a graphing calculator to calculate the amount that must be invested today to have \$8000 for a trip to Europe in 5 years' time, if the interest rate is 5.25%, compounded quarterly.

The spreadsheets in this chapter are created with Microsoft Excel. If you are using a different spreadsheet software, consult your manual. In the following Investigation, you will design a spreadsheet.

Investigation 1

Watching Money Grow

Imagine that you invest a fixed amount of money and leave it invested until retirement.

At the computer, open the spreadsheet program and start a new spreadsheet.

1. Create a spreadsheet to calculate the growth of an investment. Your spreadsheet should have the same appearance and features as shown on page 154.

UTILITY

If you need help to format a spreadsheet, refer to Utility 31.

	A	B	C	D
1	**Investing a Fixed Sum of Money**			
2				
3		Principal	$5,000	
4		Annual Interest Rate	10%	
5		Compounding Periods Per Year	2	
6				
7				
8	**Year**	**Opening Balance**	**Interest**	**Closing Balance**
9	=1/$C$5	=C3	=B9*$C$4/$C$5	=B9+C9
10	=A9+1/$C$5	=D9	=B10*$C$4/$C$5	=B10+C10

a) C3, C4, and C5 show the initial values $5000, 10%, and 2.

b) A9 and A10 will show the number of years for the investment. These values will be fractions.

c) B9 and B10 will show the balance of the investment at the beginning of each interest period.

d) C9 and C10 will show the amount of interest earned during each interest period.

e) D9 and D10 will show the balance of the investment at the end of each interest period.

2. Give complete explanations for the formulas in cells A9 to D10.

3. After the formulas in row 10 have been filled down to a suitable row number (50 or so), your spreadsheet should display the following values.

	A	B	C	D
1	**Investing a Fixed Sum of Money**			
2				
3		Principal	$5,000	
4		Annual Interest Rate	10%	
5		Compounding Periods Per Year	2	
6				
7				
8	**Year**	**Opening Balance**	**Interest**	**Closing Balance**
9	1/2	$5,000.00	$250.00	$5,250.00
10	1	$5,250.00	$262.50	$5,512.50
11	1 1/2	$5,512.50	$275.63	$5,788.13
12	2	$5,788.13	$289.41	$6,077.53

4. Test your spreadsheet by calculating the first few rows manually.

5. Using your spreadsheet, calculate the final balance if $7500 were invested at 8% per year, compounded semi-annually, for 6 years.

6. Use your spreadsheet to find how long it will take \$12 000 invested at 7.2% per year, compounded quarterly, to grow to \$15 000.
7. Save your spreadsheet with a suitable name. You will use it in *Investigation 2* and in the exercises at the end of this Tutorial.

In *Investigation 2*, you will create a *template*. A template is a pattern used to create other similar spreadsheets. For example, a sales report could be created as a spreadsheet, saved as a template, and this template is then used to create monthly sales reports. Templates can contain text and graphics, formatting and formulas, and data can be inserted as appropriate.

Investigation 2 Investing Regularly

You want to add a fixed sum to your investment every compounding period, so you must modify your spreadsheet to account for the regular additional payments. For this Investigation, assume that you want to invest \$1200 at the end of each compounding period.

1. At the computer, open your spreadsheet from *Investigation 1*. Change the title to *Investing Regularly* and save it with a different name.

2. Adjust the spreadsheet to allow a user to enter the additional payment as a variable. Your spreadsheet should look like this and have the features described below.

UTILITY

If you need help to format a spreadsheet, refer to Utility 31.

	A	B	C	D	E
1	**Investing Regularly**				
2					
3		Principal	$5,000		
4		Annual Interest Rate	10%		
5		Compounding Periods Per Year	2		
6		Additional Regular Payment	$1,200		
7					
8	**Year**	**Opening Balance**	**Interest**	**Payment**	**Closing Balance**
9	=1/$C$5	=C3	=B9*$C$4/$C$5	=$C$6	=B9+C9+D9
10	=A9+1/$C$5	=E9	=B10*$C$4/$C$5	=$C$6	=B10+C10+D10

a) A new column D has been inserted.

b) C6 has the value \$1200.

c) D9 and D10 will display the value of the regular payment.

d) E9 and E10 will display the total of the opening balance, the earned interest, and the regular payment.

3. After the formulas in row 10 have been filled down to a suitable row number (50 or so), your spreadsheet should display the following values.

	A	B	C	D	E
1	**Investing Regularly**				
2					
3		Principal	$5,000		
4		Annual Interest Rate	10%		
5		Compounding Periods Per Year	2		
6		Additional Regular Payment	$1,200		
7					
8	**Year**	**Opening Balance**	**Interest**	**Payment**	**Closing Balance**
9	1/2	$5,000.00	$250.00	$1,200.00	$6,450.00
10	1	$6,450.00	$322.50	$1,200.00	$7,972.50

4. Test your spreadsheet by calculating the first few rows manually.

5. Use your spreadsheet to calculate the final balance if $1500 were invested at 8% per year, compounded semi-annually, with additional investments of $1000 at the end of every six months for 5 years.

6. Use your spreadsheet to find how long it will take to save $35 000, if $2500 were invested at 7.2% per year, compounded quarterly, followed by an additional $400 at the end of each 3-month period.

7. Save your spreadsheet as a template called INVESTING REGULARLY.

The TVM Solver of a TI-83 calculator can be used to solve investment problems.

UTILITY

For detailed instructions on how to use the TVM Solver see Utility 22.

Example **Investing frequently**

A financial institution offers an annual interest rate of 6%, compounded monthly.

Compare $1200 invested at the end of each year to $100 invested at the end of every month.

SOLUTION

Choose a period of 2 years over which to compare the investment options.

Option 1: Invest $1200 at the end of each year.

Use the TVM Solver of the TI-83.

The number of deposits is 2, so N = 2.

The annual interest rate is 6%, so I% = 6.

The present value is the opening balance, so PV = 0.

The payment is a deposit of \$1200, so PMT = −1200.

The number of payments per year is 1, so P/Y = 1.

The number of compounding periods per year is 12, so C/Y = 12.

Enter these values into the calculator (below left) and solve for the future value, FV. This is \$2474.01.

Option 2: Invest \$100 at the end of every month.

Use the TVM Solver. Change the values of N, PMT, and P/Y to show the new investment (above right).

Deposits are made at the end of each month for 2 years, so the number of payments, N, is 24.

Each monthly deposit is \$100. So PMT = −100.

Deposits are made monthly, so P/Y = 12.

Solve for the future value, FV. This investment would result in \$2543.20 at the end of 2 years.

Since Option 2 earns more, it is the better choice for an investor, providing money is available for deposit each month.

DISCUSSING THE IDEAS

1. When you create a spreadsheet, what are the advantages of naming cells? What are the disadvantages?

2. Discuss the advantages and disadvantages of using a spreadsheet rather than a calculator. When is it better to use a spreadsheet? When is it better to use a calculator?

3. In the *Example*, explain why the value of PMT is negative. Under what circumstances would the value of PMT be positive?

EXERCISES

Checking Your Skills

1. Describe fully, a situation for which each of the following calculator screens would be a solution.

a)

```
N=24
I%=8
PV=0
PMT=-275
▪FV=7122.552807■
P/Y=12
C/Y=2
PMT:END BEGIN
```

b)

```
N=24
I%=8
PV=-5000
▪PMT=226.1364573
FV=0■
P/Y=12
C/Y=12
PMT:END BEGIN
```

2. Use the spreadsheet you created in *Investigation 1* or the TVM Solver to solve the following.

a) Consider a $5000 investment at 10%, compounded annually. How long does it take the investment to double? Does this follow the Rule of 72?

b) Suppose $12 500 is invested. Does the amount of principal affect the doubling period?

c) How is the doubling period affected if the interest is compounded monthly?

3. Shaina wishes to invest $2000 given by her grandfather. She has an option of a guaranteed investment certificate earning 8.75%, compounded quarterly, or a savings bond earning 9%, compounded semi-annually. Which investment should she choose? If each investment term is 5 years, what will be the difference in their values at the end of the term?

4. Develop a template for the following invoice that calculates the totals and requires the operator to input a minimum number of entries.

Acme Auto Parts

Parts					Labour	
Item #	**Auto Parts**	**Quantity**	**Unit Price**	**Total**	O/H Front Brakes (0.9 h @ $53.00/h)	47.70
1	Brake Pads	1	26.34	26.34		
2	Wheel Seals	2	5.25	10.50	Machined and Replaced Rotor (flat rate)	10.00
3	Rotor	1	30.16	30.16		
		Total Parts		$67.00	Total Labour	57.70

Total Parts	67.00
Total Labour	57.70
PST on Parts (8%)	5.36
GST (7%)	8.73
TOTAL	**$138.79**

5. Suppose an uncle has left you \$100 000 to invest, on condition that you give one-half of all interest earned to your local animal shelter. You decide to invest it in a savings account that pays 5% interest, compounded quarterly.

a) Create a spreadsheet to show the amount given to the shelter, and the resulting balance for each interest period.

b) How much will you have given to the animal shelter after a 10-year period?

6. Imagine that you have just won \$500 000 in a contest. You invest it at 12%, compounded semi-annually. You decide to live off the investment. Determine how much money you can withdraw each compounding period if you want the money to last 50 years.

7. Refer to exercise 6. Set up a spreadsheet to show the annual amount remaining in the investment.

a) How much money will you have left after 10 years? After 20 years?

b) Modify your spreadsheet to spend all the money within 10 years. How much money can you withdraw at the end of each 6-month period?

c) Suppose you want to leave \$250 000 from the winnings to your children in 50 years' time. How much will you be able to withdraw in each 6-month period?

Extending Your Thinking

8. Refer to exercise 6. Suggest a way of investing \$1 000 000 that would allow you to withdraw at the end of each year a payment that will increase by 3% per year with a balance of \$0 at the end of 50 years. Assume that your investment can earn 7.5% per year, compounded annually. How much will you withdraw at the end of the first year?

PROJECT

In this Tutorial, you learned to create a spreadsheet template. The skills and concepts relate to the Chapter Project, and to these projects in the Project Book.

- To Lease or Purchase?
- Landscape Gardening
- Investing in the Stock Market
- House Building Costs

COMMUNICATING THE IDEAS

Write a note to a friend explaining why, when investing, he is better off investing small amounts monthly, rather than a larger amount annually. Use examples to illustrate your explanation.

Investing for the Future

If you invest money, the interest you receive on your investment is taxed each year. However, if you qualify and invest in a registered retirement savings plan (RRSP), you can defer the tax that you must pay on both the principal and any interest earned.

In *Investigation 1*, you will compare investing in an RRSP to investing in an unregistered plan.

Investigation 1 The Power of RRSPs

Suppose you make a one-time investment of $5000 at 6%, compounded annually. Your marginal tax rate is 45%. Would it be better to invest it in an RRSP (along with the RRSP tax rebate) or in a non-registered account?

STUDENT REFERENCE

If you need help to understand marginal tax rates, look in the Student Reference.

	A	B	C	D	E	F	G	H	I	J	K	L	M
1					Comparing RRSP's with Non-Registered Investments								
2													
3			Principal	$5,000									
4			Annual Interest Rate	6%									
5			Marginal Tax Rate	45%									
6													
7				Registered Plan							Non-Registered Plan		
8	Year	Opening Balance	RSP Purchase	Subtotal	Interest	Closing Balance	Tax Rebate		Year	Opening Balance	Interest	Income Tax Payable	Closing Balance
9	1	$0	$5,000						1	$5,000			

1. Set up the spreadsheet as shown. Enter Principal, Annual Interest Rate, and Marginal Tax Rate in cells C3 to C5.
2. For the Registered Plan, the opening balance in Year 1 is $0 and the RSP purchase is $5000. Interest is calculated on the amount in the subtotal column (the total of the opening balance and the RSP purchase). When you invest in an RRSP, you will receive a tax rebate equal to the marginal tax rate times the amount of the RSP purchase. Enter appropriate formulas in cells D9 through to G9.
3. The tax rebate is invested at the beginning of the next tax-year. Therefore, the RSP purchase for Year 2 equals the tax rebate from Year 1.
4. For the non-registered plan, your opening balance is the principal. At the end of the year, you must pay income tax at your marginal tax rate on the interest received for the year, so your closing balance is reduced by the tax you must pay. Test your formulas to ensure that you obtain the following results.

	A	B	C	D	E	F	G	H	I	J	K	L	M
1				Comparing RRSP's with Non-Registered Investments									
2													
3			Principal	$5,000									
4			Annual Interest Rate	6%									
5			Marginal Tax Rate	45%									
6													
7			Registered Plan								Non-Registered Plan		
8	Year	Opening Balance	RSP Purchase	Subtotal	Interest	Closing Balance	Tax Rebate		Year	Opening Balance	Interest	Income Tax Payable	Closing Balance
9	1	$0	$5,000	$5,000	$300	$5,300	$2,250		1	$5,000	$300	$135	$5,165
10	2	$5,300	$2,250	$7,550	$453	$8,003	$1,013		2	$5,165	$310	$139	$5,335

5. Determine the formulas for row 10. Which formulas are the same as in row 9? Which ones are different?

6. Fill down the rows until you have a total of 30 years. When you withdraw any money from an RRSP, you must pay tax on the amount withdrawn. Calculate the tax payable on the closing balance after 30 years at the marginal tax rate of 45%. Calculate your net return. The net return does not include the initial investment.

7. Use the formula $\frac{\text{Net Return}}{\text{Initial Investment} \times \text{Number of Years}} \times 100\%$ to determine the average annual rate of return for each investment. How much better is the RRSP? Save your spreadsheet using a suitable name.

8. Many people borrow money to make their RRSP contribution, then use the tax rebate to pay off the loan. Calculate the net return after 30 years for an investor who borrowed the $5000 for the RSP purchase in Year 1 and used the tax rebate to repay the loan.

In *Investigation 2*, you determine the regular savings needed to achieve a financial goal.

Investigation 2 So, You Want to Be a Millionaire!

You can create a spreadsheet, or use the TVM Solver of a TI-83 calculator, to enter the results in a table.

	A	B	C	D	E	F	G	H	I
1	So, You Want to Be a Millionaire!								
2									
3		Annual Interest Rate	8%						
4		Periods Per Year	4						
5		Retirement Age	55						
6		Retirement Goal	$1,000,000						
7									
8	Present Age	Years to Retirement	Investment Per Period	Annual Investment		Total Investment		Monthly Investment	Daily Investment
9	18	37	-$1,127.24	-$4,508.97		-$166,831.91		-$375.75	-$12.35
10	20	35							
11	25	30							
12	30	25							
13	35	20							
14	40	15							
15	45	10							
16	50	5							
17	54	1							

Using a spreadsheet:

1. Create a spreadsheet similar to the one on page 161, or copy it as a table.

2. In cell B9 enter the formula "= $C$5–A9".

3. The function PMT in Excel calculates the periodic payment. It is similar to the PMT function in the TVM Solver in a TI-83 graphing calculator. You use it as follows:

   PMT(Rate, nper, PV, FV, type)
   where Rate = interest rate per interest period
   nper = total number of payments
   PV = present value
   FV = future value
   type = 0 or 1, where 0 (or omitted) means payment at the end of the period and 1 means payment at the beginning of the period.

   To calculate the Investment Per Period in cell C9, enter the formula:
   =PMT($C$3/$C$4,B9*$C$4,0,$C$6)

   This formula calculates the periodic payment required to achieve your retirement goal. The Annual Investment in cell D9 is the Investment per period multiplied by the number of periods in a year. Enter the formula "=C9*$C$4".

4. Determine the formulas for cells F9, H9, and I9.

5. Fill down the formulas in cells B9 to I9 to calculate the investment plans for different people who start to save at later times in their lives.

6. Save your spreadsheet with a suitable name.

Using a calculator:

7. Use the TVM Solver to calculate the entries in columns B to D in the table. Make appropriate entries for N, PV, FV, P/Y and C/Y, and solve for the periodic payment, PMT. Complete the table.

8. Examine your table. Notice that the older you are before you begin to invest, the more you must invest daily to reach your $1 000 000 goal. At what age do you think this goal would no longer be attainable? What would be a realistic age for you to begin to attain this dream?

9. If you could invest your money at 12%, how might this change your ability to become a millionaire?

DISCUSSING THE IDEAS

1. The advantage of investing in an RRSP is to save money during years when your marginal tax rate is high, and then withdraw some or all of the money in years when your income and your marginal tax rate are low. Discuss how creative use of RRSPs can help an individual further his or her education, or allow a parent to stay at home while he or she has pre-school children.

2. In the non-registered plan from *Investigation 1*, income tax was paid each year by withdrawing from the investment. Many investors pay the income tax without withdrawing from the investment.
   a) Describe how you would adjust the spreadsheet to reflect this.
   b) Explain why the net return on the investment would be higher if income tax were paid this way.

3. Why do you think the federal government is willing to defer tax on money invested in RRSPs? Discuss any long-range advantages there may be for the government.

4. The federal government limits the amount that you can deposit into an RRSP each year. Research this limit and describe how some people are allowed to invest more than others. Why do you think the federal government would place these limits on RRSP contributions?

EXERCISES

Checking Your Skills

1. Use the RRSP spreadsheet or the TVM Solver of your graphing calculator. Examine the effect of investing $5000 at 10%, compounded annually, for 30 years at marginal tax rates of 25%, 39%, and 44%, respectively. Assume that each year the tax rebate from the previous year is reinvested. Compare the differences in investing in a registered plan versus a non-registered plan. Copy and complete the following table.

Investing $5000 for 30 Years

	Registered Plan				Non-Registered	
Marginal Tax Rate	Closing Balance	Tax Paid	Net Return	Average Annual Rate of Return	Closing Balance	Average Annual Rate of Return
25%						
39%						
44%						

2. Many people invest their tax rebate on July 1, rather than wait until the beginning of the next tax year. Repeat exercise 1 for an investment that earns 10%, compounded semi-annually, where the tax rebate is reinvested on July 1. Make appropriate adjustments to the spreadsheet RRSP.

3. Financial analysts encourage people to invest early. Either use your TVM Solver or set up a spreadsheet to solve the following exercises.

   a) Starting at age 18, you invest $500 at the end of each 3-month period in an RRSP, at 9%, compounded quarterly. How much money will you have after 10 years?

   b) At age 28, you stop adding money to your investment, but leave it to grow until you turn 55. How much money will you have then?

4. You postpone investing in an RRSP until you are 28.

   a) What amount of money must you invest quarterly until you reach 55 (for 27 years) to have the same amount of money as in exercise 3b?

   b) How much money must you invest quarterly if you wait until you are 33, to have the same amount of money as in exercise 3b?

5. Investigate how the age at which you start to invest in an RRSP affects the total amount you must invest.

   a) Set up a spreadsheet, or use the TVM Solver of your calculator, to calculate the closing balance each year if you invest $1000 annually from ages 20 to 55, at 8%, compounded annually.

   b) If you started to invest when you were 30, how much must you invest at the end of each year in order to accumulate the same final closing balance as in part a?

   c) If you started to invest when you were 40, how much must you invest at the end of each year in order to accumulate the same final closing balance as in part a?

   d) If you started to invest when you were 50, how much must you invest at the end of each year in order to accumulate the same final closing balance as in part a?

   e) Complete the following table.

Starting Age	20	30	40	50
Annual Investment				
Balance at 55				
Total Amount Invested				

6. Repeat exercise 5 for an annual interest rate of 12%, compounded annually.

7. Compare your results in exercises 5 and 6.

a) How does the interest rate affect the amounts required to begin investing at each age?

b) At what age do you feel it would be "too late" to ever catch up to the amount of money you would have had if you had begun to invest early?

Extending Your Thinking

8. Remember to pay yourself first! If you are able to set aside 10% of every pay cheque for savings, you will be surprised how quickly your investment grows.

a) Set up a spreadsheet to calculate your future savings. Estimate your earnings for the first year you are employed full time, and enter 10% of this amount.

b) Can you expect regular increments in salary for the first few years of employment, as you gain experience? If so, estimate the average percent salary increment and adjust your investments accordingly.

c) For years when you do not expect a regular increment, assume your salary will increase approximately 2% per year to account for inflation.

d) How much money would you have after 10 years? After 20 years? After 30 years?

PROJECT

In this Tutorial, you learned the advantages of investing regularly and early. The skills and concepts relate to the Chapter Project, and to these projects in the Project Book.

- To Lease or Purchase?
- Investing in the Stock Market
- Landscape Gardening
- House Building Costs

COMMUNICATING THE IDEAS

You are a financial consultant. Prepare a memo for teachers of grade 12 students to use when explaining the benefits of investing at an early age, and why everyone should maximize RRSP contributions.

TUTORIAL 4.3

Investment Portfolios

Each type of investment has a different level of risk along with a different expected return. With "safe" investments, you have a better chance of preserving your original capital, but the rate of return will be low. Investments that offer a higher rate of return involve a greater risk of losing some or all of your money.

Investments fall under three main categories: cash investments, fixed income investments, and equity investments.

Cash investments include bank accounts, term deposits, and money market mutual funds. They are short-term investments that are easily and quickly accessible ("liquid") as the need arises. These investments are the safest, but produce the lowest returns. They are classified as "safety" investments.

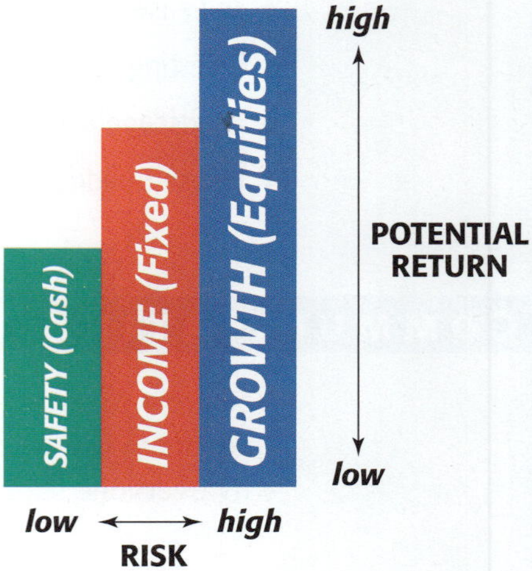

Fixed-income investments include bonds, debentures, and guaranteed investment certificates. They provide a source of regular income with limited risk, but may only be cashed in at certain times without paying an interest penalty, or losing interest. They are classified as "income" investments.

Equity investments, such as stocks and mutual funds, potentially offer a greater return on your money, but are riskier. They should be considered long-term investments, because it may take several years to realize high returns. They are classified as "growth" investments.

Your investment portfolio will be structured differently at various stages of your life. There are four main life stages in investment, as outlined below.

	Early Years 18–35	Establishment Years 35–50	Pre-Retirement Years 50–60	Retirement Years 65 +
Goals & Investment Strategy	Getting Started: Growth to Balanced	Build & Invest: Growth	Consolidate for Future: Growth to Balanced	Income & Security Safety of Principal
Investment Life Stage	• Heavy financial outlays: cash is tight • Minimal investments • Short-term financial goals • Need to avoid risk of loss • Maximize RRSPs	• More disposable income • Invest to gain maximum growth • Diversify investments, increase growth opportunity	• Minimal or no debts • Lower living expenses, mortgage paid, family grown • Able to invest more • Begin to take less risk as you plan for retirement	• Investment portfolio now will provide income, begin to withdraw money • Need to preserve capital

One Strategy:

Suggested Asset Allocation

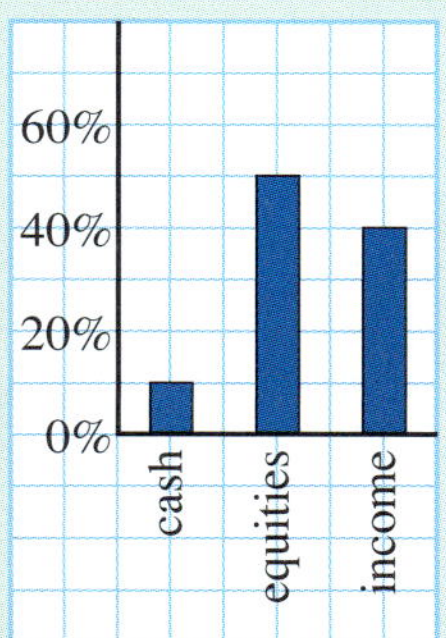

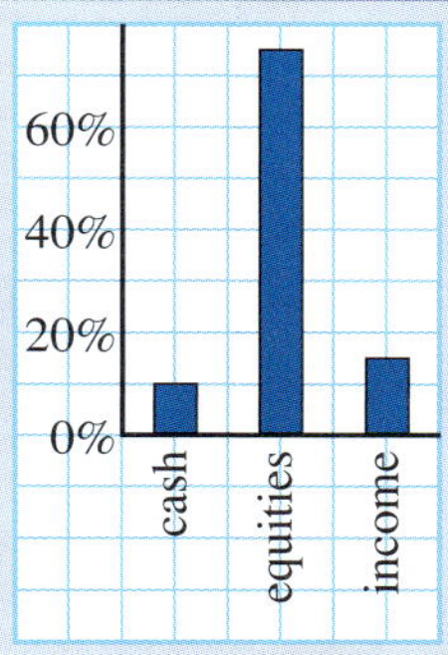

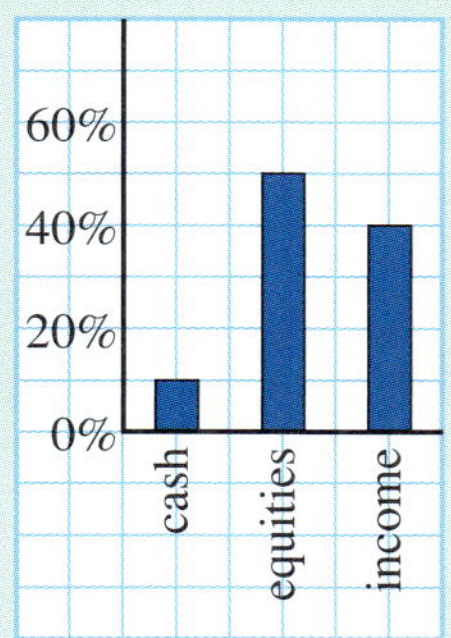

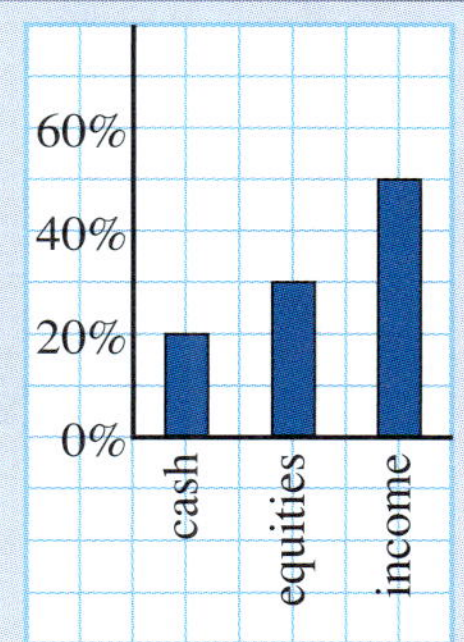

Example 1 **Allocation of resources**

Monica, a 25-year old mother, plans to invest \$2000. How much should she allocate to cash, equities, and income?

SOLUTION

A suggested amount for each type of investment in the early years is:

Cash: 10%
Income: 40%
Equities: 50%

Monica should invest as follows:

Cash: $10\% \times \$2000 = \$200$
Income: $40\% \times \$2000 = \$800$
Equities: $50\% \times \$2000 = \$1000$

When a bank advertises its rates, it refers to a "nominal rate of interest". For example, a daily interest savings account is advertised as paying 4.5% per year. In fact, your money will earn more than this. What is the actual or *effective* rate of interest? Use the TVM Solver to find the effective rate of interest. First, calculate the accumulated amount for an amount ($1 is simplest) invested at 4.5%, compounded daily, shown in the screen below left.

```
 N=1
 I%=4.5
 PV=-1
 PMT=0
▪FV=1.046024958
 P/Y=1
 C/Y=365
 PMT:END BEGIN
```

```
 N=1
▪I%=4.60249585
 PV=-1
 PMT=0
 FV=1.046024958
 P/Y=1
 C/Y=1
 PMT:END BEGIN
```

To determine the effective annual interest rate, we must find the value of I, which will leave PV and FV unchanged, but with annual compounding. Change C/Y as shown in the screen above right, then move the cursor to I, and press ALPHA ENTER. The effective annual interest rate is about 4.602%.

Example 2

Rates of return

A 40-year-old computer analyst has the following investment portfolio.

Investment	Amount	Average Rate of Return
Bank account	$1500	2%
Term deposit	$2500	$5\frac{1}{4}$%
Guaranteed investment certificate	$5000	$7\frac{1}{2}$%
Guaranteed investment certificate	$4000	8%
Mutual fund	$12 000	$14\frac{3}{4}$%

Calculate her total annual return and her average rate of return.

SOLUTION

Investment	Amount	Average Rate of Return	Return
Bank account	$1500	2%	$30.00
Term deposit	$2500	5.25%	$131.25
Guaranteed investment certificate	$5000	7.5%	$375.00
Guaranteed investment certificate	$4000	8%	$320.00
Mutual fund	$12 000	14.75%	$1770.00
TOTALS	$25 000		$2626.25

To calculate the average rate of return, use the TVM Solver. The present value is –$25 000 and the future value is $25 000 + $2626.25 or $27 626.25.

Her total annual return is $2626.25.
The average rate of return is 10.505%.

Many people who want a high return on investment and are willing to accept risk choose to deal in stocks of companies. Stocks represent shares of ownership and are traded at a stock exchange. Many exchanges use a stock market index to provide information to investors about the performance of groups of stocks. Investors may use the movements of indices as a way to assess market trends. The Toronto Stock Exchange 300 Composite Index (TSE 300) was introduced in 1977. It is a weighted average of the stock prices of 300 of the largest traded companies. The base year for all stock price indices is 1975, when the base value was set at 1000. Media comments on the performance of the stock market in Canada usually refer to the TSE 300.

Example 3 **Analyze an investment portfolio.**

An investment of $10 000 is held from December 31, 1995, to December 31, 1998. Canadian stocks that track the TSE 300 index are 75% of the portfolio. The remaining 25% of the portfolio is invested in guaranteed investment certificates (GICs) that pay 6.5%, compounded annually. On December 31, 1995, the TSE 300 index was at 4714. On December 31, 1998, the TSE 300 index was at 6486.

a) What was the value of the total portfolio on December 31, 1998?

b) What average annual rate of return does the value in part a represent?

SOLUTION

a) The value of the stock on December 31, 1995
= 75% of \$10 000
= \$7500

The value of the GIC on December 31, 1995
= \$10 000 – \$7500
= \$2500

The TSE 300 index increased from 4714 to 6486 during the investment period, so the value of the stock increased in the ratio of these values.

Value of stocks on December 31, 1998 $= \$7500 \times \frac{6486}{4714}$
$\doteq \$10\ 319.26$

The GICs earn compound interest. Use the compound interest formula, $A = P(1 + i)^n$, where $P = 2500$, $I = 0.065$ and $n = 3$.

$A = 2500(1 + 0.065)^3$
$\doteq 3019.87$

The value of the GICs on December 31, 1998 is about \$3019.87.

The total value of the portfolio on December 31, 1998 is about \$10 319.26 + \$3019.87, or \$13 339.13.

b) Use the TVM Solver.

The average annual rate of return is about 10.08%.

DISCUSSING THE IDEAS

1. Examine the table of investment strategies for various life stages. Why is the range 35 to 50 years the stage at which one should invest the most in risk-taking investments?

2. Why would a person want to invest money in a savings account, which typically pays a lower rate of interest than other investments?

3. In *Example 3*, why would it be incorrect to increase the value of the stocks by the difference of the indices, 6484 – 4714?

EXERCISES

Checking Your Skills

1. A bill for a credit card indicates an Annual Interest Rate of 18.5% and a Daily Interest Rate of 0.05054%. What is the effective rate?

2. Find the effective annual rate for each of the following.

a) 12% per year, compounded quarterly

b) 5.6% per year, compounded monthly

c) 4.8% per year, compounded daily

3. Travis, a 30-year-old contractor, has the following investment portfolio.

Investment	Amount Invested ($)	Average Rate of Return (%)	Amount of Return ($)
Savings account	2500	2.5	
Canada Savings Bonds	3600	6.0	
Guaranteed investment certificates	6300	6.75	
Mutual funds	9600	11.25	
TOTALS			

a) Ignore taxes. Complete the table. Calculate the average annual rate of return.

b) This year, Travis wants to invest $2000. Examine his portfolio and advise him on the type of investment he should consider.

4. A 55-year-old electrician has these investments.

Investment	Amount Invested	Average Rate of Return	Amount of Return ($)
Savings account	$1800	2.5%	
Money market funds	$7800	7.25%	
Canada Savings Bonds	$4200	6.5%	
Term deposits	$5600	7.0%	
Guaranteed investment certificates	$5500	8.75%	
Stock in a computer company	$3400	22.5%	
Mutual funds	$58 000	15.75%	
TOTALS			

a) Ignore taxes. Complete the table. Calculate the average annual rate of return.

b) Explain why the average annual rate of return is not simply the average of the "Rates of Return"?

c) This year, the electrician wants to invest $5000. Examine her portfolio and suggest the type of investment she should consider.

d) What percent of the total investment is in each type of investment? Should the electrician consider reallocating some investments into other types?

5. For a particular 5-year period, a mutual fund experienced the growth rates shown below. Jean invested $10 000 in 1991 in his RRSP.

Year	Opening Balance	Rate of Growth (%)	Closing Balance
1991	$10 000	8.2	
1992		11.7	
1993		14.6	
1994		(2.9)	
1995		33.1	

a) Copy the table.

b) Calculate the total value of his investment for each of the 5 years. What is his total return on the investment after 5 years? (Note: The use of brackets in 1994 indicates that the fund lost money.)

6. On her 25th birthday, Annette invested $5000 in an RRSP GIC that pays 6%, compounded semi-annually. On her 30th birthday, she placed $12 000 in an RRSP GIC that pays 8.5%, compounded annually. On her 35th birthday, she invested $8000 in an RRSP GIC that pays 7%, compounded semi-annually.

a) What will her GICs be worth when she is 65, if she keeps them invested at the rates given until her 65th birthday?

b) What is the average annual rate of return over the last 30 years of the investment?

7. Brian opened an RRSP account at his credit union on June 1, 1996, with a deposit of $2000. He added $1900 on December 1, 1997, and another $1700 on June 1, 1998. The account earns a fixed interest rate of 4%, compounded semi-annually.

a) What amount will be in his account on December 1, 2002?

b) What is the average annual rate of return over the period of the investment to December 1, 2002?

Extending Your Thinking

8. Phan deposits $3000 to open a savings account that pays 4.2% interest, compounded monthly. This account pays 4.5%, compounded monthly, for balances over $5000. At the end of 7 months, he deposits a further $3000 in the account. Three months later, he withdraws $1500 to pay some bills.

a) How much is in his account 1 year after he opened it?

b) Phan makes no further deposits to or withdrawals from the account for the next 2 years. What is his average annual rate of return if you ignore taxes?

PROJECT

In this Tutorial, you studied investment portfolios. The skills and concepts relate to the Chapter Project, and to these projects in the Project Book.

- To Lease or Purchase?
- Investing in the Stock Market

COMMUNICATING THE IDEAS

Prepare a memo for a financial advisor to give to a young client that outlines a long-term investment strategy.

TUTORIAL 4.4

Leasing

As you enter the regular workforce, you may be faced with many financial decisions. Some jobs may only be open to you if you have a fairly new vehicle. One decision you may have to make is whether to lease or buy a car. Other expensive items, such as computers, can also be leased. If you own a business, you may also consider leasing the equipment necessary to operate the business.

When you lease an item, you make monthly payments to cover the depreciation of the item over the duration of the lease, taxes on the amount of the depreciation, and interest on the outstanding balance of the full purchase price.

The price of a lease depends on three factors: the *residual value* (or lease-end value), the interest rate offered for the lease, and the length of the lease.

Some leases require a down payment. When you lease, you pay the difference between the current price of the item and the *residual value*. Lease payments are made at the beginning of the month. If you make a down payment, it is subtracted from the total amount to be paid, so monthly payments will be lower. A down payment saves on some of the interest owed.

Interest rates on some leases may be greater than the rate you would pay if you were to purchase the item and take out a loan at a bank. In this case, the total cost to lease would exceed the cost to purchase.

Investigation 1

Leases

If you have ever looked at advertisements for cars, you may have noticed that some vehicles seem to offer much better leasing prices than others. You will now investigate some of the reasons behind these offers.

The following table illustrates typical residual values for several types of vehicles, expressed as a percent of the original price.

	Length of Lease (months)				
Type of Vehicle	**12**	**24**	**36**	**48**	**60**
Luxury	67%	57%	51%	41%	35%
Sporty	55%	49%	42%	36%	32%
Economy	57%	47%	41%	31%	26%
Truck 1/4 ton	59%	51%	46%	38%	35%
Sport Utility Vehicle	70%	64%	60%	52%	48%

1. Copy and complete this table for a sporty car with a purchase price \$22 000.

	Length of Lease (months)				
	12	**24**	**36**	**48**	**60**
Residual Values (\$)					
Depreciation (\$)					
Average Annual Cost (before interest and taxes)					

2. Examine the entries in the table. Explain why a long-term lease is more cost effective than a short-term lease.

3. One car dealership offers a 24-month lease for \$395 each month, with a down payment of \$2350 and a residual value of \$8500. The advertised annual interest rate is 10.9%, compounded monthly.

a) Use the TVM Solver to determine the present value of the lease.

b) What is the total price charged for the car?

c) How much has been paid in total at lease end?

d) Suppose the car is then purchased. How much has been paid for the car without taxes?

Many people, who shop for a new vehicle, are faced with the dilemma of whether to lease or to buy. Various circumstances may dictate which option is preferable. In *Investigation 2*, you will use a spreadsheet to examine the financial implications of each option.

Investigation 2 To Lease or to Buy?

Open the LEASEBUY template provided by your teacher.

	A	B	C	D	E
1	**Purchasing vs. Leasing an Automobile**				
2					
3	Make and Model of Vehicle				
4					
5	Purchase Price	*\$18,500.00*			
6	Down Payment	*\$0.00*			
7	Finance Rate	*1.90%*			
8	Provincial Sales Tax Rate	*7%*			
9					
10					
11	**Leasing Information**			**Purchasing Information**	
12	Purchase Option Price	*\$9,805.00*		Purchase Price	\$18,500.00
13	Lease Term (months)	*48*		GST	\$1,295.00
14	Base Monthly Lease Payment	\$203.46		PST	\$1,295.00
15	Monthly Payment (includes taxes)	\$231.95		**Total Purchase Price**	\$21,090.00

1. The values that are inputted into the spreadsheet include: Make and Model of Vehicle (B3); Purchase Price (B5); Down Payment ($0, if none) (B6); Finance Rate (for all 3 sections) (B7), (E18), and (B31); Purchase Option Price (the residual value) (B12); Lease Term (B13); the Filing Fee (B16); the Option Finance term (B30); and Provincial Sales Tax rate (B8). Determine and enter formulas for the appropriate cells.

For a lease, payments are due at the beginning of the month. Recall that the PMT function has the following format: =PMT(Rate per period, Number of periods, Present value (negative if you are calculating loan payments), Future value, Type (0 or1))

a) To calculate the lease costs:
- Base Monthly Lease Payment: use PMT function
- Monthly Payment (Base monthly lease payment + PST + GST)
- Refundable Security Deposit: The security deposit is equal to one monthly payment, rounded up to the nearest $25. To calculate this, use the function "=CEILING(B15,25)".
- Total Due upon Signing: the security deposit, the first month's lease payment, and the filing fee.
- Total Lease Cost (less security deposit): the total of all lease payments plus the filing fee.

b) To calculate the cost to purchase the vehicle:
- Purchase Price: copied from the entered information
- GST
- PST
- Total Purchase Price
- Less: Down Payment: copied from the entered information
- Amount to Finance
- Filing Fee
- Finance Term (months): either enter this value or copy from the term for the lease
- Monthly Payment: use PMT function
- Total Cost

c) To calculate the cost to purchase at the end of the lease:
- Purchase Price: the residual value of the vehicle: copy from the Purchase Option Price
- Taxes on Purchase Option Price
- Total Cost to Purchase
- Total Cost of Vehicle
- Option Finance Term (months)
- Finance Rate
- Monthly Payment: use PMT function
- Total Cost of Vehicle: cost of lease plus cost to purchase at end of lease

2. Use your spreadsheet to determine the following.

a) The total leasing cost

b) The total cost of the vehicle if purchased at the end of the lease

c) The purchase cost if you purchase the vehicle through financing at the beginning under these conditions:
Purchase Price: $18 500, Down Payment: $0; Finance Rate for all situations: 1.90%, Provincial Sales Tax: 7%, Purchase Option Price: $9805; Lease Term: 36 months; Purchase Finance Term, and Purchase Finance Option Term both 36 months.

3. a) Refer to exercise 2, part c. What is the difference between the monthly cost to lease and the monthly cost to purchase?

b) What is the total cost difference if you purchase the car through financing to begin with or purchase the car through financing at the end of the lease?

4. a) Refer to exercise 2, part c. What is the cost of financing if you purchase the vehicle?

b) What is the cost of financing if you lease the vehicle, and then purchase it with financing at the end of the lease?

5. Save your spreadsheet with a suitable name.

Example **Lease-end value**

Elvin sees an advertisement for a car he is interested in leasing. Its purchase price is $16 750. The monthly lease cost is $229 plus taxes, for 48 months. The finance rate is 3.7%, compounded monthly. Elvin wants to know the purchase price of the car at the end of the lease, but this information is not given. Calculate the approximate purchase option price.

SOLUTION

The purchase option price is approximately the value of the car at the end of the lease.

Use the TVM Solver. Enter the values shown in the screen.

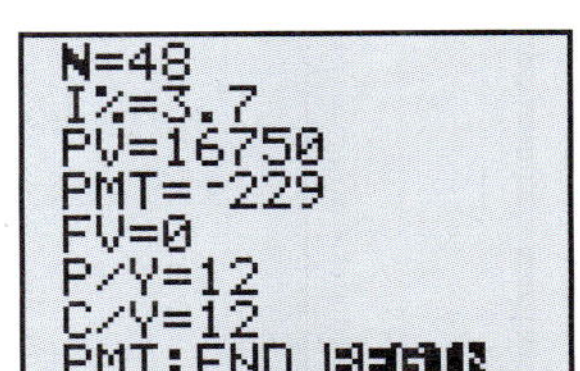

Note that the value of PV is positive (Elvin receives this amount) while the value of PMT is negative (Elvin pays this amount).

Solve for FV as shown to the right.

The future value, and therefore the approximate purchase option price, is $7553.49. This value appears as a negative number, because it represents a payment that Elvin is going to have to make.

Discussing the Ideas

1. Discuss why people lease vehicles. Make a class list of these reasons.

2. What other items might a consumer lease?

3. Name items that might be leased by a business.

4. What are the advantages of leasing? What are the disadvantages?

Exercises

Checking Your Skills

1. A new luxury car has a purchase price of $24 500. It can be leased for 36 months with no down payment for $345 per month, plus taxes. After the 36 months, the car has a purchase option price of $14 500 plus taxes. GST and PST are both 7%.

a) Use the TVM Solver to find the finance rate.

b) Compare the purchase option price with the residual value of the car using the table from *Investigation 1*.

c) What down payment would you need to make in order for the purchase option price to be equal to the residual value of the car?

2. A new car has a purchase price of $25 000. The finance rate is 2.9%, compounded monthly. It can be leased for 24 months, with no down payment and a purchase option price of 49% of the purchase price. Use the spreadsheet you created in *Investigation 2*. Assume that PST is 8% and GST is 7%.

a) Compare the total cost of the car if purchased and financed over a 5-year period, to leasing for 2 years then purchasing over a 3-year period.

b) Compare the monthly lease payment to the monthly loan payment if the vehicle is purchased. Why is the lease payment larger?

c) For this vehicle, the residual value for a 2-year lease is 42%. Change your spreadsheet to calculate values for a 36-month lease, followed by the purchase financed over 2 years. What are the monthly lease payments and the monthly loan payments when the vehicle is purchased?

d) Compare the total cost of purchasing the car to leasing for 3 years and then purchasing. Explain which option will be cheaper and why.

3. Vanessa wants to lease a new $45 000 luxury car for 3 years. She can also lease a one-year-old or a two-year-old car. The residual values of the car are the same as shown in the Residual Values table from *Investigation 1*. GST and PST are both 7%.

a) Use your Residual Values table to determine the purchase price of both the one-year-old and the two-year-old cars.

b) The interest rate for the lease is 2.9%. Calculate the monthly payments for each of the 3 vehicles, then the total cost of leasing each vehicle. How much less are the monthly payments for one-year-old and two-year-old cars compared to new cars? How much do you save over the cost of the lease?

4. Many drivers do not wish to drive a car that is more than four years old. Assume that rates and purchase prices remain the same. Would Vanessa, from exercise 3, be better off leasing 3 new vehicles over a 12-year period (4-year leases), or 4 one-year-old vehicles (3-year leases)?

5. Tyler wants to lease either a $\frac{1}{2}$-t truck or a sport utility vehicle for 3 years. The purchase price of each vehicle is $38 000. The residual value of the truck is 53%, while it is 60% for the sport utility vehicle. Assume that GST and PST are both 7%.

a) If the interest rate is 3.9%, compare the lease costs of each vehicle. Which one should Tyler lease? Explain why.

b) Tyler wants to purchase the vehicle at the end of the lease, financing it at 4.9% for 3 years. Compare the total cost for each vehicle. Which is less expensive?

c) If Tyler were to purchase the vehicle over 5 years at 3.9%, compounded monthly, what would be the monthly payment? What would be the total cost?

6. A car lease runs for 36 months at $305 per month, taxes included, with a down payment of $1105, a lease-end value of $7105, and an interest rate of 11.6%, compounded monthly. Maintenance is the purchaser's responsibility. Set up a spreadsheet to include the monthly values of the opening balance, interest paid, lease payment, and closing balance. Use the spreadsheet to solve the following exercises. Assume GST and PST are 7%.

a) What part of the $305 is used to pay the interest on the $7105?

b) What total price is being charged for the car?

c) What is the change in the monthly lease payment, if the lease-end value is reduced to $5700?

d) What is the monthly payment for a straight purchase over 36 months with a 20% down payment?

e) What is the annual percentage depreciation rate assumed with the $7105 lease-end value?

7. A financial planner wants a new computer. The purchase price is $3294. Financing is available at 18.99%, compounded monthly. Assume GST and PST are both 7%.

a) Calculate the monthly payment to purchase this computer over 48 months. What is the total amount paid for the computer? What are the payments and the total paid if the planner finances it for 36 months? For 24 months?

b) The same computer can be leased. Lease rates, before taxes, are as follows: 24 months for $182.88/month, 36 months for $132.32/month, and 48 months for $106.92/month. All lease rates include taxes. Calculate the total paid for each of the lease terms.

c) For each of the three leasing/purchasing periods described above, compare the total paid to lease the computer to the total paid to finance it. Comment on the cost of leasing versus buying.

8. Nina is entering university and needs a computer. She sees a computer advertised for $1399. Alternatively, she can lease the same computer for 36 months for $52.98 a month, including taxes.

a) Calculate the total purchase price of the computer.

b) The computer company offers to finance purchases over 36 months at 18.99%, compounded monthly. Calculate the monthly payments using the TVM Solver.

c) If the computer has no residual value at lease end, calculate the interest rate charged on the lease. Calculate the total payment for the lease.

d) Compare the cost to lease to the cost to purchase the computer. Is one option better than the other? Under what circumstances might Nina decide to lease?

Extending Your Thinking

9. For many people, one determining factor in the decision to lease or buy is income tax. If you use a piece of equipment for business purposes, you can claim an income tax deduction. If you own the equipment, you can depreciate it at a certain rate every year and claim the depreciation. However, if you lease the equipment, you can claim the total lease payments in a tax year. Complete the following tables for the computer obtained over a 3-year period, by the financial planner in exercise 7. The financial planner's marginal tax rate is 44%.

Purchase Computer				
Year	**Value at the Beginning of the Year ($)**	**Depreciation (30%)**	**Value at the End of the Year ($)**	**Tax Savings ($)**
1	3294	988.20	2305.80	434.81
2				
3				
			Total	

a) What is the total net cost of purchasing the computer?

b) What is the total net cost of leasing the computer?

c) What are the advantages of leasing for business purposes? What are the advantages of purchasing the computer?

Lease Computer		
Year	**Total Annual Payments ($)**	**Tax Savings ($)**
1	1587.84	698.65
2		
3		
Totals		

PROJECT

In this Tutorial, you learned about leasing an item. The skills and concepts relate to the Chapter Project, and to these projects in the Project Book.

- To Lease or Purchase?
- Landscape Gardening
- House Building Costs

COMMUNICATING THE IDEAS

Suppose you own a car dealership. Prepare a pamphlet that explains to potential customers what a lease is and the benefits of leasing.

Lease or Buy a Car?

The time has come to get a car. Use your knowledge of the financial differences between leasing and buying a vehicle to decide what vehicle you want to drive. Choose one vehicle described in the advertisements you collected earlier.

1. What leasing information do you already have? What leasing information do you still need?

2. If you know the monthly payments for the lease, use the TVM Solver to find the car's residual value. Does the amount quoted for the monthly payment include taxes? If so, you must subtract these taxes before you solve for the residual value. If PST = 7% and GST = 7%, the total tax is 14%, so:

$$\text{Total Monthly Payment} = \text{Regular Payment} \times 1.14$$

To find the Regular Payment, divide both sides by 1.14.

$$\text{Regular Payment} = \frac{\text{Total Monthly Payment}}{1.14}$$

3. If you do not know the regular monthly lease payments, or the residual value, you can estimate the residual value *for the purpose of this project only* using the Residual Value table from *Investigation 1* of *Tutorial 4.4*. (Residual values will vary from car to car and company to company, depending on the current market conditions.)

4. Use a spreadsheet. Open the LEASEBUY template and enter the information about the vehicle you have chosen. Analyze the information. You may adjust the term of the lease, if you wish, but be sure to adjust the residual value accordingly.

CHAPTER PROJECT

To complete this Chapter Project you must learn about purchasing or renting a home. Complete *Tutorials 4.5* and *4.6* to develop an understanding of the financial implications for each option. You will need this knowledge when you return to the Chapter Project on page 198 and for the following projects in the Project Book.

- To Lease or Purchase?
- Landscape Gardening
- Home Building Costs

5. Write a summary.
   - Describe your vehicle. Is it new or used? If possible, include a picture of the vehicle. Why did you make this choice?
   - Will you lease the vehicle or purchase it? Give reasons for your choice.
   - If you lease the vehicle, do you think you will purchase the vehicle at the end of the lease? Why or why not?
   - What are your monthly payments? With your project, include a printout of the spreadsheet analysis for the vehicle.

   Save your summary.

Career Skills

To analyze the financial implications of options may require complex calculations. The ability to use technology appropriately and to interpret results are skills highly valued by employers.

TUTORIAL 4.5

Mortgages

Very few people who buy a house can afford to pay cash for it. Most people must borrow the money. This loan usually takes the form of a mortgage for which the house acts as security. Canadian law permits interest rates for mortgages to be compounded, at most, semi-annually.

Before you start to look for a home, you should know the maximum mortgage amount for which you qualify. A lender will not lend you more money than it feels you can comfortably pay back. You must also have a down payment for your home. The standard minimum down payment is 10% of the purchase price, but some home buyers may qualify for a 5% down payment. If your down payment is less than 25% of the property value, a premium for mortgage default insurance will be added to your regular payment. The *amortization period* is the number of years it will take you to pay off the mortgage loan.

Practise Your Prior Skills

You will calculate annual and monthly income. Use these exercises to review your knowledge.

1. Samantha is a computer programmer who earns $22.40 an hour. If she works 37.5 h each week, calculate her annual income, then her average monthly income.
2. Paul is a car salesman with a base salary of $200 per week plus 1.5% commission on all sales. If his average weekly sales total $35 000, determine his average annual income.

Investigation

Which House Can You Afford?

Lars and Freya want to buy their first home. They need to know what price range they can afford. In this Investigation you will determine the maximum allowable mortgage.

1. Lars works as a waiter in an upscale restaurant and earns $7.50 an hour. He works 35 h a week. His tips average $425 per week. Freya works as a dental assistant and earns $1050 biweekly. Calculate the gross monthly income for the couple.
2. The total monthly house payment should not exceed 27% to 32% of total gross monthly income. Determine the maximum house payment that Lars and Freya should consider.

3. Lenders limit the amount you can borrow. The total monthly payments for your house and other obligations, such as car loans and credit card payments, may not exceed 37% to 40% of your gross monthly income. If the couple has an outstanding car loan with monthly payments of \$335 and credit card payments averaging \$250 a month, calculate the maximum monthly house payment Lars and Freya are allowed.

4. The amount of income available for a house payment is the lesser of your answers for exercises 2 and 3. This monthly house payment includes heat and property taxes. Suppose Lars and Freya estimate their heating costs to be \$75 per month and property taxes to be \$220 per month. Calculate the maximum amount they may spend on mortgage payments.

5. The monthly principal and interest payment for a mortgage is:

$$\text{Payment} = \frac{\text{Mortgage Amount}}{\$1000} \times \text{Cost per } \$1000$$

$$\text{or Mortgage amount} = \frac{\text{Payment} \times \$1000}{\text{Cost per } \$1000}$$

The cost per \$1000 depends on the amortization period and the interest rate. The following table illustrates a few rates and periods for \$1000.

	Amortization Periods			
Fixed Interest Rate	**10 years**	**15 years**	**20 years**	**25 years**
6.00%	\$11.065099	\$8.3988279	\$7.1218844	\$6.3980662
6.25%	\$11.187683	\$8.5307643	\$7.2628110	\$6.5474164
6.50%	\$11.310930	\$8.6636948	\$7.4050041	\$6.6982379
6.75%	\$11.434837	\$8.7976086	\$7.5484429	\$6.8504964
7.00%	\$11.559399	\$8.9324944	\$7.6931063	\$7.0041575
7.25%	\$11.684609	\$9.0683408	\$7.8389728	\$7.1591867
7.50%	\$11.810465	\$9.2051366	\$7.9860210	\$7.3155491
7.75%	\$11.936960	\$9.3428700	\$8.1342289	\$7.4732100
8.00%	\$12.064090	\$9.4815294	\$8.2835748	\$7.6321346
8.25%	\$12.26526	\$9.70140	\$8.52066	\$7.88450
8.50%	\$12.39857	\$9.84740	\$8.67823	\$8.05227

Suppose the current interest rate is 8.50%, compounded semi-annually, and Lars and Freya decide on a 25-year amortization period. Determine the maximum amount the bank will lend them.

6. Suppose Lars and Freya plan to make a 10% down payment on a house. Determine the maximum price they can afford if the bank offers a mortgage at 8.5%, compounded semi-annually, with an amortization period of 25 years.

There are several strategies that a borrower might adopt in order to reduce the total cost of the mortgage. One way to reduce this cost is to reduce the amortization period.

Unless otherwise stated, assume that all mortgages are compounded semi-annually.

Example 1 The effect of changing amortization periods

A couple has a mortgage of \$100 000 at 8%, amortized over 25 years. They make monthly payments on the mortgage.

a) How much will they save if they reduce the amortization period to 15 years?

b) How much will they save if they keep the amortization period at 25 years, but make bi-weekly payments?

SOLUTION

Use the TVM Solver to first calculate the monthly payment.

Enter the values shown, then solve for PMT. This is \$763.21.

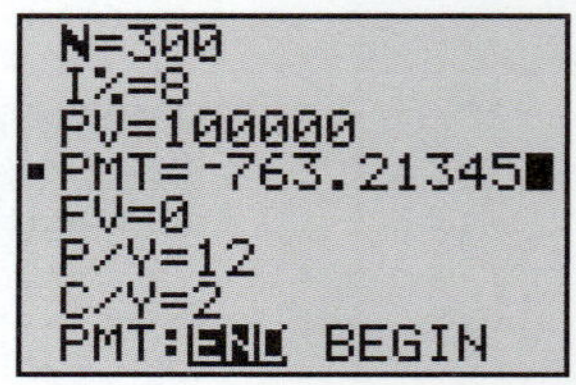

The couple will make 300 monthly payments of \$763.21.

The total amount paid for the mortgage is \$763.21 × 25 × 12, or \$228 963.

a) Repeat the calculation for an amortization period of 15 years. The couple will make 180 monthly payments of \$948.15.

The total amount paid for this mortgage is \$948.15 × 180, or \$170 667.

By reducing the amortization period from 25 years to 15 years, the couple will save \$228 963 – \$170 667, or \$58 296.

b) The couple will make 26 bi-weekly payments each year, so N becomes 25 × 26 = 650, and P/Y becomes 26.

The total amount paid for this mortgage is \$351.63 × 650, or \$228 559.50.

By paying the mortgage bi-weekly, the couple will save \$228 963 – \$228 559.50, or \$403.50.

Example 2 The outstanding balance on a mortgage

The couple in the previous example agreed to a mortgage of $100 000 at 8%, amortized over 25 years. How much will they still owe on the mortgage after 5 years?

SOLUTION

From *Example 1*, the monthly payments are $763.21. Use the TVM Solver to calculate the outstanding balance after 5 years.

Change the value of N to 60 and use the TVM Solver to calculate the value of FV, the Future Value.

The outstanding balance on the mortgage after 5 years is $92 136.01.

DISCUSSING THE IDEAS

1. Explain why there is a range for the maximum amount of mortgage allowed. Why might there be a problem for some people to pay the maximum amount?

2. Income tax takes a large part of a person's gross pay. Can you suggest a better method to determine the maximum mortgage a family could afford?

3. The balance of a mortgage shrinks slowly during the first few years. Why do you think this is?

EXERCISES

Checking Your Skills

1. Anita can afford monthly mortgage payments of $650. If interest rates are 7.5%, what size mortgage can Anita afford with each of the following amortization periods?

   a) 25 years **b)** 20 years **c)** 15 years

2. Which would costs less: a $120 000 mortgage at 8.2% over 20 years; or a $110 000 mortgage at 7.5% over 25 years? By how much?

3. Grant has \$20 000 saved for the down payment on a house. He works as a systems analyst and earns \$65 000 a year. He has a personal loan with monthly payments of \$339. He estimates that he will have property taxes of \$4000 a year and heat bills of \$60 a month.

a) Determine the maximum mortgage payment allowed.

b) If interest rates are 7%, determine the maximum mortgage for which Grant qualifies, given a 25-year amortization period.

c) Grant wants to pay off his house in less than 25 years. Calculate the maximum mortgage available to Grant for 20, 15, and 10 years, respectively.

d) Calculate the maximum price for a house Grant can consider for each of the 4 amortization periods.

4. Jessica and Daniel have a combined income of \$54 000 per year and wish to buy a house. They have a car lease for \$199 a month, and a line of credit with monthly payments of \$245. They are applying for an \$85 000 mortgage. The taxes on the house are \$250 a month and the heating bill is \$48.

a) Determine the maximum mortgage payment allowed.

b) If interest rates are currently 7.5%, what is the lowest amortization period for which Jessica and Daniel qualify?

c) The couple decides to obtain a mortgage with a 15-year amortization period. What will their monthly payment be?

d) Suggest reasons why the couple might have chosen a 15-year period, rather than a lower or higher one.

5. Victoria and Steven are purchasing a home and require a mortgage for \$100 000. They need to decide between a 15-year, a 20-year, or a 25-year amortization period.

a) Complete the following tables, if the interest rate is 6.5%.

Amortization Period (years)	Monthly Payments	Monthly Payments
15		
20		
25		

Comparing Amortization Period (years)	Total Saving
20 vs. 25	
15 vs. 25	
15 vs. 20	

b) Complete the above tables for an interest rate of 8.25%.

c) Comment on the savings for different amortization periods as interest rates increase.

6. Drew and Janine currently have a 5-year mortgage for \$125 000, at 8%, amortized over 25 years. This means that after 5 years, they will have to either re-negotiate the terms of the mortgage or pay off the mortgage.

a) How much is owed on the mortgage at the end of the 5 years? How much interest have they paid?

b) At the end of the 5 years, they re-negotiate their mortgage and obtain a 5-year mortgage at 5.75%, amortized over 15 years. Calculate their new payment. How much will they still owe at the end of the 5-year term? How much interest will they have paid during this 5-year period?

7. In the late 1970s many people purchased homes at 10.25%, with a 30-year amortization period and a 5-year term. Interest rates then rose in the early 80s, and these people had to refinance at 18.5%. Explain why a couple who purchased a home with a \$45 000 mortgage might have to sell under these circumstances. Use the TVM Solver to compare the original mortgage payment to the new one.

Extending Your Thinking

8. People generally try to pay off their mortgage as quickly as possible. Is this always the best idea? Suppose you have an extra \$2000 a year. Decide whether to use it to reduce the amount you owe on your mortgage or to invest in an RRSP that earns 6%. Assume you have a marginal tax rate of 39%. Your mortgage is amortized over 20 years, at 7.5%, and the balance you owe is \$65 000.

a) What is the total amount you will pay for your mortgage if you do not make extra payments?

b) Set up a spreadsheet to show the amortization of the mortgage over the 20 years. Adapt the spreadsheet to allow for an extra payment of \$2000 at the end of each year. In how many years will your mortgage be paid off? How much will you pay in total interest? How much will you save if you make these extra payments?

c) If you invest in an RRSP, how much can you invest each year? Remember to factor in your marginal tax rate. This exercise assumes that you are allowed to make these extra RRSP contributions. If you have already reached your maximum allowable contribution, this scheme will not be available to you.

d) Expand the spreadsheet that you developed in part b to show the growth of your RRSP. What will be the value of your RRSP the same year that your mortgage is paid off?

e) Are you better off to invest in an RRSP or to pay off your mortgage as quickly as possible?

PROJECT

In this Tutorial, you learned about mortgages. The skills and concepts relate to the Chapter Project, and to these projects in the Project Book.

- To Lease or Purchase?
- House Building Costs
- Landscape Gardening

COMMUNICATING THE IDEAS

Assume that you are a loans officer in a bank. A couple has asked your advice on a mortgage they would like to re-negotiate. Their combined annual salaries total \$95 000. Their only other debt is a monthly credit card payment of \$200. Taxes on their home are \$265 per month and heating is \$62 per month. The outstanding balance on their mortgage is \$65 000, and interest rates are currently at 7.25%. Write a letter to the couple to suggest an amortization period they should consider. Explain why you think this is their best option.

TUTORIAL 4.6

Renting versus Buying

Home ownership is not for everyone. Many different circumstances will influence one's decision to rent or buy a home.

Many costs are associated with home ownership other than the mortgage. Utilities (generally hydro, gas, and water) must be paid. Homeowner's insurance is mandatory. There are many maintenance costs that range from paint, carpeting, and landscaping to replacing items such as the roof or the furnace. Although any one of these costs will not occur every year, they must be included in a household budget.

STUDENT REFERENCE

If you need to review property taxes, look in the Student Reference.

Practise Your Prior Skills

In an earlier grade you studied budgets. Use these exercises to review your skills.

1. Grant and Laura own a home valued at $130 000.

a) For tax purposes, the assessed value is 75% of the fair market value, and the tax rate is 32.8 mills. Calculate their annual property tax bill. What is their monthly property tax payment?

b) Their estimated expenses for utilities, insurance, and maintenance are listed below. Calculate their average monthly expenses.

Item	Cost
Hydro	$105.00 a month
Gas	$48.00 a month
Water	$116.25 quarterly
Insurance	$57.25 a month
Maintenance	$900.00 annually

Investigation

Is It Better to Rent or to Buy?

This question is a complex one, for there are so many variables. In this *Investigation*, you will examine the financial considerations of this dilemma.

If you purchase a home, you must first have a down payment. It is likely that, for the first several years at least, your mortgage payments and other expenses will exceed the amount you would pay if you were to rent.

What if you were to invest the money you saved for a down payment and the annual difference between the cost of home ownership and renting?

1. Open the spreadsheet RENTBUY. Enter the values for the titles that are indicated in italics.

2. Maura and John are considering purchasing a home for $145 000. They have $25 000 for a down payment. Use the TVM Solver on your calculator to calculate the regular monthly payment for the mortgage at 7.5%, compounded semi-annually, amortized over 25 years. Complete the entries in the *Buying a Home* section of the spreadsheet.

3. Maura and John anticipate these monthly expenses:
   - Taxes – $275.00
   - Hydro – $90.00
   - Gas – $55.00
   - Water – $40.00
   - Insurance – $42.50
   - Maintenance – $100.00

   Enter these values in the *Monthly Expenses of Owning a Home* section of the spreadsheet.

4. Maura and John are currently renting a home. If they continue to rent, their monthly expenses will be:
   - Rent – $975.00
   - Hydro – $90.00
   - Gas – $55.00
   - Water – $40.00
   - Insurance – $25.50

   Enter these values in the *Monthly Expenses of Renting* section of the spreadsheet.

5. The interest rate on investments is currently 6.0%, the average annual appreciation in Real Estate is 1.5%, and the average annual inflation in rent, utilities, and maintenance is 2.0%. Enter these values in the appropriate cells on the spreadsheet.

6. Examine the buying of a home as an investment. You need to consider the annual maintenance costs, the costs of carrying the mortgage as well as your *equity* in the home. Equity is the difference between the market value and the amount still owing on the house. The amounts still owing on the house can be found in column L. The Market Value in cell Q4 is the purchase price plus the annual increase in the value of real estate.

7. Examine what happens when the difference between the money that would have been spent buying a home and the money spent renting a home is invested.

8. To decide whether you want to buy or rent, there are many calculations that still have to be made. These may include the following:
   - The total cost of buying a home
   - The total cost of renting a home
   - The final value of a house as an investment (equity)
   - The final value of the investment fund

   a) Are there other considerations that you might make? Are there any non-financial considerations?

   b) What would be the role of income taxes in these calculations?

9. Financially, is it better for Maura and John to buy a home or continue to rent? Explain your answer.

DISCUSSING THE IDEAS

1. List maintenance items that occur around the home. You may wish to ask an adult to review your list. Estimate, where possible, the average cost of these items. Produce a complete class list.

2. Estimate the average monthly utility expenses for your community. Ask your family what they spend monthly on utilities. Bring the figures to class, and determine class averages for each utility. What are some factors that might account for variations in utility costs for different families?

3. A telephone is an expense whether you rent or own a home. What other expenses are common to both renters and homeowners? Any expenses common to both would not be considered in a cost comparison.

4. Determine the method of assessing property in your area, and the mill rate. Examine your tax bill, or one provided to you by a neighbour or your teacher. Are education taxes included on the tax bill? How are they calculated?

EXERCISES

Checking Your Skills

1. Justin and Penny own a home worth \$165 000. For tax purposes, the assessed value is 80% of the fair market value, and the tax rate is 36.4 mills. They have a mortgage of \$120 000, amortized over 15 years, at 8%. Their other expenses are shown in the table. Calculate Justin and Penny's average monthly cost for their home.

Item	Cost
Hydro	\$90 monthly
Gas	\$62 monthly
Water	\$110 quarterly
Insurance	\$550 annually
Maintenance	\$1150 annually

2. Stephanie and Steven must decide whether to buy a condominium or rent a garden apartment. The condo costs \$110 000. They have a \$15 000 down payment, and will finance the balance at 6.5% for 20 years. Taxes on the condominium are \$160 a month, utilities are \$120, maintenance fees are \$225, and insurance is \$46. The rent on the garden apartment is \$1045 a month and all utilities are included. Insurance is \$32 a month. If they rent the apartment, they will invest the \$15 000 in an RRSP at 7.5%, compounded annually, and will invest the difference in the price they would pay for the condo and the rent. They estimate the value of the condo to appreciate 3% a year, while the inflation rate for rent and utilities is 2% annually. They plan to move in 10 years' time. Use the RENTBUY spreadsheet to analyze both options, then advise Stephanie and Steven on which is the better option.

3. Michaela and Dustin plan to move to a new city. They must choose whether to buy a home for \$180 000 with a \$25 000 down payment or rent a similar house for \$1025 a month. They have four options:

- Buy the house with a 15-year mortgage and continue investing at the same rate after the mortgage is paid.
- Buy the house with a 25-year mortgage.
- Rent the house and invest both the \$25 000 and the difference each month between the rent and the mortgage payment.

They can invest the money in an RRSP at an interest rate of 8.5%, compounded annually. The current mortgage rate is 7.75%, compounded semi-annually. The utilities for each property are: hydro \$110, gas \$62, and water \$45. Maintenance on the house they would purchase is estimated to be \$120 a month. Insurance is \$56 if they purchase a home and \$34 if they rent. Taxes on the home are \$260 a month.

Assume that annual inflation for expenses and rent averages 2.5%, and that the annual appreciation on the house is 3%.

a) Use the RENTBUY spreadsheet to examine each of these options, then advise Michaela and Dustin on their best option.

b) If the appreciation on the house were 5% per year, would your advice change?

Extending Your Thinking

4. One dilemma many people face is whether to buy the "ideal" home they want as their first home, or to buy a less expensive home, pay off part of the mortgage, then sell the home and purchase a second one. They repeat this buy-sell routine until they have the home of their dreams.

Assume you qualify for a mortgage with approximately \$1100 monthly payments. Interest rates are currently 7.5%, compounded annually. You have \$15 000 saved for a down payment. Assume property appreciates at 3% per year.

- **a)** First, consider purchasing a home for \$165 000. Calculate the monthly payment on a mortgage of \$150 000, amortized over 25 years. What is the total cost of the mortgage? How much is paid in interest?
- **b)** Now consider a less expensive home for \$100 000. How much will you have to borrow? If you were to make payments of approximately \$1100, what would your amortization period be?
- **c)** You decide to sell this home in 5 years. Use the Selling Your Home spreadsheet from your teacher to calculate the proceeds from the sale.
- **d)** Use the proceeds as a down payment for the purchase of a home priced at \$150 000. If you keep your payments at approximately \$1100 and interest rates are still 7.5%, what will be your amortization period?
- **e)** Suppose you sell this home after 5 years. Calculate the proceeds.
- **f)** What would the original \$165 000 home be worth now? For your third purchase, you will select a home with the equivalent current value to this home. Use the proceeds from the sale of the second home as the down payment. If you keep your payments at approximately \$1100 and rates are still 7.5%, what will be your amortization period? How much will you pay in total for this mortgage, if you do not sell this home?
- **g)** How much do you pay in total if you use the buy-sell-buy routine? Is the third home paid off any sooner than the original 25 years?
- **h)** Which is less expensive: to purchase the \$165 000 home at the beginning or to buy a total of 3 homes? What is the cost difference?
- **i)** Are there any costs associated with moving? Estimate what these costs might be and re-examine your results.
- **j)** Is it worth it to buy several homes and follow this process? Why or why not? What possible impacts would this scheme have on various members of your family?

PROJECT

In this Tutorial, you learned to compare renting a home with buying a home. The skills and concepts relate to the Chapter Project, and to these projects in the Project Book.

- To Lease or Purchase?
- Landscape Gardening
- House Building Costs

COMMUNICATING THE IDEAS

A person's decision to rent or buy is frequently not based solely on financial considerations. Describe other factors that enter into the decision of whether home ownership is right for you.

Major Decision: Rent or Buy a Home?

Use your knowledge of renting and buying a house to complete your project. You may use a graphing calculator, or any of the spreadsheets and templates you have created and used in the preceding Tutorials.

1. Examine your expected income. Recall that you have just either leased or purchased a vehicle, and have regular payments. Estimate the property taxes you will pay for your city, town, or municipality. Assume that your heating costs will be $60 a month.

2. What is the largest mortgage payment you can afford?

3. Assume you have $12 000 saved for the down payment. Determine the highest-priced home you can purchase.

4. Use the information that you gathered at the beginning of this chapter. Cut out or print the picture of your house, and include a written description of the house and its location, and the reasons you chose this particular house. How much is the house?

5. Assemble similar information about a comparable house to rent. Include the listing and the price.

6. Use current interest rates. Decide on the amortization period you will choose for your mortgage. Remember, you cannot exceed your maximum allowed payment. Give reasons for your decision.

7. Assume that three years have now passed. Your mortgage is up for renewal. Your combined salaries have increased: 3% in year 1, 2% in year 2, and 4% in year 3. Calculate your new total monthly income. You still have a car payment. Determine the maximum mortgage payment you can now afford.

8. You decide to increase your payments and shorten your amortization. Note how much you still owe and determine your new amortization period and monthly payments. Give reasons for your choice.

9. Suppose real estate has increased in value by 3% each year. How much equity would you have in your home?

10. Suppose rents have increased 2% each year. How much rent would you be paying now, if you had not bought a house?

11. Given these figures, do you think it would be worthwhile for you to buy a house? Why or why not?

CONSOLIDATING YOUR SKILLS

What Do I Need To Know?

Tutorial	Skills and Concepts	Important Results and Formulas
4.1	• Design a spreadsheet template	• $A = P(1 + i)^n$
4.2	• Investigate savings sheltered with an RRSP	
4.3	• Investment portfolios	
4.4	• Leasing items	
4.5	• Investigate mortgages	• Total cost of a mortgage = regular payment × number of payments
4.6	• Analyze the benefits of renting vs. buying an increasing asset	

What Should I Be Able To Do?

Part A

1. Assume you invest a total of $40 000 throughout your working years, up to age 65, by contributing an equal amount each year.

a) To complete the table either set up a spreadsheet or use your graphing calculator.

Age at Initial Investment	Number of Years to Invest	Investment per Year	Future Value, if Growth Rate Is 6%	Future Value, if Growth Rate Is 10%	Future Value, if Growth Rate Is 14%
25	40				
30	35				
35	30				
40	25				
45	20				
50	15				
55	10				

b) For each growth rate, how much more money will you have if you begin to invest at age 25 instead of age 30?

c) For each growth rate, how much more money will you have if you begin to invest at age 30 instead of age 40?

d) What is the effect of the growth rate on the differences in the final amount of the investment?

Part B

2. Nelson invested $1000 in a mutual fund.

a) To complete the table, calculate the compound amount for each of the first 5 years, given the rates of return for each year.

Year	Opening Balance	Rate of Return	Closing Balance
1	$1000	12.1%	
2		8.3%	
3		(6.8%)	
4		19.8%	
5		55.1%	

b) When mutual fund companies report the rate of return for their funds, they first indicate the return for the previous year, then the previous 2 years, 3 years, and 5 years. Calculate the average rate of return for years 4 and 5, years 3, 4, and 5, and then for all 5 years.

3. Design a spreadsheet to allow a user to calculate his or her net worth.
Net worth = Total assets − Total liabilities.
The debt/equity ratio is calculated as:

$$\frac{\text{Total liabilities} - \text{Mortgages}}{\text{Net worth}}$$

Your spreadsheet will have the following entries.

Statement of Net Worth

ASSETS (WHAT YOU OWN)

1. Liquid/Current Assets	
i. Bank Accounts	______
ii. Cash	______
Total Liquid Assets	______
2. Semi-Liquid Assets	
i. Mutual Funds	______
ii. Stocks/Bonds	______
iii. RRSPs	______
iv. Registered Pension Plan	______
v. Life Insurance/Cash Value	______
Total Semi-Liquid Assets	______
3. Non-Liquid Assets	
i. Principal Residence	______
ii. Vehicles	______
iii. Other	______
Total Non-Liquid Assets	______
TOTAL ASSETS	______

LIABILITIES (WHAT YOU OWE)

4. Short-Term Debt	
i. Credit Card Debt	______
ii. Short-Term Loan	______
Total Short-Term Debt	______
5. Long-Term Debt	
i. Mortgage	______
ii. Line of Credit	______
iii. Other	______
Total Long-Term Debt	______
TOTAL LIABILITIES	______
NET WORTH	______
DEBT/EQUITY RATIO	______

4. Use the spreadsheet you created in exercise 3 to determine the net worth and the debt/equity ratio for Evan:

- His home is assessed at \$112 000 with a mortgage balance of \$72 500.
- He has a car valued at \$13 600, but the outstanding loan is \$6400.
- He has \$1900 in a savings account, \$3800 in Canada Savings Bonds, and \$1450 in mutual funds.
- He has a life insurance policy with a cash surrender value of \$5300.
- His RRSPs are worth \$16 100 and his registered pension plan is worth \$45 000.
- His credit card debt is \$2554 and furniture loan is \$3650.

5. Use the spreadsheet RRSP to examine a one-time investment of \$1000 that earns 12%, compounded semi-annually, if the investor has a marginal tax rate of 25%.

a) Calculate the net interest earned per year, after taxes, if the investment is not registered.

b) Use your results from part a and a graphing calculator to determine how to calculate the net amount of an unregistered investment.

c) Use your graphing calculator to complete the following table.

Investment Amount	Interest Rate	Marginal Tax Rate	Net Interest Rate	Number of Years of Investment	Total Value of Investment
\$1000	12%	25%		10	
\$2500	8.5%	39%		15	
\$4000	14.5%	44%		25	

6. Cole, a 35-year-old lawyer, has an investment portfolio.

Investment	Amount	Average Annual Return
Bank account	\$2400	3%
Term deposit	\$4800	6.25%
Guaranteed investment certificate	\$12 500	8.5%
Guaranteed investment certificate	\$8000	9.25%
Mutual Fund	\$42 000	28.75%

Calculate his total annual return and his average rate of return.

7. Two cars, each priced at \$16 900, can be leased for a 3-year term. One lease costs \$199 monthly before taxes; the other one costs \$239 monthly before taxes.

a) Suggest two reasons why the lease prices might be different.

b) If interest rates for each vehicle are set at 3.9% and no down payment is required, what is the residual value at the end of the lease for each car?

8. Eric wants to lease a sport utility vehicle with a purchase price of \$45 000. He lives in Manitoba, where the PST and GST are both 7%.

a) Use the rates for residual values found in the table in *Investigation 1* of *Tutorial 4.4* to determine the residual value at the end of 2 years, 3 years, and 4 years.

b) Determine the after-tax lease payment for a 2-year, a 3-year, and a 4-year term, if the interest rate is 2.4%, compounded monthly.

c) Eric plans to purchase the vehicle at the end of the lease. Suggest an appropriate lease term if:

i) he has no other debts.

ii) he has to take out a bank loan at 9.8%, compounded monthly.

d) If the interest rate on the lease is 13.6%, compounded monthly, and Eric's bank loan is at 8.9%, compounded monthly, suggest an appropriate lease term.

9. Heather has \$15 000 saved for the down payment on a house. She wants to buy a house priced at \$88 000. As a dental hygienist, she earns \$38 000 per year. Her monthly car lease payments are \$189. She estimates property taxes to be \$2200 per year and heating bills to be \$60 per month.

a) What is the maximum mortgage payment Heather can afford?

b) Heather wants to pay off her house as soon as possible. What is the shortest amortization period Heather can obtain if interest rates are 8%?

10. Jelena and Scott are trying to decide whether to buy a condominium or rent an apartment. The condo costs \$42 500. They have a \$5000 down payment, and will finance the rest at 7.5% for 15 years. Taxes on the condo are \$85 per month, utilities are \$88, and insurance is \$42. Rent on the apartment is \$645 per month and all utilities are included. Insurance is \$29 per month. If they rent the apartment, they will invest the \$5000 in an RRSP, at 6.75%. They estimate that the value of the condo will appreciate 1% a year, while the inflation rate for rent and utilities will be 2%. They plan to move in 10 years. Use the RENTBUY spreadsheet to analyze both options and advise Jelena and Scott on the better option.

Part C

The Chapter Project combines many skills and concepts in a single problem. Here is a sampling of a student's work from different stages of the project. Review these samples to gain insights into your own project.

11. Here is a sample of one student's work comparing buying and leasing a car. Are there any errors in the calculations? Do you agree with the student?

Financing Your Car

The car I have choosen is a 2000 LS1 from Saturn.

1. The leasing information given is listed in the following table.

Purchase Price	\$23 670
Down Payment	\$3 150
Monthly Payment	\$254
Finance Rate	2.4%
Residual	\$12 545
Secuity Deposit	\$300
Kilometers Allowed	60 000 km
Extra Charge	\$0.12/km

All of the neccesary leasing information is given in the ad.

2. The monthly payment, before taxes, is \$254.

The amount of the residual was calculated using the TVM solver on the calculator.

```
N=36
I%=2.4
PV=20520
PMT=-252
FV=-12634.71101
P/Y=12
C/Y=12
PMT:END BEGIN
```

Therefore the calculated residual is \$12 634, slighlty higher then advertised by Saturn.

3. The spreadsheet is enclosed.

4. I have choosen to purchase a new Saturn LS1. I like the idea of the company, Saturn, and the way they treat their customers. Also the car looks good, is practical and affordable.

I would choose to purchase the car. The monthly payment was calculated to be \$426 for a 60 month loan, considerable more then the lease at \$292 after taxes. However at the end of the loan the car is mine.

At the end of the lease the option to purchase would result in a new 3 year loan with monthly payments of \$415. This would be in addition to any other charges incurred for distances. I plan on doing a lot of driving, and these charges could be significant.

In conclusion I would rather pay a known amount every month for 5 years, rather then differing amounts for 6 years.

Purchasing vs. Leasing an Autgmobile

Make and Model of Vehicle

Purchase Price	\$23,670.00
Down Payment	\$3,150.00
Finance Rate	2.40%
Provincial Sales Tax Rate	8%

Leasing Information	
Purchase Option Price	\$12,545.00
Lease Term (months)	36
Base Monthly Lease Payment	\$254.00
Monthly Payment (includes taxes)	\$292.10
Filing Fee	\$22.00
refundable security deposit	\$300.00
Total Due upon Signing	\$614.10
Total Lease Cost (less security deposit)	**\$10,537.60**

Purchasing Information	
Purchase Price	\$23,670.00
GST	\$1,656.90
PST	\$1,893.60
Total Purchase Price	\$27,220.50
Less: Down Payment	\$3,150.00
Amount to Finance	\$24,070.50
Finance Rate	2.40%
Finance Term (months)	60
Monthly Payment	\$426.13
Total Cost	**\$25,567.64**

Option to Purchase at End of Lease	
Purchase Price	\$12,545.00
Taxes on Purchase Option Price	\$1,881.75
Total Cost to Purchase	\$14,426.75
Total Cost of Vehicle	**\$24,964.35**

Financing Purchase at End of Lease	
Option Finance Term (months)	36
Finance Rate	2.40%
Monthly Payment	\$415.74
Total Cost of Vehicle	**\$25,504.36**

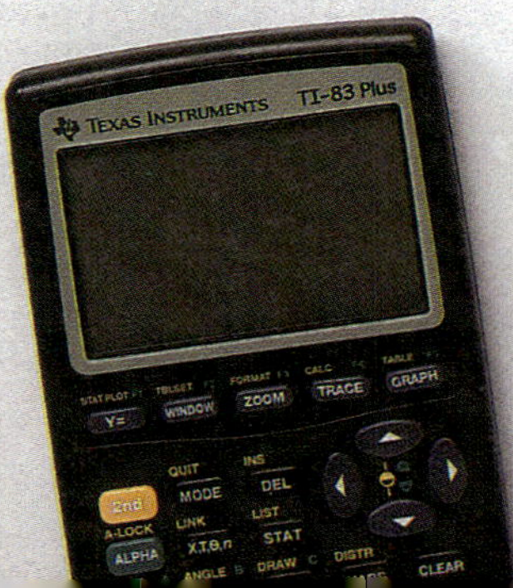

12. Here is a sample of a student's work on buying a home. Are there any errors in the calculations?

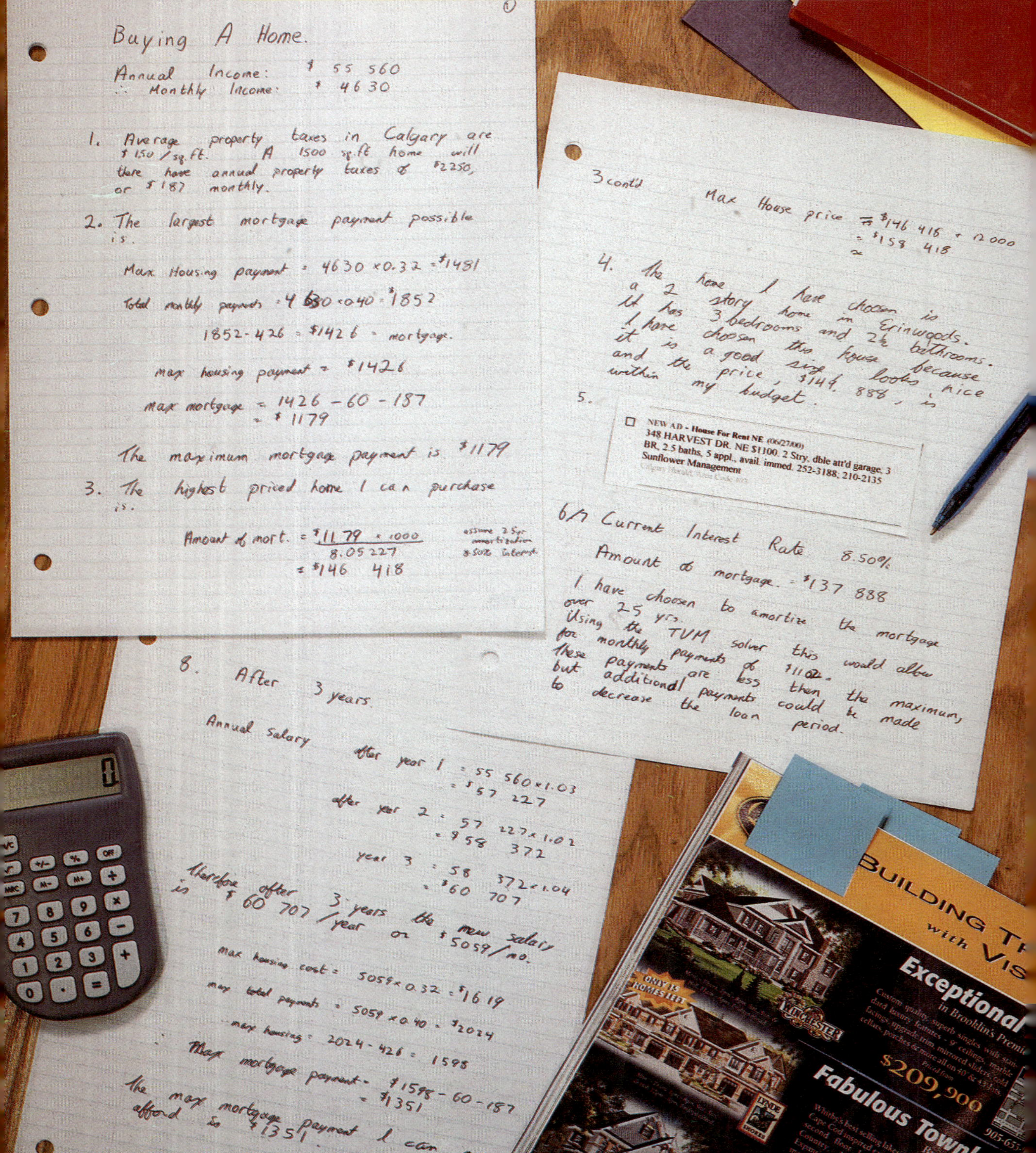

5 Sinusoidal Data

Overview

Chapter Section	Skills and Concepts
Chapter Project: Analyzing Periodic Data, page 208	Research periodic data
Tutorial 5.1: Collecting and Plotting Periodic Data, page 210	Examine the characteristics of periodic data
Chapter Project: Angle of Elevation of the Sun, page 220	Collect, plot, and analyze periodic data
Tutorial 5.2: Radian Measure and Sine Curves, page 222	Define radians; graph $y = \sin(x)$
Tutorial 5.3: Fitting Sine Curves to Data, page 225	Determine a sinusoidal curve that best fits periodic data
Tutorial 5.4: The Characteristics of $y = a\sin(bx + c) + d$, page 233	Study the effects of the parameters a, b, c, and d on the characteristics of $y = a\sin(bx + c) + d$
Tutorial 5.5: Applications, page 241	Work with periodic data and make predictions
Chapter Project: Regression and Prediction, page 248	Use a sinusoidal regression equation to make predictions
Consolidating Your Skills, page 250	Review the work of the chapter

PROJECT

You should do at least one project that uses the skills you learn in this chapter. The recommended projects in the Project Book are:

- Surviving on the Ocean's Shore
- Sun, Moon, and Stars

Analyzing Periodic Data

Mathematicians, scientists, engineers, economists, meteorologists, and many other professionals make predictions based on data.

In this Chapter Project, you will collect and analyze periodic data. You will then use your analysis to make predictions.

Here are some questions to consider.

- Why are some sets of data easier to work with than others?
- What makes data periodic?
- What are some sources of periodic data?
- How are predictions made from periodic data?

Research periodic data. The Internet, your library, magazines, and science teachers are possible sources of information.

CHAPTER PROJECT

To complete this Chapter Project, you will need to know how to collect and plot periodic data. Complete *Tutorial 5.1* to develop your skills. You will return to the Chapter Project on page 220.

FYI Visit: www.awl.ca/school/connections

You can start your research from our Internet site: Click on MATHLINKS, followed by the *Applied Mathematics 12* logo, then select a topic under Analyzing Periodic Data.

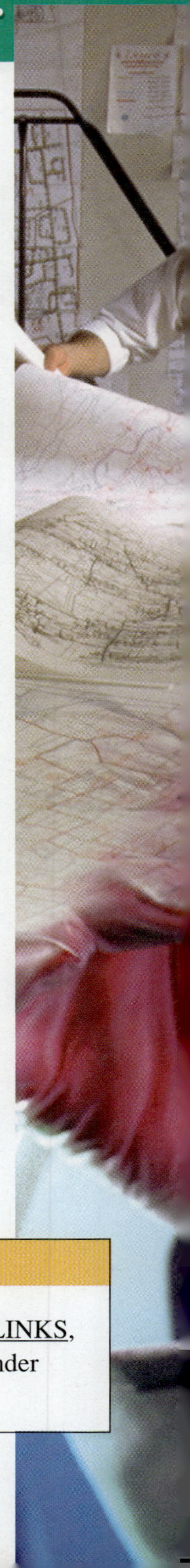

TUTORIAL

5.1 Collecting and Plotting Periodic Data

Consider the following graphs.

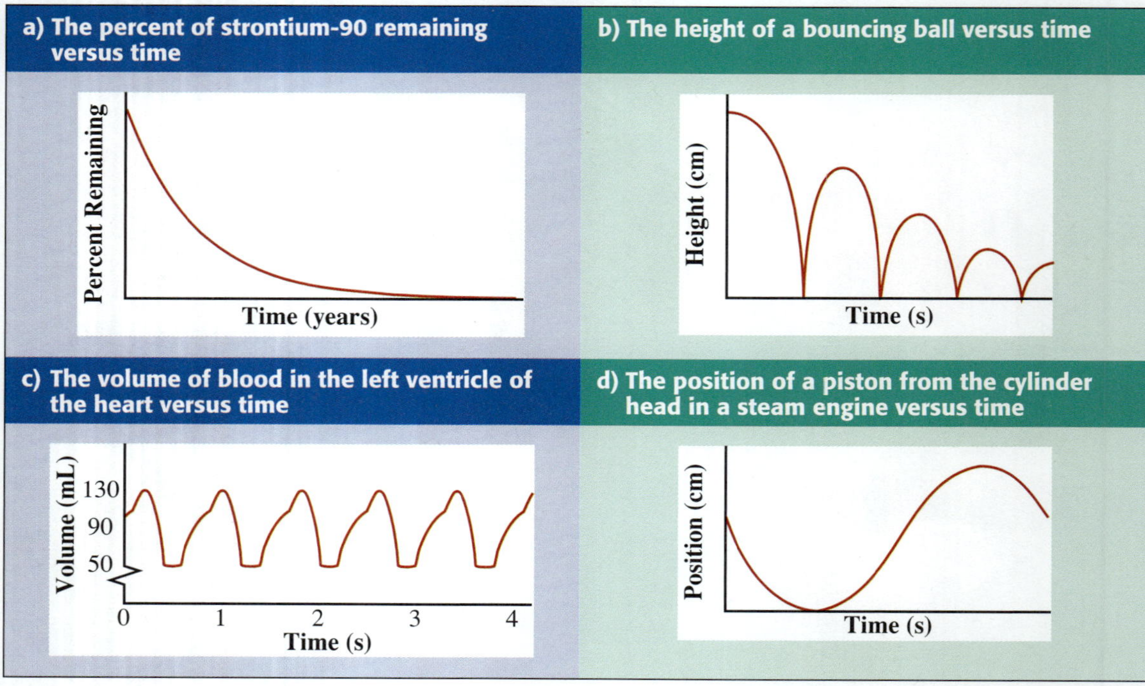

In graph a, the percent of strontium-90 decreases gradually. There is no repeating pattern. In graph b, the height of the ball increases and decreases, but there is still no repeating pattern. In graph c, there is a definite pattern that repeats, but it is not wave-like. In graph d the position of the piston follows a smooth wave-like pattern.

Data often contain patterns that can be found when you graph them. In this Tutorial, there are three Investigations in which you can collect and analyze data from a variety of sources.

You should complete at least one of these Investigations. You will need grid paper for each Investigation.

Investigation 1 Patterns in a Machine: Clock Time

You will investigate the relationship between the hours elapsed and the position of the small hand on a clock.

1. Construct a circle with a radius of 5 cm. Use it to draw a standard 12-h clock face.

2. During one complete revolution, the hour hand passes through 12 h. It also passes through 360°. Through how many degrees does the hour hand pass in 1 h?

3. If the position-angle of the hour hand is set to 0° at noon, what is the position-angle at each time?

a) 1:00 P.M. **b)** 2:00 P.M. **c)** 4:00 P.M.

d) 5:30 P.M. **e)** 10:12 P.M.

4. Complete the following table that lists the time and position-angle of the hour hand on a 12-h clock face, every hour for 24 h, starting at noon. The position-angle is always measured in the clockwise direction.

Time	Position-angle
12:00 noon	0°

5. Graph the data on grid paper. Should the points be connected? Explain.

6. Describe the pattern in this graph.

Investigation 2 Patterns in Nature: Times of Sunrise

The following data record the time of sunrise in Central Standard Time for the 15th of each month for Brandon, Manitoba for 1998 and 1999. Times are expressed in hours and decimal fractions of an hour.

1998	Day Number	Time of Sunrise (A.M.)	1999	Day Number	Time of Sunrise (A.M.)
Jan. 15	15	8.53	Jan. 15	380	8.53
Feb. 15	46	7.83	Feb. 15	411	7.83
Mar. 15	74	6.88	Mar. 15	439	6.90
Apr. 15	105	5.78	Apr. 15	470	5.80
May 15	135	4.90	May 15	500	4.92
June 15	167	4.52	June 15	531	4.52
July 15	197	4.80	July 15	561	4.80
Aug. 15	228	5.50	Aug. 15	592	5.48
Sept. 15	259	6.27	Sept. 15	613	6.25
Oct. 15	289	7.02	Oct. 15	653	7.02
Nov. 15	320	7.87	Nov. 15	683	7.87
Dec. 15	349	8.52	Dec. 15	714	8.52

1. Plot the Day Number versus the Time of Sunrise.
2. Should the points be connected? Explain.
3. Describe the pattern in this graph.
4. Suppose this pattern continues for an 850-d period. Use your graph to predict the time of sunrise on March 15, 2000.
5. Use the table to predict the time of sunrise on March 15, 2000. How close is this estimate to your estimate from exercise 4?
6. Find the actual time of sunrise on March 15, 2000. How close were your estimates?

Investigation 3 Patterns in Physics: Light and Power Sources

You will need a CBL with a light probe, an incandescent light source, and a fluorescent light source. Work in pairs.

1. Connect the CBL to your calculator.
2. Point the light probe at the incandescent light source for about 10 s.
3. Graph the data that the CBL gathers.
4. Describe the pattern in this graph.
5. Point the light probe at the fluorescent light source for about 10 s.
6. Graph the data that the CBL gathers.
7. Describe the pattern in this graph.
8. Why are the graphs different?
9. Are both graphs periodic? Explain.
10. Would you expect to get a similar graph if you use a candle as a light source? Explain.

As you have seen in these Investigations, many mathematical applications involve quantities that change in a regular way. Data that contain cycles, which repeat at regular intervals, are called *periodic data*. The length of the shortest interval that repeats is called the *period*.

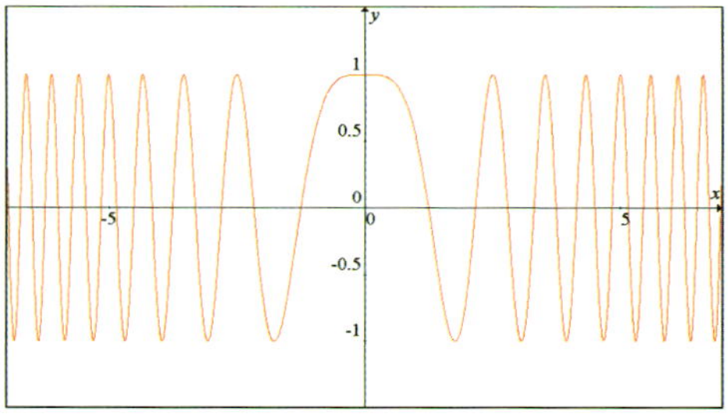

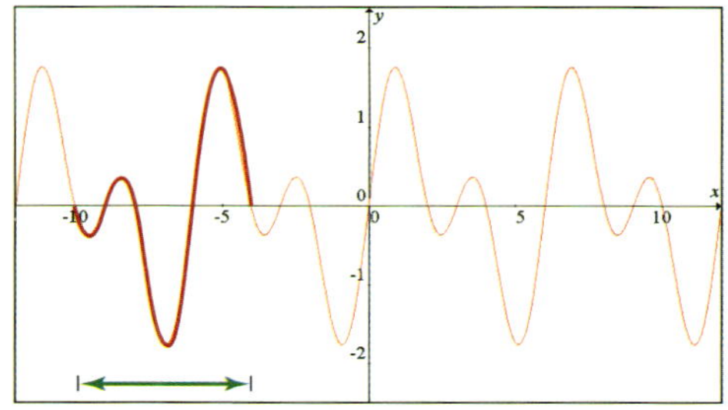

Period:
$-4 - (-10) = 6$

The graph (above left) illustrates data that are not periodic.

A *cycle* in the graph of periodic data is a part of the graph from any point to the first point where the graph begins to repeat the same pattern. The *period* is the difference in the values of x for the points at the ends of a cycle. In the graph (above right) the data are periodic. One cycle is highlighted on the graph. These data have a period of 6 units.

Example 1 **Temperature**

The maximum temperature was recorded every day during January 1997, at Waseca, Saskatchewan. At the end of the month, these values were averaged to obtain –14.1°C. This procedure was then repeated every month, for two years.

Month in 1997	Number of Month	Mean Maximum Temperature (°C)	Month in 1998	Number of Month	Mean Maximum Temperature (°C)
Jan.	1	−14.1	Jan.	13	−15.5
Feb.	2	−5.0	Feb.	14	−3.2
Mar.	3	−3.5	Mar.	15	−2.3
Apr.	4	5.9	Apr.	16	13.7
May	5	16.3	May	17	21.7
June	6	20.7	June	18	21.1
July	7	24.2	July	19	24.5
Aug.	8	24.1	Aug.	20	26.5
Sept.	9	19.8	Sept.	21	18.6
Oct.	10	7.8	Oct.	22	9.6
Nov.	11	0.6	Nov.	23	−2.1
Dec.	12	−0.7	Dec.	24	−8.7

a) Construct a graph to illustrate these data.

b) Describe the period and range of this graph.

c) Assuming this pattern continues, what is your prediction for the mean monthly maximum temperature for June 1999?

SOLUTION

a)

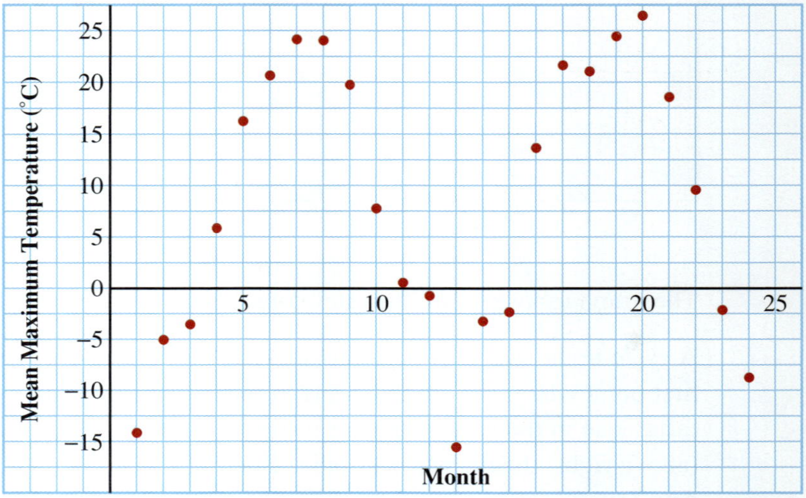

b) This graph represents the continuous variation in the monthly mean maximum temperatures at Waseca. The data seem to have a cycle that repeats. The length of the period is approximately 12 months. The overall average temperature is about 5°C. The temperature drops about 20°C below the average, to about –15°C in January. The temperature rises about 20°C above the average, to about +25°C in August.

c) Extend the pattern in the graph. The mean maximum temperature for June 1999 is approximately 21°C.

The graph of the function $y = \sin x$ (shown below) is called a *sine curve*.

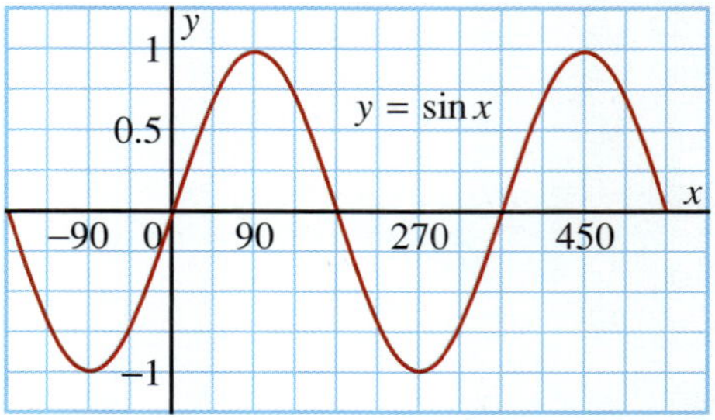

When we plot periodic data, the resulting graph looks like a sine curve. Data that fit this pattern are called *sinusoidal data*. The data on times of sunrise from *Investigation 2* are sinusoidal.

Example 2 Ferris wheel

The following data describe the height of a seat on a Ferris wheel, as the wheel rotates.

Time (s)	0	4	8	12	16	20	24	28	32	36	40	44	48
Height (m)	0.8	8.0	25.4	42.8	50.0	42.8	25.4	8.0	0.8	8.0	25.4	42.8	50.0

a) Sketch the graph of the data in your notebook. Should you connect the points? Explain.

b) Describe the pattern in this graph. What is the period of these data?

c) How many complete revolutions will the Ferris wheel make in 2 min of operation?

d) Haviar got on the Ferris wheel. How high was Haviar above the ground after 60 s?

SOLUTION

a) The points should be connected because the height of the seat is a continuous variable.

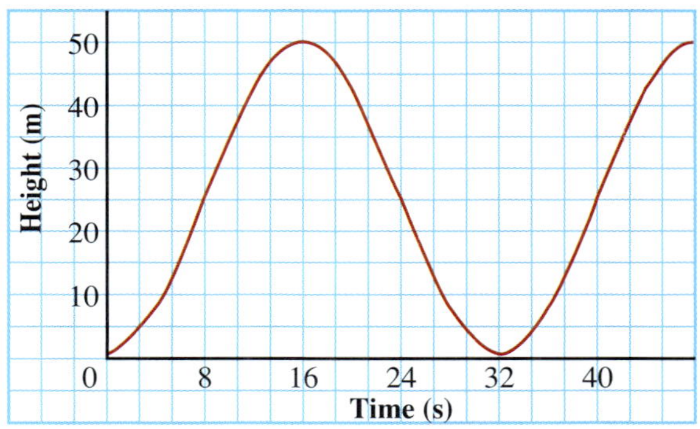

b) The graph is periodic and sinusoidal. The Ferris wheel makes one complete revolution every 32 s. Thus, the period is 32 s.

c) It takes 32 s to complete one revolution. In two minutes, or 120 s, it will complete $\frac{120}{32}$ or 3.75 revolutions.

d) Extend the graph.

Haviar will be about 8 m above the ground after 60 s.

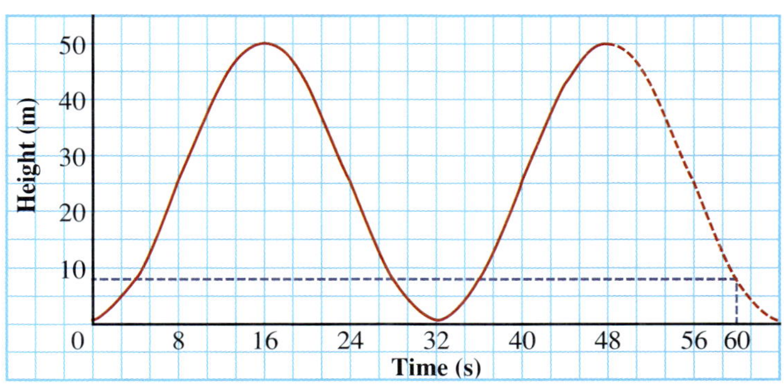

DISCUSSING THE IDEAS

1. Why are patterns in a set of data important?
2. Are all periodic data sinusoidal?
3. Are all sinusoidal data periodic?
4. In *Example 2*, explain why the graph of the height of the seat versus time has no values below 0.8 m.

EXERCISES

Checking Your Skills

1. For each of the following functions, state whether it is periodic or not. If it is periodic, estimate its period.

   a) This is the graph produced by an oscilloscope when a violin is played.

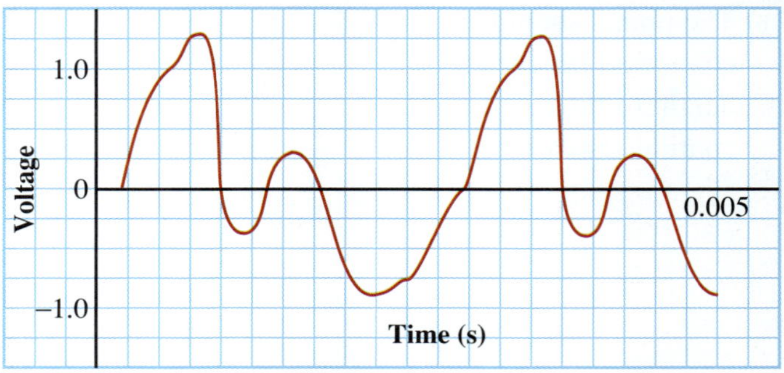

   b) This graph shows the vertical motion of a cork that floats on a pond after a pebble has been dropped into the water.

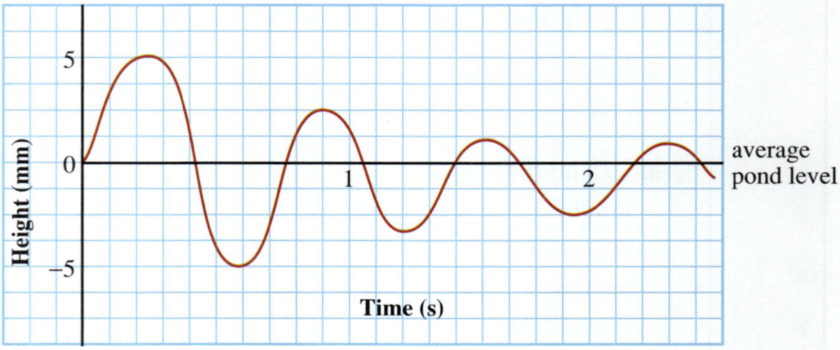

c) Sunspots are dark spots that appear from time to time on the surface of the sun. This graph shows how the number of sunspots varied from 1944 to 1998.

Monthly mean sunspot numbers, 1944–1998

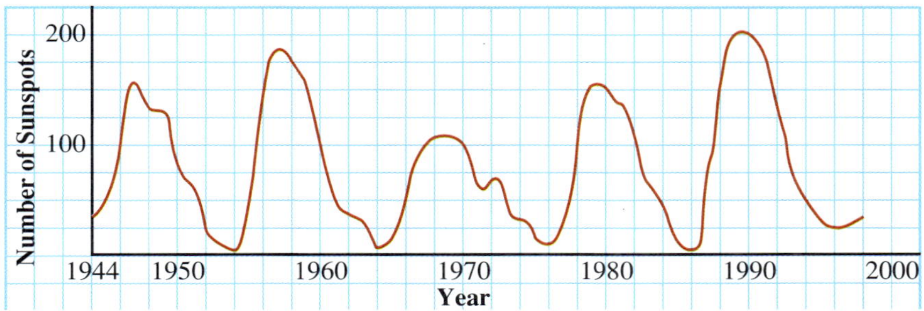

2. An electrocardiogram records the electrical activity of the heart over time. The output from an electrocardiogram (ECG) is shown. Describe the graph and suggest reasons for the changes in the graph.

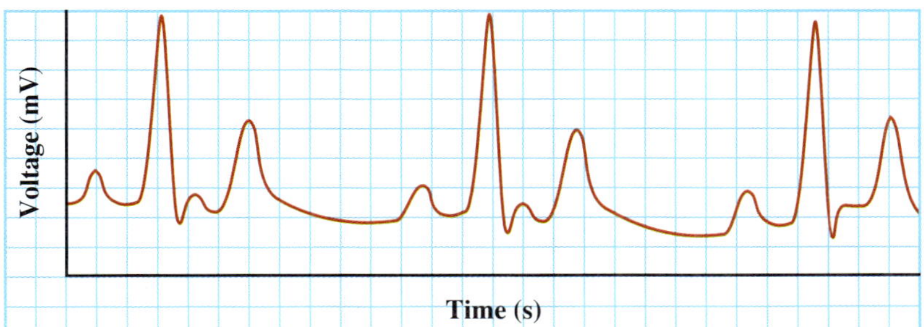

3. This graph shows how the blood pressure in the left ventricle of the heart varies during 5 consecutive heartbeats.

Pressure of blood in left ventricle

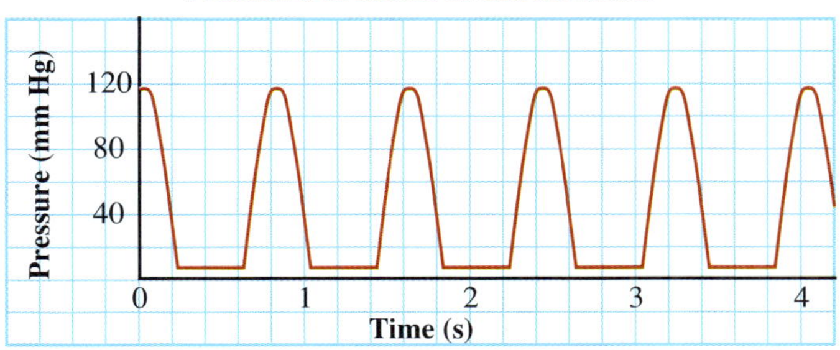

During intense physical activity, the heart beats faster to satisfy the body's demand for more oxygen. Suppose a graph was drawn to show the variation in blood pressure in this situation.

a) How would this graph differ from the one above?

b) How would this graph be similar to the one above?

4. The volume of air in your lungs is a periodic function of time. This graph shows how the volume of air in the lungs varies during normal breathing.

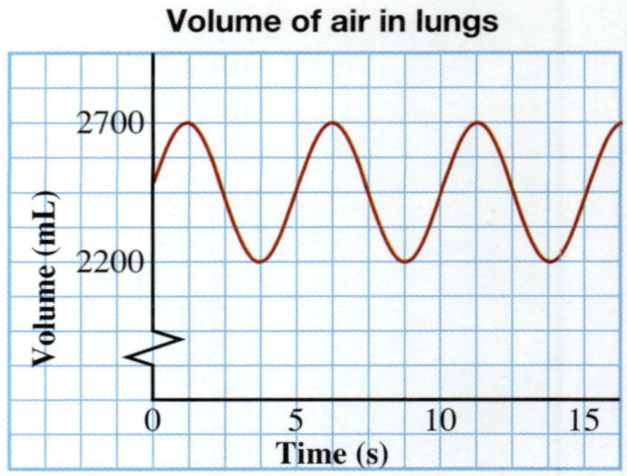

a) How long does a person take to inhale and exhale once?

b) What information can you determine about the person's level of physical activity from this graph?

5. In the northern hemisphere, the sun sets at its latest on June 21st, and at its earliest on December 21st.

a) The times in the table are expressed in hours on the 24-h clock. On the same grid, sketch graphs that show the times of sunset at Fairbanks, Alaska, and Mexico City from January to December.

Approximate Times of Sunset

	March 21	June 21	September 21	December 21
Fairbanks	19.2 h	23.8 h	19.2 h	14.6 h
Mexico City	18.7 h	19.4 h	18.7 h	18.0 h

b) Calgary lies between Fairbanks and Mexico City. On the same grid, sketch another graph to estimate the times of sunset at Calgary.

6. This graph shows how the length of the shadow of a 100-m building varies during a 3-day period.

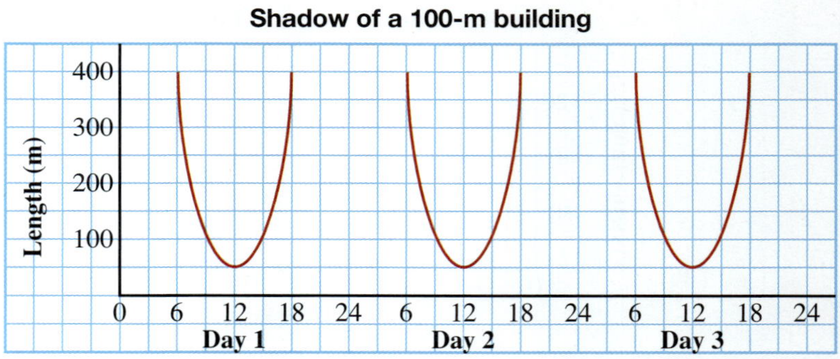

a) How long is the shadow at 8:00 A.M. and at 2:00 P.M.?

b) For about how many hours during the day is the shadow longer than 100 m?

c) In some cities, such as Singapore, the sun is directly overhead at noon during some days of the year. What changes would there be in the graph if it were drawn for a building in Singapore on one of those days?

d) A building of the same size and shape as this one is located in a city where the sun rises earlier and sets later. Describe how the graph would change to reflect this difference.

7. This table provides the times of sunrise for the 1st of each month at Medicine Hat, Alberta, for 2000.

Date	Day Number	Time of Sunrise	Date	Day Number	Time of Sunrise
Jan. 1	1	8.37	July 1	183	4.30
Feb. 1	32	7.95	Aug. 1	214	4.87
Mar. 1	61	7.08	Sept. 1	245	5.63
Apr. 1	92	5.97	Oct. 1	275	6.38
May 1	122	4.97	Nov. 1	306	7.22
June 1	153	4.30	Dec. 1	336	8.00

a) Graph the data.

b) Use your graph to predict the time of sunrise on June 27, 2000.

c) Tela lives in Medicine Hat and wants to see the first sunrise at Medicine Hat in 2001. At what time can she expect sunrise to occur?

Extending Your Thinking

8. To create a pendulum, hang a weight at the end of a long string. Pull the weight to one side and let it go to set the pendulum in motion. Suppose that you graphed the horizontal displacement of the pendulum versus the time for a short period of time (less than 30 s).

a) What type of graph would you expect to find?

b) What characteristic of the pendulum's swing would correspond to one cycle of the graph?

c) What characteristic of the pendulum's swing would correspond to the maximum horizontal displacement?

d) Collect data to support your prediction.

PROJECT

In this Tutorial, you collected and graphed periodic data. You will use these skills to complete the Chapter Project, and in the following projects in the Project Book.

- Surviving on the Ocean's Shore
- Sun, Moon, and Stars

COMMUNICATING THE IDEAS

Write to explain the characteristics of periodic data and sinusoidal data. Explain how to find the period of periodic data.

Angle of Elevation of the Sun

In this part of the project, you will make a series of observations that can be used to determine the angle of elevation of the sun at different times during the day.

> **WARNING:** Never look directly at the sun. It is so bright that its light can permanently damage your eyes.

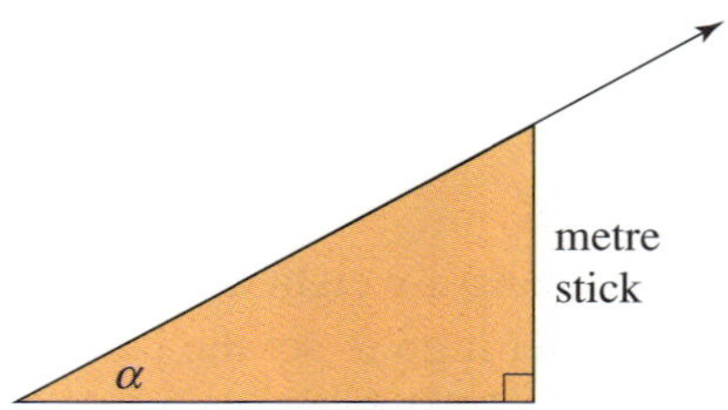

To complete this part of the project you will need a metre stick and a tape measure.

1. Create a table with four columns and at least 8 rows to record: time (h:min), time (h), shadow length (cm), and angle of elevation (°).
2. Devise some technique to hold a metre stick vertically. (You might tape the metre stick to a vertical edge of a cardboard box.)
3. On a sunny day, place the metre stick in an outdoor location where the ground is level and there are no obstructions.
4. Record the time and measure the length of the shadow of the metre stick, to the nearest 0.5 cm.
5. Repeat step 4 at least 8 times, with no more than 1 h between observations. (Note that 15 or more observations will provide a better picture of the sun's elevation throughout the day.)

CHAPTER PROJECT

In the next stage of this Chapter Project, you will use a regression equation to estimate the angle of elevation of the sun. Complete *Tutorials 5.2* to *5.5* to develop an understanding of the related skills. You will return to the Chapter Project on page 248.

6. Convert the times in hours and minutes to hours and decimal fractions of hours in a 24-h system.

7. Use trigonometry to convert each shadow length into an angle of elevation, α, of the sun.

8. Graph the data for the angle of elevation versus time.

9. From this graph, estimate the period of the curve and the maximum angle of elevation of the sun during the day.

10. Estimate the times for sunrise and sunset. Explain how you obtained these from your graph.

Career Skills

Often, ingenuity must be used to create solutions to problems such as that of ensuring the metre stick is vertical. Taking and recording accurate measurements are essential skills in industry.

TUTORIAL 5.2

Radian Measure and Sine Curves

Physical quantities can be measured in various units and systems of measure. For example, length can be measured in metres, yards, or cubits. Volume can be measured in litres, gallons, bushels, or barrels. Mass can be measured in kilograms, pounds, or carats. Some systems of measure, such as the International Metric System, were created to standardize and simplify measurements.

Thus far, we have used degrees to measure angles. The division of a complete rotation into 360° can be traced back at least 4000 years, to ancient Babylon. When the metric system was first devised in the late 18th century, the *gradian* was created as a new unit for angle measure. Four hundred grads correspond to one complete rotation, and a right angle corresponds to 100 grads. Mathematicians, scientists, and engineers have developed another unit for angle measurement, the *radian*.

You may have noticed that a scientific calculator has a key labelled DRG. This key switches the units used to measure angles from degrees to radians to gradians, and back again. As you press this key, one of the symbols DEG, RAD, or GRAD will appear in the calculator display. On a graphing calculator, the MODE menu allows you to switch between Radian and Degree measures.

Investigation 1 What is a radian?

You will need compasses, a ruler, a protractor, string or thread, and scissors.

1. Draw a circle with a radius of at least 5 cm; smaller circles are more difficult to manipulate.
2. Label the centre of the circle C, then draw a radius and label the end A.
3. Cut a piece of string equal to the length of the radius.
4. Place one end of the string on point A. Carefully lay the string along the circumference of the circle and label the far endpoint of the string B. The path AB along the circle is an arc. In this case, the arc length AB equals the length of the radius.

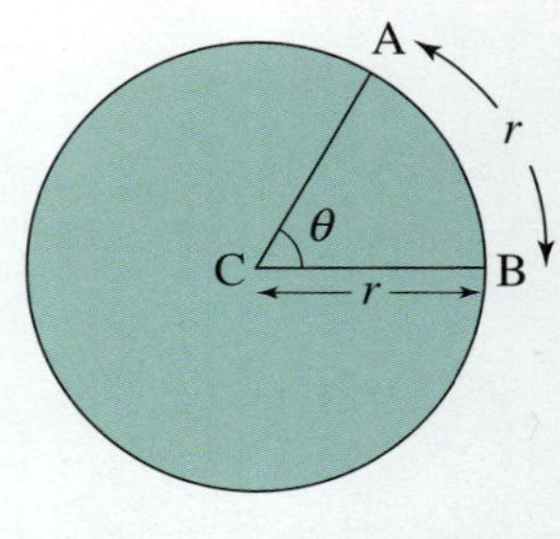

5. Connect B to C. Measure the sector angle ∠BCA. Compare your answer to that of at least two other students. What do you notice?

6. Did the size of the circle affect the results of this Investigation?

7. One radian is the measure of a central angle of a sector with arc length equal to the radius. The measure of ∠BCA is 1 radian. Approximately how many degrees are in 1 radian?

8. If a circle has radius r, what is its circumference?

9. How many arc lengths, with length r, will fit into the circumference?

10. How many radians will fit into one complete revolution?

One radian is the measure of the central angle of a sector with arc length of the sector equal to the radius.

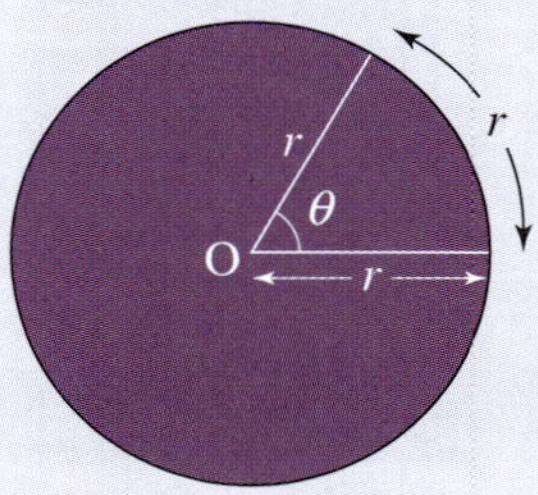

$\theta = 1$ radian

2π radians $\doteq 6.28$ radians $= 360°$

1 radian $\doteq 57.295\,78°$

The TI-83 uses a raised r, ^{r}, as the symbol to indicate that an angle measure is expressed in radians. By convention, if no units are shown for an angle, then it is assumed to be in radians.

When we use the TI-83 to fit a sine curve to periodic data, the output is **always** in radians.

Investigation 2 The Characteristics of $y = \sin x$

1. Set your calculator to radian mode.

2. Enter the equation Y1 = $\sin x$.

3. Use window settings Xmin $= -2\pi$, Xmax $= 2\pi$, Xscl $= 1$, Ymin $= -1$, Ymax $= 1$, and Yscl $= 0.5$.

4. Graph the function and sketch it carefully in your notebook. Label the axes.

5. With x in radians, predict the x-intercepts of the curve $y = \sin x$. Use the ZERO feature in the CALC menu to check your predictions.

6. Predict the coordinates of the maximum and minimum points of $y = \sin x$. Use the MAXIMUM and MINIMUM features in the CALC menu to check.

7. The *amplitude* of a sinusoidal function is half the vertical distance between the maximum and minimum points. It can be found using the formula $\frac{\text{maximum value} - \text{minimum value}}{2}$. What is the amplitude of $y = \sin x$?

8. What is the period of $y = \sin x$?

Properties of $y = \sin x$

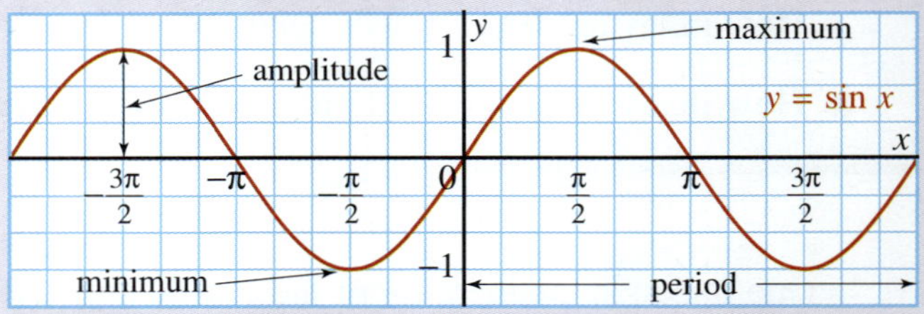

Amplitude = 1
Period = 2π radians

Maximum value = 1
Minimum value = −1

DISCUSSING THE IDEAS

1. Explain why a radian is the same size, regardless of the length of the radius of the circle.

2. On the graph of $y = \sin x$, the amplitude and the maximum value are equal. Can you draw a sinusoidal function for which the amplitude and the maximum value are different?

3. A sine curve oscillates up and down. Why is the amplitude always given as a positive number?

4. Could the period of a sine curve be defined as the horizontal distance between two consecutive maximum points? Explain.

PROJECT

In this Tutorial, you worked with radian measure and determined the basic properties of a sine curve. You will use these skills to complete the Chapter Project, and in the following projects in the Project Book.

- Surviving on the Ocean's Shore
- Sun, Moon, and Stars

COMMUNICATING THE IDEAS

Write to explain what a radian is and how it is related to degrees.

TUTORIAL 5.3

Fitting Sine Curves to Data

When you build a mathematical model of a process, a key step is to represent a collection of data with an equation. One powerful method by which to obtain such an equation is statistical regression. A best-fit equation can then be used to analyze the original process and to make predictions.

In this Tutorial, you will use a graphing calculator to fit sine curves to periodic data, and use the resulting equations to make predictions. First, you have to determine if a set of data is periodic in nature. Then, you need to make a rough estimate of the period. With this information you are ready to determine the sinusoidal regression.

Investigation 1

Times of Sunrise

The data in the following table list the time of sunrise for Brandon, Manitoba for 1998 and 1999 presented in *Investigation 2* of *Tutorial 5.1*. You will fit a regression equation to these data then predict the time of sunrise.

1998	Day Number	Time of Sunrise (A.M.)	1999	Day Number	Time of Sunrise (A.M.)
Jan. 15	15	8.53	Jan. 15	380	8.53
Feb. 15	46	7.83	Feb. 15	411	7.83
Mar. 15	74	6.88	Mar. 15	439	6.90
Apr. 15	105	5.78	Apr. 15	470	5.80
May 15	135	4.90	May 15	500	4.92
June 15	167	4.52	June 15	531	4.52
July 15	197	4.80	July 15	561	4.80
Aug. 15	228	5.50	Aug. 15	592	5.48
Sept. 15	259	6.27	Sept. 15	613	6.25
Oct. 15	289	7.02	Oct. 15	653	7.02
Nov. 15	320	7.87	Nov. 15	683	7.87
Dec. 15	349	8.52	Dec. 15	714	8.52

1. Enter the data for Day Number in list L1 of the TI-83 calculator and the data for Time of Sunrise in list L2. Draw a scatterplot of the data.

2. Use your graphing calculator to find the sinusoidal regression equation for these data.

3. Use the regression equation to estimate the maximum and minimum values, the amplitude, and the period.

UTILITY

If you need help to perform a sinusoidal regression using the TI-83, Utility 32 shows you how. If you use a different graphing calculator, consult the user's manual that came with it.

4. What is the value of a in the regression equation? How is this related to the amplitude?

5. Recall that the period of a basic sine curve is 2π. What is the value of b in the regression equation? Calculate $\frac{2\pi}{b}$. How is this related to the period?

6. The median value is found using this formula:

$$\text{median value} = \frac{\text{maximum value} + \text{minimum value}}{2}$$

Find the median time of sunrise.

7. What is the value of d in the regression equation? How is this related to the median time of sunrise?

8. Use the regression equation to predict the time of sunset on December 31, 2000. Convert the time back to hours and minutes before you compare your time to published values.

You will need a CBL with a Vernier microphone/amplifier and a tuning fork to investigate the frequency of a musical note.

Investigation 2 **Determine the Frequency of a Musical Note.**

Work with a partner.

1. Gently tap the tuning fork and hold it toward the microphone for a few seconds.

2. From the data gathered by the CBL, select an interval of 0.05 s.

3. Graph the data generated in this interval. Estimate the period of the vibration.

4. Use the sinusoidal regression feature to find an equation in the form $y = a\sin(bx + c) + d$. Graph the regression equation with the data.

5. What are the maximum, minimum, and median value, the amplitude, and the period of this curve? How do the values of a, $\frac{2\pi}{b}$, and d relate to the amplitude, period, and median?

6. Use the period derived from the regression equation to predict the frequency of the tuning fork. Recall that frequency (Hz) $= \dfrac{1}{\text{period(s)}}$.

7. Which do you think is more accurate — the frequency obtained from the regression equation or the frequency stamped on the tuning fork? Explain.

The periodic rise and fall of the water along ocean shores is the result of gravitational forces of the moon and the sun. Despite the complex nature of tides, the depth of the water versus time, at a given location, can be approximated by a sinusoidal function.

Example

Sine regression and tides

The following data are water levels recorded on November 18 and 19, 2000 at the Oak Bay station in British Columbia.

Time (h)	Depth (m)	Time (h)	Depth (m)	Time (h)	Depth (m)	Time (h)	Depth (m)
0	1.0	13	2.8	26	0.9	38	2.6
2	0.6	14	2.6	27	0.9	39	2.4
3	0.7	15	2.5	28	1.0	40	2.3
4	0.9	16	2.5	29	1.2	41	2.2
5	1.3	17	2.4	30	1.5	42	2.1
6	1.6	18	2.4	31	1.8	43	2.1
7	2.0	19	2.3	32	2.2	44	2.1
8	2.4	20	2.3	33	2.5	45	2.1
9	2.7	21	2.2	34	2.8	46	2.1
10	2.9	22	2.0	35	2.9	47	2.0
11	3.0	23	1.8	36	3.0		
12	2.9	24	1.5	37	2.9		

a) Use the sinusoidal regression feature to find the equation of the sinusoidal curve of best fit. Graph the regression equation.

b) Use the graph of the regression equation to estimate the maximum, minimum, and median values, the amplitude, and the period. How do the values of a, $\frac{2\pi}{b}$, and d relate to the amplitude, period, and median?

c) Use your equation to estimate the depth of the water at 6:30 P.M. on November 18. Is the tide coming in or going out at this time?

d) Between what times is the depth of the water more than 2.5 m?

SOLUTION

a) 

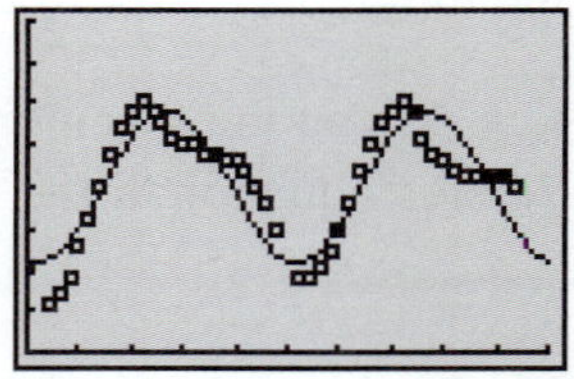

$D \doteq 0.91 \sin (0.25t - 1.81) + 2$, where D is the depth of water in metres and t is the time in hours.

b) Use the CALC feature to find the maximum and minimum values. The maximum depth is approximately 2.91 m and the minimum depth is approximately 1.09 m.

The median is approximately $\frac{2.91 + 1.09}{2}$ or 2. This is the same as the value of d.

The amplitude is approximately $\frac{2.91 - 1.09}{2}$ or 0.91. This is the same as the value of a.

On the graph, the first maximum occurs at 13.52 h, and the second maximum occurs at 38.66 h. The period is approximately 38.66 – 13.52 or 25.14 h. The value of $\frac{2\pi}{b}$ is 25.13. This is very close to the estimate of the period.

c) At 6:30 P.M. or 18.5 h, $D = 0.91 \sin [0.25(18.5) - 1.81] + 2$

$\doteq 2.29$ m

From your graph you can see that the depth of the water was decreasing at 6:00 P.M., so the tide must have been going out.

d) Set Y2 to 2.5 and press GRAPH.

Method 1: The times when $D > 2.5$ can be estimated from the graphs to occur between 10:00 and 17:30, and between 10:15 and 18:00.

Method 2: The CALC-Intersect feature of the TI-83 can be used to find the points of intersection more accurately.

UTILITY

Refer to Utility 19 for instructions on how to use the CALC-Intersect feature of the TI-83.

The water is more than 2.5 m deep from 9.92 h to 17.44 h, and from 34.36 h to 41.89 h. After converting these times into hours and minutes, the times are from 9:55 to 17:26 on November 18, and from 10:22 to 17:53 on November 19.

DISCUSSING THE IDEAS

1. What pattern must be evident in the data before you decide to use SinReg?

2. Why should you provide an estimate of the period when you use SinReg?

3. Why do you need to interpret the results after you use SinReg?

4. In the *Example*, how well does the regression equation fit the data? Explain what this means in terms of the accuracy of predictions made using the regression equation.

EXERCISES

Checking Your Skills

1. The table shows the times of sunset in Central Standard Time at Prince Albert, Saskatchewan for the 1st of each month for the years 1999 and 2000. Times are shown as decimal fractions of an hour on the 24-h clock.

Date (1999)	Day Number	Time of Sunset	Date (2000)	Day Number	Time of Sunset
Jan. 1	1	16.93	Jan. 1	366	16.93
Feb. 1	32	17.78	Feb. 1	397	17.77
Mar. 1	60	18.68	Mar. 1	426	18.70
Apr. 1	91	19.63	Apr. 1	457	19.65
May 1	121	20.53	May 1	487	20.55
June 1	152	21.33	June 1	518	21.35
July 1	182	21.55	July 1	548	21.55
Aug. 1	213	20.97	Aug. 1	579	20.95
Sept. 1	244	19.88	Sept. 1	610	19.85
Oct. 1	274	18.68	Oct. 1	640	18.65
Nov. 1	305	17.53	Nov. 1	671	17.50
Dec. 1	335	16.83	Dec. 1	701	16.82

a) Enter the data in your calculator and use the sinusoidal regression feature to find a best-fit equation to represent the data.

b) Predict the time of sunset for March 1, 2001.

c) What is the median time of sunset?

d) How much does the maximum time of sunset vary from the median time of sunset?

2. The following data are water levels recorded on November 18 and 19, 2000 at Sidney, British Columbia.

Time (h)	Depth (m)
0	1.3
2	0.7
3	0.6
4	0.7
5	1.0
6	1.4
7	1.9
8	2.4
9	2.8
10	3.1
11	3.3
12	3.3

Time (h)	Depth (m)
13	3.1
14	2.9
15	2.7
16	2.5
17	2.3
18	2.3
19	2.2
20	2.3
21	2.3
22	2.2
23	2.0
24	1.8

Time (h)	Depth (m)
26	1.2
27	1.0
28	0.9
29	1.0
30	1.3
31	1.6
32	2.1
33	2.5
34	2.9
35	3.1
36	3.3
37	3.2

Time (h)	Depth (m)
38	3.0
39	2.7
40	2.4
41	2.2
42	2.0
43	1.9
44	1.9
45	2.0
46	2.1
47	2.1

a) Use the sinusoidal regression feature to find the equation of the sinusoidal curve of best fit. Graph the regression equation.

b) Find the median depth.

c) Predict the depth of the water at 11:45 A.M. on November 19.

d) Between what times is the depth of the water more than 3 m?

Sailboats near Sidney, British Columbia

3. The depth of the water at Egeria Bay, Langara Island, BC was measured seven times during one day.

Time (h)	1.88	5.20	8.60	11.88	15.08	17.73	20.48
Depth (m)	4.4	2.7	0.7	2.3	3.9	3.0	2.0

Times are for a 24-h clock, and have been expressed in hours and decimal fractions of an hour.

a) Enter the data in your calculator and use the sinusoidal regression feature to find a regression equation to represent the data.

b) Predict the depth of the water at 14:58.

c) What is the median depth of the water?

d) How much does the maximum depth of water vary from its median depth?

4. The following table shows the height of a seat on a Ferris wheel, at specific times as the wheel rotates.

Time (s)	Height (m)	Time (s)	Height (m)
0	1.5	25	22.8
5	4.8	30	15.8
10	12.3	35	7.3
15	20.2	40	2.0
20	24.5	45	3.8

a) Enter the data in your calculator and use the sinusoidal regression feature to find a regression equation to represent the data.

b) How long does the Ferris wheel take to make one revolution?

c) What is the maximum height of the seat above the ground?

d) What is the diameter of the Ferris wheel?

e) How high will the seat be after 64 s?

5. The table lists the maximum daily temperature for the 1st of each month during 1999 at Fort Smith, NWT.

Date (1999)	Day Number	Max. Temp (°C)
Jan. 1	1	−28.8
Feb. 1	32	−13.8
Mar. 1	60	−7.2
Apr. 1	91	3.2
May 1	121	12.8
June 1	153	20.4
July 1	183	17.8
Aug. 1	214	22.0
Sept. 1	245	13.0
Oct. 1	275	1.1
Nov. 1	306	0.6
Dec. 1	335	−13.2

a) Enter the data in your calculator, and obtain a scatterplot of the data. Estimate the period.

b) Use the sinusoidal regression feature to obtain a best-fit equation for the data.

c) Use the graph of this equation to predict the highest maximum temperature, the lowest maximum temperature, and the dates when both of these will occur.

d) What is the median yearly temperature?

e) How much do the maximum temperatures vary from the median yearly temperature?

f) How well does the regression equation fit the data? How accurate are your answers from parts c, d, and e?

Extending Your Thinking

6. The following are the times of sunrise and sunset at Medicine Hat, Alberta, for the 2nd of each month during 2000.

Date	Day Number	Time of Sunrise	Time of Sunset
Jan. 2	2	8.35	16.53
Feb. 2	33	7.93	17.30
Mar. 2	62	7.05	18.12
Apr. 2	93	5.93	18.95
May 2	123	4.93	19.73
June 2	154	4.28	20.42
July 2	184	4.30	20.58
Aug. 2	215	4.90	20.05
Sept. 2	246	5.67	19.07
Oct. 2	276	6.42	17.97
Nov. 2	307	7.25	16.95
Dec. 2	337	8.03	16.38

a) Analyze the data. Enter appropriate values in your calculator. Find regression equations for each of the following.

i) Sunrise time versus the day of the year

ii) Sunset time versus the day of the year

iii) The number of hours of sunlight versus the day of the year

b) Some students in Medicine Hat and the surrounding area, catch the bus as early as 7:15 A.M., and do not get home until 4:30 P.M. School starts at 8:30 A.M., and ends at 3:15 P.M. each day. During how many days will students ride the bus in the dark? List any assumptions that you make when you answer this question.

PROJECT

In this Tutorial, you fitted a regression equation to periodic data. You will use this skill to complete the Chapter Project, and in the following projects in the Project Book.

- Surviving on the Ocean's Shore
- Sun, Moon, and Stars

COMMUNICATING THE IDEAS

Explain the advantages and disadvantages of using the curve of best fit to make predictions.

TUTORIAL

5.4 The Characteristics of $y = a\sin(bx + c) + d$

Many situations that involve periodic data can be studied by fitting a sine curve of the form $y = a\sin(bx + c) + d$ to the data so that the maximum values, minimum values, and period match the given data closely. In this Tutorial, you will study how the characteristics of a sinusoidal pattern are related to the equation and the graph of the pattern.

Parameters are the constants in the equation of a function. In the Investigation you will examine how the parameters in the equation of a sinusoidal function are related to the characteristics of the graph of the function.

Investigation

Exploring the Effect of the Parameters *a*, *b*, *c*, and *d*.

The Jones' have a water wheel on their farm. The water wheel has radius 1 m, turns clockwise, and makes one complete revolution every 6.28 s. The waterline is at the middle of the wheel.

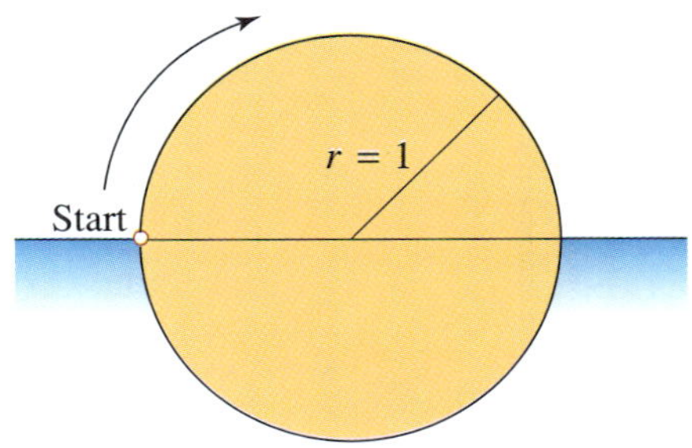

Set the window on the TI-83 calculator to Xmin = 0, Xmax = 25.5, Xscl = 1, Ymin = –2, Ymax = 2, Yscl = 0.5.

Part A: The Basic Sine Curve

1. Use the point where the wheel exits the water as the start point. Copy and complete the table.

Revolution	0	$\frac{1}{4}$	$\frac{1}{2}$	$\frac{3}{4}$	1	$1\frac{1}{4}$	$1\frac{1}{2}$	$1\frac{3}{4}$	2
Time (s)	0	1.57	3.14	4.71	6.28	7.85	9.42	10.99	12.56
Height (m)	0								

2. Enter the data for Time in list L1 and the data for Height in list L2 in your calculator. Find the sinusoidal regression equation of best fit. Store the regression equation in Y1. Round the value of each of the parameters to two decimal places.

3. Graph the regression equation and print the results or sketch it into your notebook.

Part B: Shifting the Sine Curve Horizontally

Change the start point to the point where the wheel enters the water.

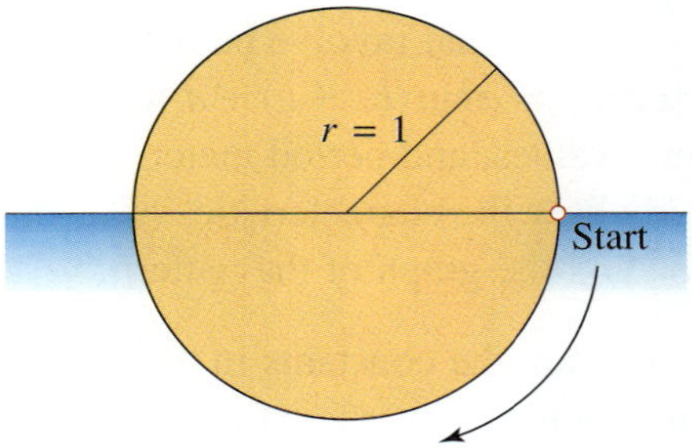

4. Copy and complete the table.

Revolution	0	$\frac{1}{4}$	$\frac{1}{2}$	$\frac{3}{4}$	1	$1\frac{1}{4}$	$1\frac{1}{2}$	$1\frac{3}{4}$	2
Time (s)	0	1.57	3.14	4.71	6.28	7.85	9.42	10.99	12.56
Height (m)	0								

5. Enter the data in your graphing calculator, find the sinusoidal regression equation of best fit and store the equation in Y2. Round the value of each of the parameters to two decimal places.

6. Graph this equation and the equation from exercise 2 on the same screen. Print this graph or copy it into your notebook.

7. Use the graph of the equation from exercise 5. Find each value.

a) amplitude

b) period

c) median value

d) maximum and minimum values

e) the x-coordinate of the point on the increasing part of the curve closest to the y-axis whose y-coordinate is the median value.

8. Use the parameters in the regression equation to find each value. How do these values compare to your results in exercise 7?

a) a **b)** $\frac{2\pi}{b}$ **c)** d

d) $d + a$ and $d - a$ **e)** $-\frac{c}{b}$

9. Compare the equations from exercise 2 and exercise 5. How are they different? Compare the graphs of these equations. How are they different?

Part C: Changing the Amplitude

The water wheel on the Colyn's farm has radius 2 m. It makes one complete revolution every 6.28 s.

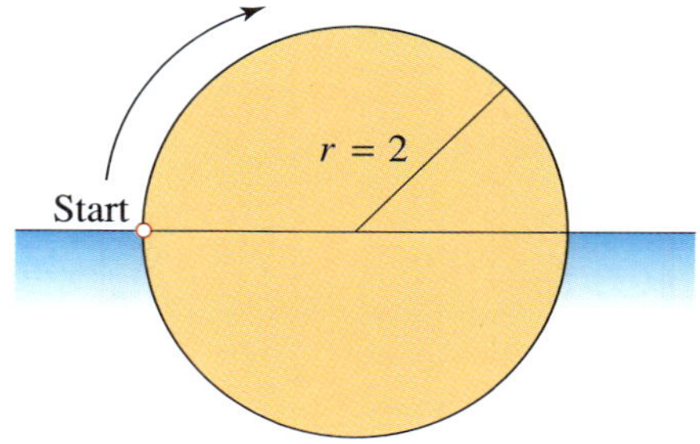

10. Use the point where the wheel exits the water as the start point. Copy and complete the table.

Revolution	0	$\frac{1}{4}$	$\frac{1}{2}$	$\frac{3}{4}$	1	$1\frac{1}{4}$	$1\frac{1}{2}$	$1\frac{3}{4}$	2
Time (s)	0	1.57	3.14	4.71	6.28	7.85	9.42	10.99	12.56
Height (m)	0								

11. Repeat exercises 5 to 9 for the Colyn's water wheel.

Part D: Changing the Period

The Jones' adjust the rate of water flow so the wheel makes one complete revolution every 12.56 s.

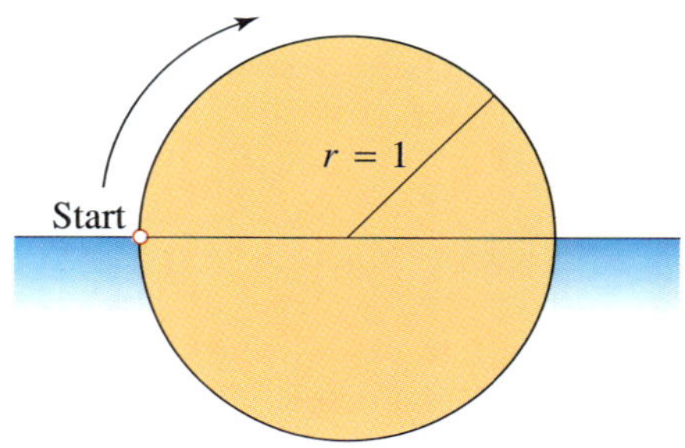

12. Use the point where the wheel comes out of the water as the start point. Copy and complete the table. Explain why the times are different in the table below.

Revolution	0	$\frac{1}{4}$	$\frac{1}{2}$	$\frac{3}{4}$	1	$1\frac{1}{4}$	$1\frac{1}{2}$	$1\frac{3}{4}$	2
Time (s)	0	3.14	6.28	9.42	12.56	15.70	18.84	21.98	25.12
Height (m)	0								

13. Repeat exercises 5 to 9.

Part E: Changing the Median Value

The Jones' re-adjust the water flow so the wheel again makes one complete revolution every 6.28 s. The water level drops so the water just touches the bottom of the wheel.

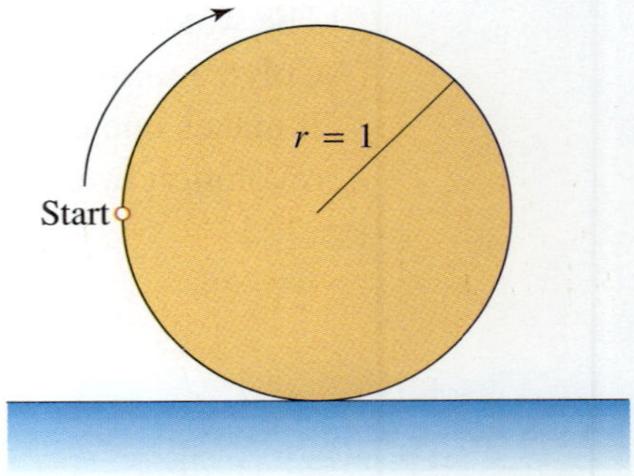

14. Use the same starting point as in exercise 1. Copy and complete the table.

Revolution	0	$\frac{1}{4}$	$\frac{1}{2}$	$\frac{3}{4}$	1	$1\frac{1}{4}$	$1\frac{1}{2}$	$1\frac{3}{4}$	2
Time (s)	0	1.57	3.14	4.71	6.28	7.85	9.42	10.99	12.56
Height (m)	1								

15. Repeat exercises 5 to 9.

Part F: Changing More Than One Parameter

Use the scenario from part E, but move the starting point to the point where the wheel touches the water.

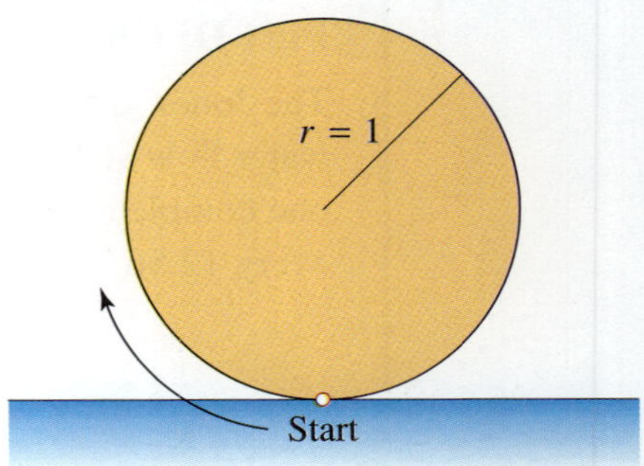

16. Copy and complete the table.

Revolution	0	$\frac{1}{4}$	$\frac{1}{2}$	$\frac{3}{4}$	1	$1\frac{1}{4}$	$1\frac{1}{2}$	$1\frac{3}{4}$	2
Time (s)	0	1.57	3.14	4.71	6.28	7.85	9.42	10.99	12.56
Height (m)	0								

17. Repeat exercises 5 to 9.

A sinusoidal function is continuous in both directions, so there is no start or endpoint. For the sake of simplicity, we define the **start point** as (0, 0) on $y = \sin x$. This is the middle point on the increasing portion of the sine curve that is closest to the y-axis. Here is a summary of the results of the Investigation.

$$y = a \sin(bx + c) + d$$

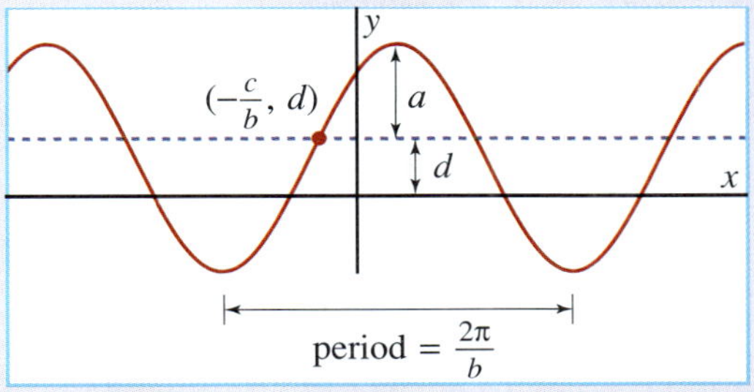

Amplitude: a

Period: $\frac{2\pi}{b}$

Middle or Median value: d

Maximum value: $d + a$

Minimum value: $d - a$

Example 1 **Determine the characteristics from an equation.**

Given $y = 5 \sin 3x + 2$.

a) Find the characteristics of the curve.

b) Graph the equation.

SOLUTION

a) Compare the equation $y = 5 \sin 3x + 2$ to $y = a \sin(bx + c) + d$, then use the results of the Investigation.

$a = 5$	so	amplitude = 5
$b = 3$	so	period = $\frac{2\pi}{3}$
$c = 0$	so	start point = (0, 2)
$d = 2$	so	median value = +2
		maximum value = 2 + 5 = 7
		minimum value = 2 − 5 = −3

b) Use a graphing calculator with window settings Xmin = $-\pi$, Xmax = π, Xscl = $\frac{\pi}{6}$, Ymin = −4, Ymax = 8, Yscl = 1 to obtain this graph.

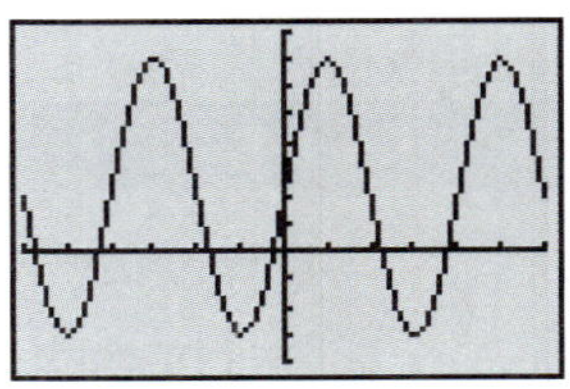

Example 2 Determine the characteristics from a graph.

The graph of $y = a\sin(bx + c) + d$ is shown.

```
WINDOW
 Xmin=-6.5
 Xmax=6.5
 Xscl=1
 Ymin=-5
 Ymax=3
 Yscl=1
 Xres=1
```

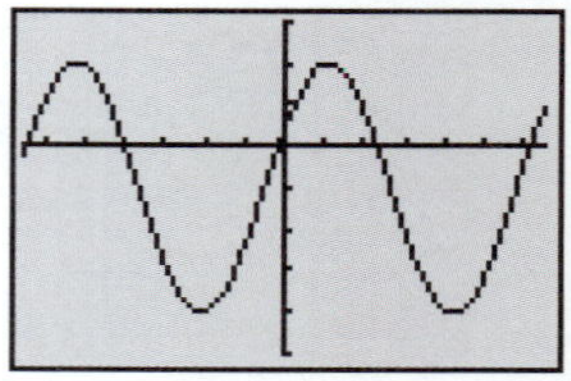

Use the graph to answer the following questions.

a) What is the amplitude?

b) What is the middle value?

c) What is the period?

d) What is the start point?

SOLUTION

a) From the graph, the maximum is +2 and the minimum is –4. The amplitude is 3.

b) The middle value is –1.

c) One period extends from about –4 to 2.2, or approximately 6.2.

d) The start point is about (–0.5, –1).

DISCUSSING THE IDEAS

1. Summarize how each parameter in the equation $y = a\sin(bx + c) + d$ can be estimated from its graph.

2. Explain how the amplitude and the median value of a sinusoidal function are related.

EXERCISES

Checking Your Skills

1. For each of the sinusoidal functions state the following.

i) maximum value
ii) minimum value
iii) amplitude
iv) period
v) median value
vi) start point

a) $y = 2\sin x - 1$ **b)** $y = \sin(2x + \frac{\pi}{3})$ **c)** $y = 3\sin 4x$

d) $y = 0.66\sin 2x - 0.5$ **e)** $y = \sin(3x - 2) + 1$

2. Determine the characteristics for each of the sinusoids shown in the following graphs.

a)

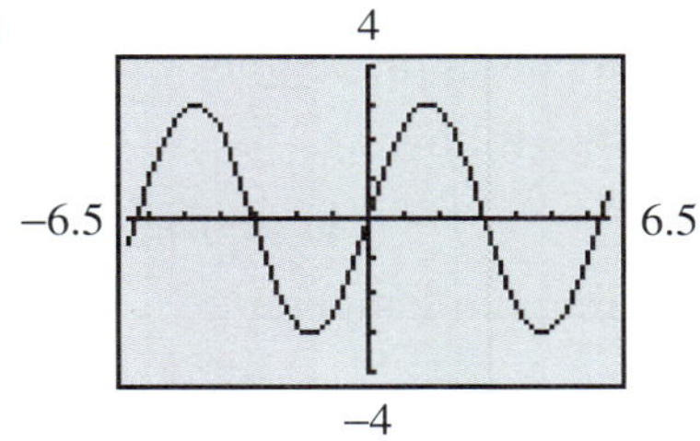

b)

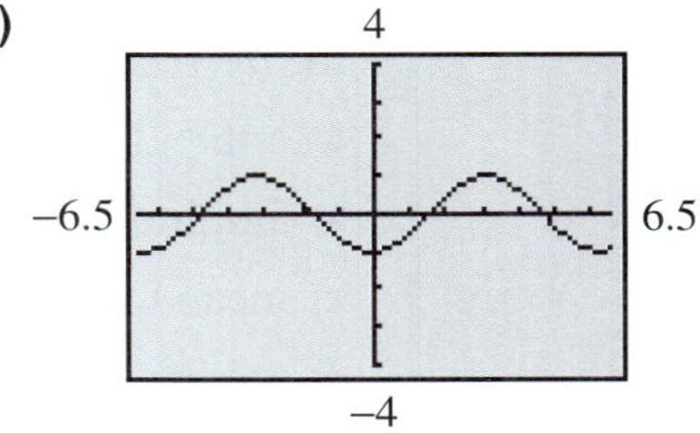

c)

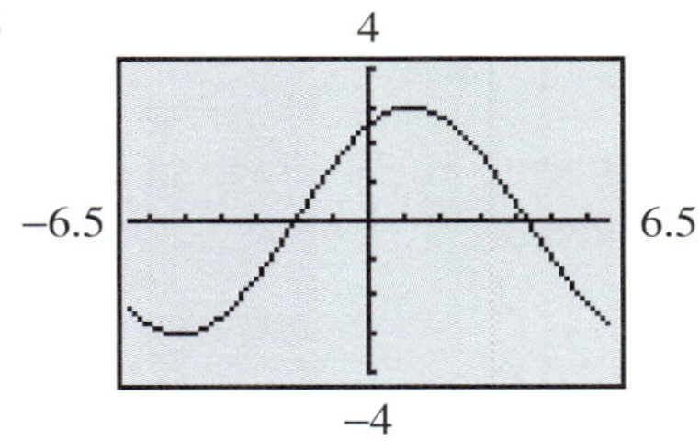

d)

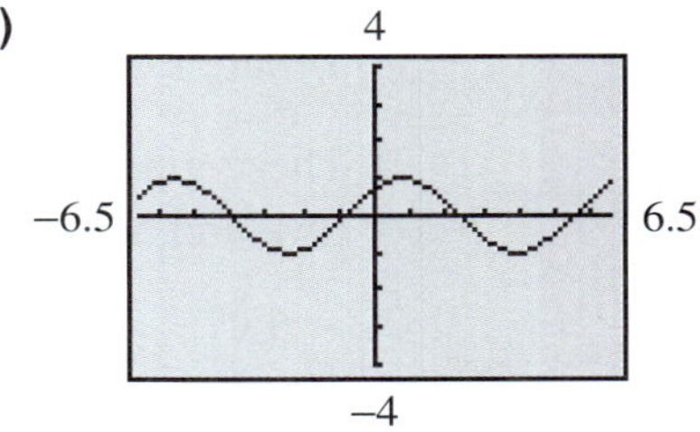

3. Start with the graph $y = \sin x$ and describe in words how it could be changed to look like the graph $y = 3\sin 2x - 1$.

4. In music, each note has related notes in higher and lower octaves. The following graphs show the wavelengths of the note A played on three tuning forks at three consecutive octaves.

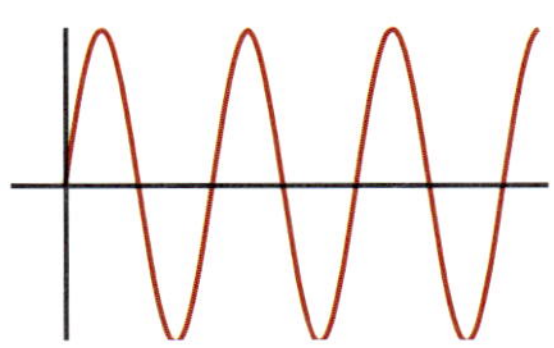

Frequency = 880 Hz

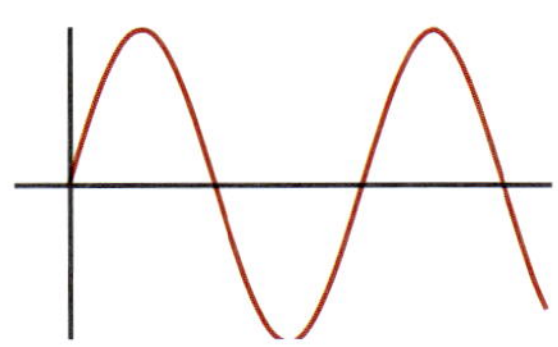

Frequency = 440 Hz

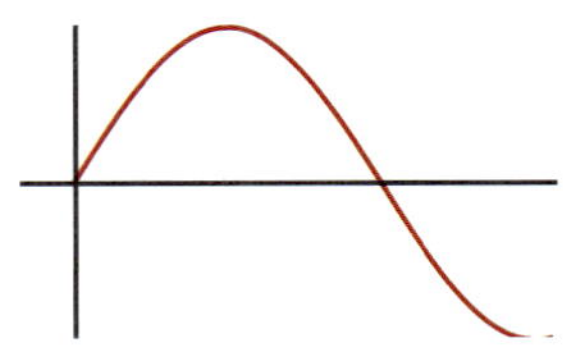

Frequency = 220 Hz

a) What can you conclude about octaves?

b) Describe the characteristics of the three graphs.

5. The graph illustrates the vibration of a guitar string after it has been struck. Describe how the amplitude and period change over time. How would you be able to tell that this was happening just by listening to the guitar?

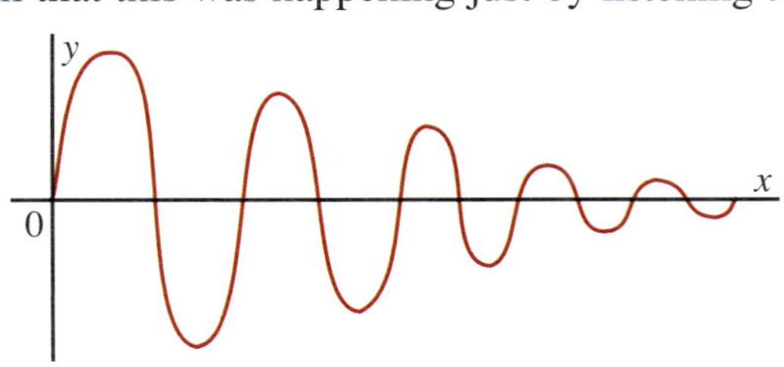

6. A temperature-time graph is shown for Fort Chipewyan, Alberta. It was created with long-term mean data for each date during a calendar year. The variable plotted on the horizontal axis is the number of days after January 1st. The variable plotted on the vertical axis is the temperature in degrees Celsius.

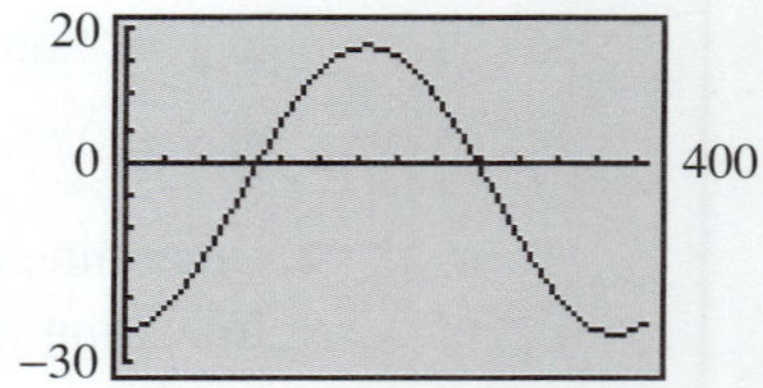

Use the graph to estimate the following.

a) amplitude
b) period
c) mean maximum and minimum temperature
d) median temperature for the year
e) date for the maximum temperature
f) date for the minimum temperature

7. These graphs show the patterns produced on an oscilloscope when a tuning fork and a clarinet were played.

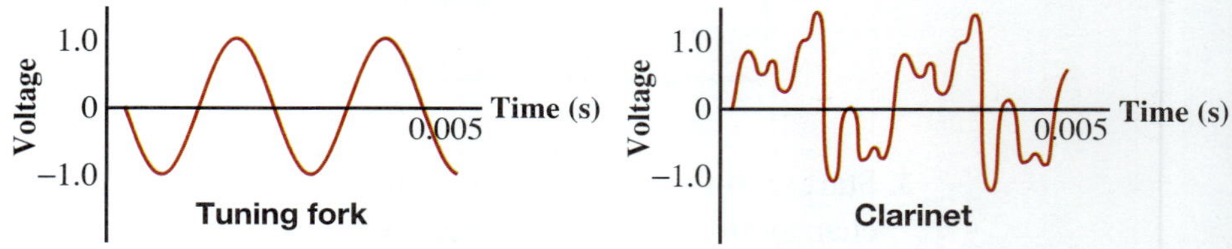

a) For each instrument, use the graph to:
   i) estimate the amplitude
   ii) estimate the period
b) Describe how each graph would change if the instrument were played louder.

8. The twin towers of the World Trade Center in New York were once the tallest buildings in the world. During a strong wind, the top of each tower can swing back and forth as much as 80 cm, with a period of 10 s. Sketch a graph to show the movement of the top of one of the buildings from the rest position as a function of time, for a 20-s interval.

PROJECT

In this Tutorial, you examined the characteristics of sinusoidal functions. You will use this skill to complete the Chapter Project, and in the following projects in the Project Book.

- Surviving on the Ocean's Shore
- Sun, Moon, and Stars

COMMUNICATING THE IDEAS

Write to explain how the value of the parameters a, b, c, and d are related to the graph of $y = a\sin(bx + c) + d$.

Applications

Many applications involve sinusoidal data. In this Tutorial, you will have more opportunities to develop your skills as you work with, and analyze situations with sinusoidal data.

Example 1

AC voltage

Voltage is generated between the ends of a wire when it is moved through a magnetic field. When a wire is coiled around an armature, and the armature is rotated in a magnetic field, then a periodic voltage is generated. A graph of voltage versus time is shown.

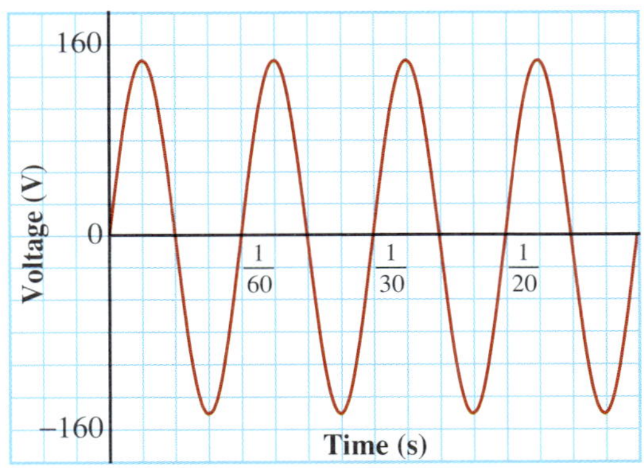

a) What is the median value of the voltage?

b) What is the amplitude of this graph?

c) What is the period of this graph?

d) What is the start point of this curve?

SOLUTION

a) The median voltage through one cycle is zero.

b) The amplitude is 160. This means that the voltage varies from +160 V to −160 V.

c) The period is $\frac{1}{60}$ s, which is the time it takes the armature to complete one rotation.

d) The start point of this curve is (0, 0).

Example 2 **Average daily temperatures at Eureka**

The following equation provides an estimate of the average monthly temperatures at Eureka, Nunavut.

$$T = 22.39 \sin(0.53M - 1.96) - 16.78,$$

where T is the temperature (°C) and M is the number of the month of the year (January is 1, February is 2, etc.)

Compare this equation to $y = a \sin(bx + c) + d$.

a) What is the value of d? What does this number represent in terms of the temperature data?

b) What is the value of a? What does this number represent in terms of the temperature data?

c) What is the value of b? What is the period? What does this number represent?

d) What is the value of c? What information can we get from this?

e) What is the daily mean temperature for July?

SOLUTION

a) The value of d is –16.78, so the median temperature is –16.78°C. This means that the average temperature for the whole year is –16.78°C.

b) The value of a is 22.39, so the amplitude is 22.39. Combined with the value of d, this means that the average monthly temperature varies from $-16.78 + 22.39$ or +5.61°C to $-16.78 - 22.39$ or –39.17°C.

c) The value of b is 0.53, so the period is $\frac{2\pi}{0.53} \doteq 11.86$ months. This means that the temperature cycle repeats about every 11.86 months.

d) The value of c is –1.96, so the x-coordinate of the start point is $\frac{1.96}{0.53} \doteq 3.7$. This means that the median temperature of –16.78°C is first reached during the fourth month of the year.

e) Substitute $M = 7$ and calculate to find that the daily mean temperature for July is approximately 5.3°C.

The sun appears to move further south in the sky during the winter and to return toward the north during summer. The declination of the sun is equal to its angle above or below the equator, as measured by a person on the equator at noon.

Example 3 The apparent annual motion of the sun

The following table provides the angle of declination of the sun on different days during the year. Positive angles indicate that the sun is north of the equator.

Date	Day Number	Angle of Declination	Date	Day Number	Angle of Declination
Jan. 4	4	−22.8	July 4	186	22.9
Jan. 18	18	−20.7	July 18	200	21.1
Feb. 4	35	−16.5	Aug. 4	217	17.4
Feb. 18	49	−11.9	Aug. 18	231	13.3
Mar. 4	63	−6.7	Sept. 4	248	7.4
Mar. 18	77	−1.2	Sept. 18	262	2.1
Apr. 4	94	5.5	Oct. 4	278	−4.1
Apr. 18	108	10.6	Oct. 18	282	−9.4
May 4	124	15.8	Nov. 4	309	−15.2
May 18	138	19.4	Nov. 18	323	−19.1
June 4	156	22.4	Dec. 4	338	−22.2
June 18	170	23.4	Dec. 18	352	−23.4

a) Find the equation of the sinusoidal curve of best fit.

b) Use your equation to predict the angle of declination of the sun on December 7.

SOLUTION

a) Enter the data in a graphing calculator. Use the sinusoidal regression feature. The equation of the sinusoidal curve of best fit is $A = 23.553\sin(0.017D - 1.364) - 0.003$, where A is the angle of declination in degrees and D is the number of the day in the year.

b) Counting from Jan. 1, Dec. 7 would be day 341. Substitute $D = 341$ into the equation in part a.
$A = 23.553\sin[0.017(341) - 1.364] - 0.003 \doteq -22.64$. On December 7, the declination of the sun would be about $-22.64°$.

Discussing the Ideas

1. Refer to *Example 1*. Where else does voltage vary with a period of $\frac{1}{60}$ s?

2. Given the period in *Example 2*, is it possible for the winters to get warmer and the summers colder given enough time? Explain.

3. In *Example 3*, why does the sun's declination vary from −23.4° to +23.4° during a year?

Exercises

Checking Your Skills

1. A mass is supported by a spring so that it rests 0.5 m above a tabletop. The mass is pulled down 0.4 m and released at time $t = 0$. This creates a periodic up-and-down motion, called simple harmonic motion. It takes 1.2 s for the mass to return to the low position each time.

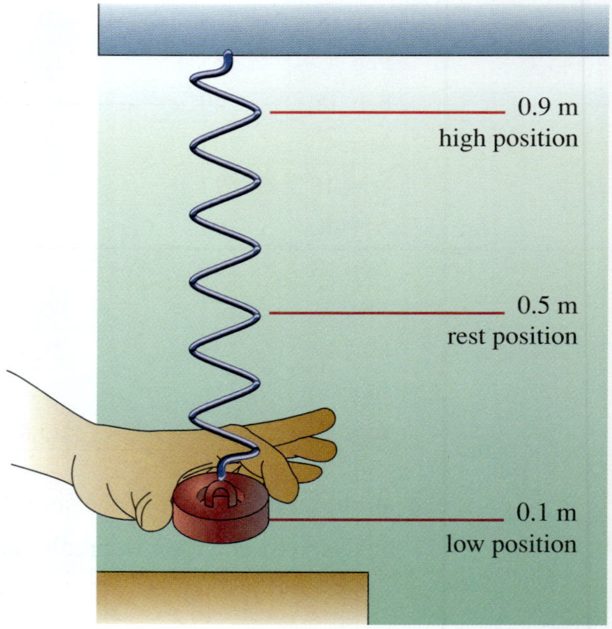

   a) From the given data, identify the amplitude, period, and vertical translation for the motion of the oscillating mass. Identify the horizontal translation that would apply to a graph of displacement versus time.

   b) Sketch a graph that describes the motion of the mass.

   c) Use your graph to find the height of the mass above the table after 0.3 s and 1.5 s.

   d) Determine the time when the mass is 0.75 m above the tabletop for the first time.

2. The graph shows how the depth of water, in metres, at Copper Island, Hectate Strait, BC, varies during a typical day.

a) What are the characteristics of this graph?

b) Use the graph to find the depth of the water at 7:30 P.M.

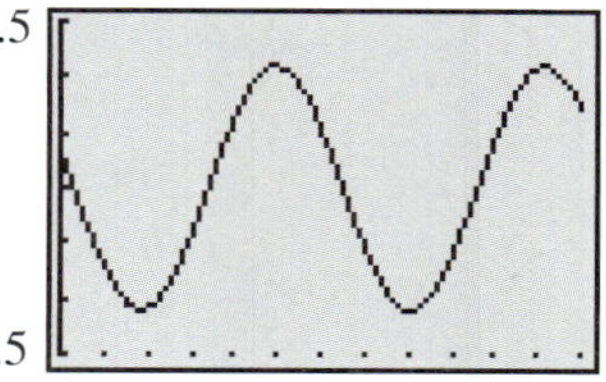

3. The following table provides the mean minimum monthly temperature during two years at Winnipeg, Manitoba.

	1997	T min (°C)		1998	T min (°C)
1	Jan.	−24.5	13	Jan.	−18.4
2	Feb.	−20.0	14	Feb.	−7.6
3	Mar.	−15.8	15	Mar.	−9.7
4	Apr.	−5.7	16	Apr.	0.2
5	May	0.7	17	May	4.7
6	June	11.4	18	June	10.7
7	July	13.9	19	July	12.4
8	Aug.	11.1	20	Aug.	13.3
9	Sept.	7.1	21	Sept.	6.9
10	Oct.	0.1	22	Oct.	1.5
11	Nov.	−9.0	23	Nov.	−7.1
12	Dec.	−9.0	24	Dec.	−17.4

a) Find the equation of best fit for these data.

b) Use your equation to predict the mean minimum monthly temperature for January, 1999.

c) Find the actual mean minimum monthly temperature for January, 1999. How does the value compare to your predicted value? Why might these values be different?

Aerial view of downtown Winnipeg and the Assiniboine River.

d) Suppose that in 1999 Winnipeg was, on average, 5°C warmer than in 1998. How would you adjust your equation of best fit to better predict the temperatures for 2000?

4. A piston in an engine moves up and down in its cylinder in a sinusoidal fashion. The motion of the piston is described by the equation $h = 20\sin(125.66t) + 20$, where h is the height in centimetres and t is the time in seconds.

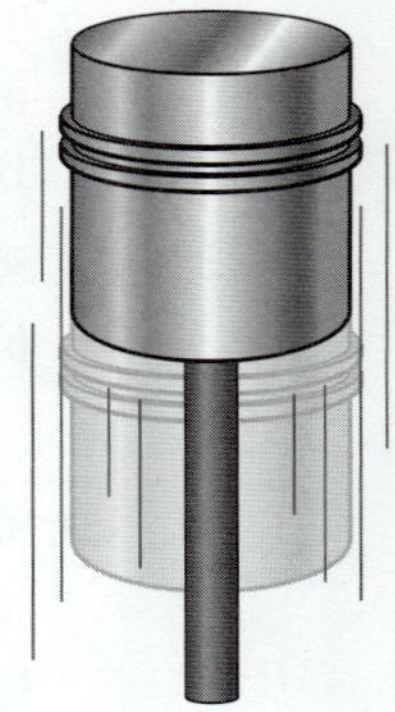

a) State the piston's maximum height, minimum height, and period.

b) Suppose the piston operates for 1 h. How many complete cycles does it make?

5. The mean daily maximum temperatures at Komakuk Beach, in the Yukon, were recorded for each month from 1958 to 1990. The data are represented by the formula $T = 17.11\sin(0.60M - 2.83) - 5.38$, where T is the temperature (°C) and M is the number of the month in the year (1 to 12).

a) What is the mean daily temperature for May?

b) In a typical year, during how many months will the mean daily temperatures be higher than 0°C?

6. The following table lists the angle of elevation of the centre of the sun during one day, at Kamloops, BC, on January 23, 2000. Positive values indicate that the sun is above the horizon, negative values, below. The sun has an angular diameter of 0.5°. Sunrise occurs when the upper limb of the sun first appears above the horizon. Sunset occurs when the upper limb of the sun just disappears below the horizon.

Time (h)	Elev. (°)	Time (h)	Elev. (°)	Time (h)	Elev. (°)
6.67	−11.1	10.67	16.8	14.67	12.9
7.00	−8.1	11.00	17.9	15.00	11.0
7.33	−5.2	11.33	18.8	15.33	9.0
7.67	−2.3	11.67	19.4	15.67	6.7
8.00	0.8	12.00	19.8	16.00	4.4
8.33	3.2	12.33	19.8	16.33	1.9
8.67	5.6	12.67	19.6	16.67	−1.0
9.00	7.9	13.00	19.1	17.00	−3.8
9.33	10.1	13.33	18.4	17.33	−6.7
9.67	12.1	13.67	17.4	17.67	−9.7
10.00	13.9	14.00	16.1		
10.33	15.5	14.33	14.6		

a) Find the equation of the sinusoidal curve of best fit.

b) What are the sunrise and sunset times for Kamloops on January 23, 2000?

c) For how many minutes of the day will the sun be above the horizon?

d) Is the sun ever directly overhead on January 23, 2000?

Extending Your Thinking

7. Botanical Beach Provincial Park on Vancouver Island is noted for its striking geological features and abundant shore life. When the tide is out, visitors can see a wide variety of live sea creatures in tidal pools. The following data on water depth were collected on one day.

Time (h)	5.15	11.32	17.67	23.70
Depth (m)	1.19	2.86	1.11	2.71

a) Graph the data on grid paper.

b) Connect the points to make a smooth sinusoidal curve.

c) Use the graph to estimate values for the parameters a, b, c, and d.

d) Substitute your values into the equation $D = a\sin(bT + c) + d$.

e) Plot the points using a graphing calculator and enter the equation in part d.

f) By trial and error, adjust the values of the parameters until your equation touches all four points.

g) Using the final version of your equation, estimate the depth of the water at 5:15 P.M.

h) Now use the sinusoidal regression feature to find a regression equation for the data. Use this equation to estimate the depth of the water at 5:15 P.M.

i) What are the advantages and disadvantages of using trial and error, versus sinusoidal regression?

PROJECT

In this Tutorial, you worked with periodic data and made predictions. You will use these skills to complete the Chapter Project, and in the following projects in the Project Book.

- Surviving on the Ocean's Shore
- Sun, Moon, and Stars

COMMUNICATING THE IDEAS

Search for a situation in which scientists model equations to predict future events. Write a short report on what you find. For example, you might research the topic of "global warming" to find how scientists try to predict the weather 50 years from now, or you might research "binary stars" to find how astronomers predict the positions of orbiting stars.

Regression and Prediction

In your previous work on the Chapter Project, you collected and plotted data on the angle of elevation of the sun.

Now, use a graphing calculator to determine the sinusoidal regression for this data.

1. Enter the data into your calculator.
2. Use SinReg to find the regression equation. Draw the graph of the regression equation.
3. What was the maximum angle of elevation of the sun that day? This angle can be estimated from the graph, or found using the TRACE function on the TI-83 calculator.
4. From your graph, estimate the times of sunrise, sunset, and maximum elevation of the sun.
5. What was the maximum angle of elevation of the sun that day? This angle can be estimated from the graph, or found using the TRACE function on the TI-83.

Your latitude, the declination of the sun, and the angle of the elevation of the sun are related by the following equation:

$\varphi = \theta - \alpha + 90°$

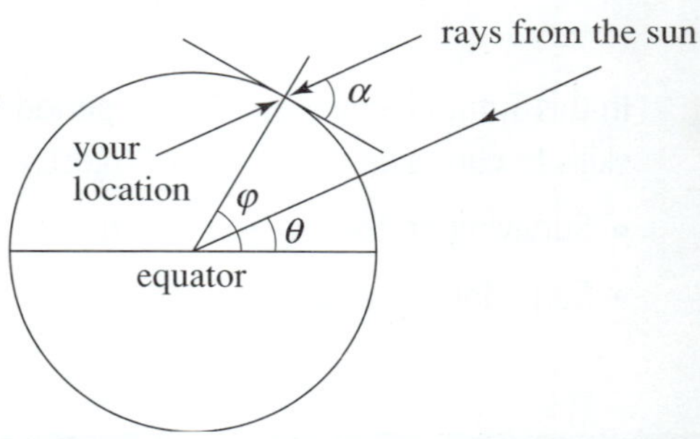

where

φ is the latitude of the observer
θ is the declination of the sun that day, and
α is the maximum angle of elevation of the sun on that day.

Marine biologist taking readings with a refractometer.

6. Use the equation $\varphi = \theta - \alpha + 90^\circ$ to estimate your latitude. You may want to review *Example 3* in *Tutorial 5.5* to find the declination of the sun on the day of your observation.

7. How would your data change if you did this experiment at a different location?

8. How would your data change if you did this experiment at a different time of year?

9. What is the period of the graph of the regression equation? How does this compare to the predicted period?

Your final project will consist of the following items.

- Your table of data
- Your hand drawn graph, with the estimates of the period and amplitude
- The equation of the curve of best fit
- Your interpretation of this curve
- Your estimates of the times of sunrise, sunset, and maximum elevation of the sun

Career Skills

The ability to use data to predict future events is a valuable skill in today's workplace. Scientists, meteorologist, seismologist, oceanographers, and land developers, to mention a few, often develop a formula based on data to determine a pattern in the data and to make future predictions from that pattern.

CONSOLIDATING YOUR SKILLS

What Do I Need To Know?

Tutorial	Skills and Concepts	Important Results and Formulas
5.1	• Examine the characteristics of periodic data	• Periodic data have cycles that repeat at regular intervals • Sinusoidal data are periodic data that, when graphed, resemble a sine curve
5.2	• Define and work with radian measure	• 1 radian = $\frac{180}{\pi}$ degrees • 1 degree = $\frac{\pi}{180}$ radians
5.3	• Fit a sinusoidal curve to sinusoidal data	
5.4	• Learn the vocabulary associated with periodic data • Determine the effects of changing the parameters $y = a\sin(bx + c) + d$	$y = a\sin(bx + c) + d$ • Amplitude $= \frac{\text{maximum value} - \text{minimum value}}{2} = a$ • Median value $= d$ • Period = $\frac{2\pi}{b}$ radians • Start point $(-\frac{c}{b}, d)$ • Maximum value $= d + a$ • Minimum value $= d - a$
5.5	• Work with applications of periodic data	

What Should I Be Able To Do?

After you have completed the work in Chapter 5, you should be able to solve the problems that follow. Part A is investigative in nature.

As you work through the exercises, refer to the *Student Reference*, the *Utilities*, and the *Tutorials* as necessary.

Part A

1. The following data represent the angle of elevation of the sun in Nanisivik, Nunavut, on February 12, 2000.

Time	Angle (°)	Time	Angle (°)
8:00	−8.4	13:30	3.2
8:30	−6.4	14:00	2.8
9:00	−4.5	14:30	2.1
9:30	−2.8	15:00	1.2
10:00	−1.2	15:30	0.1
10:30	0.6	16:00	−1.9
11:00	1.6	16:30	−3.5
11:30	2.4	17:00	−5.3
12:00	3.0	17:30	−7.3
12:30	3.4	18:00	−9.3
13:00	3.4		

a) Draw a scatterplot of the data for February 12, 2000. Predict the times of sunrise and sunset in Nanisivik for February 12.

b) Enter the data in the calculator and determine the equation of the sinusoidal curve of best fit for each day.

c) Use the graph to determine the sunrise and sunset times for February 12, 2000. How close were your predictions from part a.

d) Choose a location near the equator. Research the times of sunrise and sunset at this location on February 12, 2000. How do these times compare to the times in Nanisivik? Explain.

Part B

2. Tall buildings are built so that they flex with the wind. For one tall building, the top is designed to sway up to 40 cm from its rest position in a 80 km/h wind. The swaying motion can be described by an equation of the form $y = a\sin(bx + c) + d$. Which parameter would correspond to the 40-cm distance?

3. The following data show how the depth of water at Parksville, BC, varies during one particular day due to tides.

Time (h:min)	6:02	8:50	11:36	14:12	16:45	20:00	23:36
Depth (m)	5.0	3.9	3.1	3.8	4.6	2.6	0.8

a) Find the equation of the sinusoidal curve of best fit.

b) Use your equation to predict the length of time the depth of the water is below 1.3 m.

4. Variable stars change in brightness over time. The star Mira (Omircon Ceti) is a variable star. It has a period of about 11 months. A light curve, summarizing Mira's changes in brightness over time, is shown to the right.

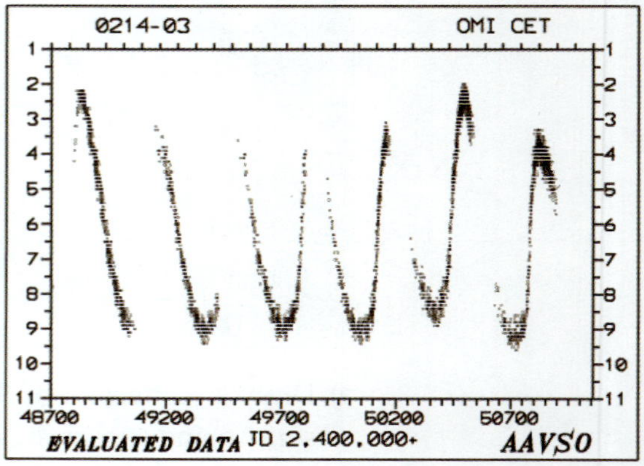

The vertical scale indicates the brightness of Mira in stellar magnitudes. This is a scale used by astronomers to describe the brightness of a star. The horizontal scale is in Julian days. A Julian Date is the number of the day, in a continuous sequence beginning January 1, 4713 BC. On the graph, the Julian Date of 2 450 700 corresponds to September 8, 1997.

a) Use the graph to estimate the amplitude, period, and median value of the light curve.

b) Predict the brightness of Mira on Julian day 2 452 000.

5. The following table lists the maximum daily temperature in degrees Celsius, for the 1st of each month during 1999 at the international airport in Vancouver, BC.

Date (1999)	Day Number	Max. Temp (°C)
Jan. 1	1	6.9
Feb. 1	32	7.2
Mar. 1	60	8.6
Apr. 1	91	10.3
May 1	121	13.3
June 1	153	12.8
July 1	183	17.4
Aug. 1	214	21.2
Sept. 1	245	19.1
Oct. 1	275	14.7
Nov. 1	306	9.2
Dec. 1	335	7.5

a) Graph the data on grid paper.

b) From your graph, estimate the daily maximum temperature for August 11.

c) Enter these data into a graphing calculator and determine the equation of the sinusoidal curve of best fit.

d) Use the regression equation to estimate the daily maximum temperature for August 11.

e) Comment on the usefulness of the regression procedure to predict specific values.

6. The table provides the mean monthly temperature at Thompson, Manitoba, for two time periods. The second column contains the temperatures for 1997. The third column contains the average of the temperatures for the years 1967 to 1990.

Month	Mean Temperature (°C) 1997	Mean Temperature (°C) 1967–1990
Jan.	−25.7	−25.0
Feb.	−20.4	−21.4
Mar.	−15.7	−13.3
Apr.	−4.7	−2.4
May	3.3	6.3
June	13.2	12.3
July	17.4	15.7
Aug.	14.0	13.8
Sept.	9.4	7.2
Oct.	−2.1	0.0
Nov.	−12.1	−11.8
Dec.	−14.8	−22.1

a) Graph the data on grid paper.

b) Enter the data for 1997 into a graphing calculator and determine the equation of the sinusoidal curve of best fit.

c) Enter the data for 1967 to 1990 into a graphing calculator and determine the equation of the sinusoidal curve of best fit.

d) Compare and contrast the two equations. Was 1997 an unusually warm, or cold, year at Thompson?

Pisew Falls, near Thompson, Manitoba

Part C

Here is a sample of a student's work on the Chapter Project.

7. Here is the table of observations and calculations. What are the greatest sources of error?

ANGLE OF ELEVATION OF THE SUN

JUNE 3

Time (h:min)	Time (h)	Shadow Length (cm)	Angle of Elevation (°)
10:06	10.10	116.0	40.8
10:55	10.92	91.5	47.5
11:45	11.75	78.0	52.0
12:52	12.87	56.0	60.8
1:59	13.98	62.0	58.2
2:54	14.90	74.0	53.5
3:45	15.75	96.0	46.1
4:29	16.48	112.0	41.8

8. Here is the student's hand drawn graph. Why is the time for the maximum angle of elevation approximately 13 h instead of 12 h? Is the student's work accurate? Explain.

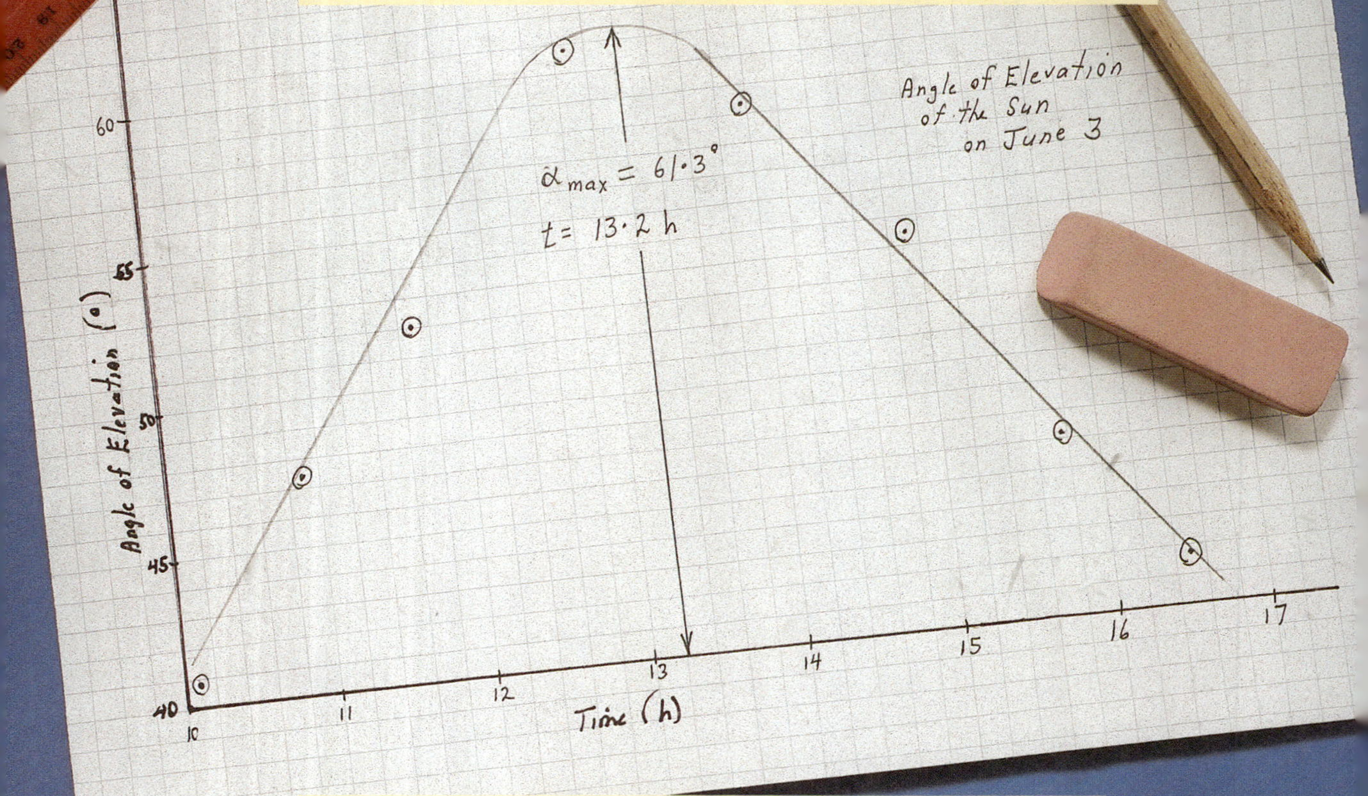

9. Here is the determination of latitude. Estimate the percentage error in this number. Is the student's work accurate? Explain.

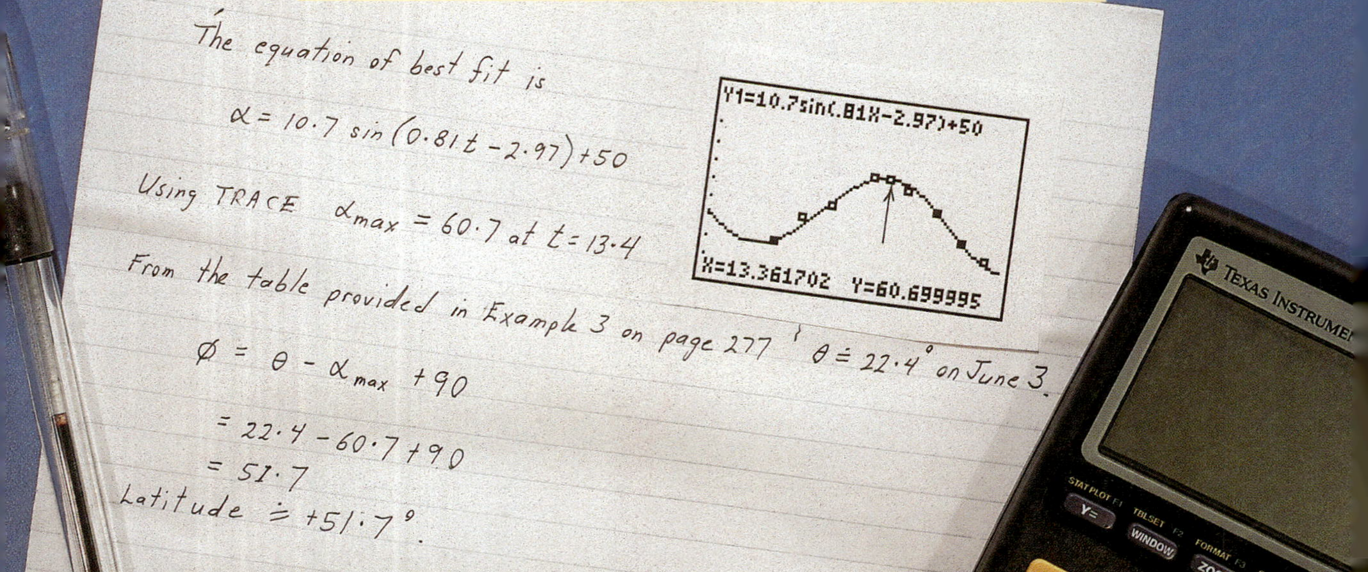

6 Patterns

Overview

Chapter Section	Skills and Concepts
Chapter Project: Caffeine and Your Body, page 258	Research foods that contain caffeine
Tutorial 6.1: Sequences, page 260	Investigate arithmetic, geometric, and recursively generated sequences
Tutorial 6.2: Modelling Using Sequences, page 266	Solve problems that involve sequences and number patterns
Chapter Project: How Long Does Caffeine Remain in Your Body?, page 274	Analyze the rate at which caffeine is removed from the body
Tutorial 6.3: Introduction to Fractals, page 276	Investigate and create fractals using self-similarity and recursion
Tutorial 6.4: Some Properties of Fractals, page 283	Analyze how the perimeter and area of fractals change as the number of iterations of the fractal increases
Chapter Project: How Does Caffeine Affect Your Body?, page 292	Create a poster to illustrate the effects of caffeine on the human body
Consolidating Your Skills, page 294	Review the work of the chapter

PROJECT

You should do at least one project that uses the skills you learn in this chapter. **To cover the curriculum, you should complete Building the Best Soft-Drink Container in the Project Book.**

Other recommended projects in the Project Book are:

- The Ponzi Scheme
- Threatened and Endangered Species

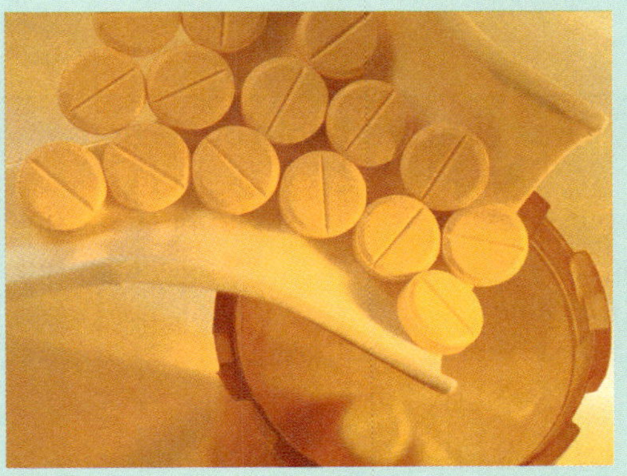

Caffeine and Your Body

Many foods and medications that we consume contain caffeine. Instead of being digested to provide nourishment, caffeine is absorbed directly into the bloodstream, where it acts as a stimulant. Once in the bloodstream, caffeine is gradually removed from your blood and eliminated. A similar process occurs when you take medicine: as soon as it is absorbed, the process of elimination begins. This is why a prescription comes with specific directions as to how much medicine to take, and when to take it. If you do not follow the directions carefully, you may not obtain the full benefit of the medication.

You may be surprised which substances contain caffeine. Here are some questions to consider.

- What substances, besides coffee, contain caffeine?
- How much of each of these foods do you consume daily?
- Caffeine is a stimulant. How does a stimulant affect your body?
- What are the side effects of caffeine?

Research the sources and the effects of caffeine on the Internet or at the library. A biology teacher or a pharmacist may be able to suggest additional sources of information.

CHAPTER PROJECT

To complete this Chapter Project, you must be able to work with sequences. Complete *Tutorials 6.1* to *6.2* to develop the skills you require. You will return to the Chapter Project on page 274.

FYI Visit: www.awl.ca/school/connections

You can start your research from our Internet site: Click on MATHLINKS, followed by the *Applied Mathematics 12* logo, then select a topic under The Effects of Caffeine.

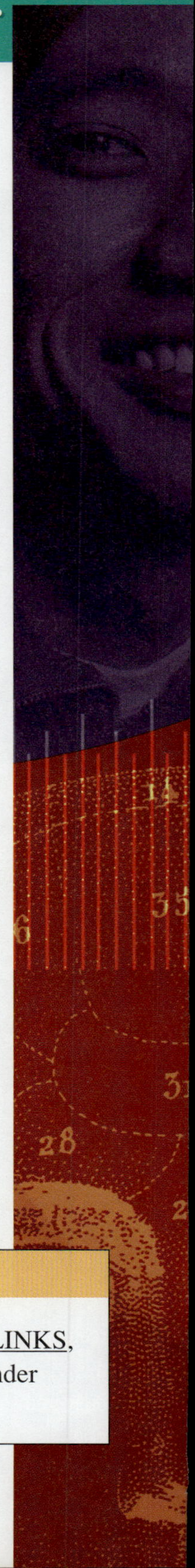

CHOCOLAT
MILK CHOC

TUTORIAL 6.1

Sequences

Many events produce data that form patterns. In this chapter, you will use technology to investigate patterns in numbers and in geometrical figures.

In 1705, the astronomer Edmund Halley noticed similarities in the records of major comets that had appeared in 1531, 1607, and 1682. He also noticed that those dates were separated by a little more than 75 years. He concluded that it was really the same comet returning on a regular basis. It last appeared in 1986 and records have been found for every appearance of it since 239 B.C.

The pattern described above is an arithmetic sequence. Any term in an arithmetic sequence, after the first, can be found by adding a constant to the preceding term.

STUDENT REFERENCE

To review "arithmetic sequence," look in the Student Reference.

A potato stored in a warm, dark room starts to grow new shoots capable of producing whole new plants. Potato farmers typically divide 1 seed potato to start 6 new plants. Over several generations of potatoes, the number of potato plants descended from a single plant forms a sequence 1, 6, 36, 216, 1296, 7776, 46 656, 279 936, ... In each successive generation, the number of potato plants is 6 times greater than the number of potato plants in the preceding generation. These numbers form a *geometric sequence.*

In a geometric sequence, to calculate each term after the first, you multiply the preceding term by a constant number, called the *common ratio*.

In both arithmetic and geometric sequences, each term, after the first, is generated by using the terms that come before it. These are both *recursively-generated sequences*. The Fibonacci sequence 1, 1, 2, 3, 5, 8, 13, ... is a recursively generated sequence that is neither arithmetic nor geometric. It can be found in population growth, the number of petals of common flowers, or the shell of a Chambered Nautilus mollusk. To calculate a term of a Fibonacci sequence, after the second term, add the two preceding terms.

A spreadsheet program can be used to generate a sequence, particularly when the sequence is recursively generated.

Investigation 1 The Growth of Bacteria

You will use a spreadsheet to study the growth of a bacteria population.

Bacteria are single-celled animals that divide in two to reproduce. Suppose you start with a single bacterium in a Petri dish. Under the right conditions, it divides every 30 min.

1. Describe the number of bacteria as a sequence and list the first 10 terms.
2. What type of sequence is it? How do you know?
3. Create a spreadsheet to describe the situation. Include columns for the number of generations of bacteria, the number of bacteria, and the time elapsed in hours. Use **Fill Down** to create data for 110 generations.
4. After how many generations will the number of bacteria equal the population of your school? Your town? Your province?
5. After how many hours will the number of bacteria equal the population of the Canada?
6. Suppose a single bacterium has a mass of 0.001 g. Add a fourth column to your spreadsheet to calculate the total mass of the bacteria.
7. After how many generations will the total mass of bacteria be equal to your mass?
8. Why is the world not consumed by bacteria?

In *Investigation 2*, you will use a CBR, or a CBL, with a motion detector to collect data.

Investigation 2 A Bouncing Ball

For this Investigation you should work in groups of two or three. You will need the following materials:

- CBR (or CBL)
- TI-83 graphing calculator
- Large ball (a volleyball works well)

The CBR should be held steadily and pointed down toward the bouncing ball. Place the CBR on top of a bookcase or table and point it downward, or hold it next to a wall and drop the ball along the wall. Some type of containment is also needed to prevent the ball from bouncing away, so you may choose to drop the ball against a flat surface.

1. Begin with a few test bounces of the ball. Drop the ball, rather than throw it downward.

2. Run the RANGER program on your calculator.

3. From the MAIN MENU choose 3:APPLICATIONS

4. Select 1:METERS, then 3:BALL BOUNCE.

5. Follow the directions on your calculator screen. Hold the ball under the CBR, and then release the ball. Press the TRIGGER key on the CBR **as the ball first strikes the ground**.

6. Your graph should have a minimum of 6 bounces. It should look similar to this screen. If you are not satisfied with the results of your experiment, press ENTER and select 5: REPEAT SAMPLE.

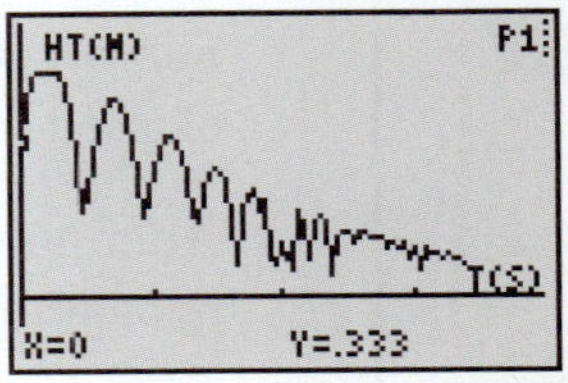

7. Copy this table.

Bounce	Maximum Height	Difference in Heights	Ratio of Heights
1			
2			
3			
4			
5			

8. Complete the table.
   a) Use the TRACE feature of the calculator. Record the maximum height reached by the ball after each bounce.
   b) To calculate the difference between bounces, subtract successive maximum heights. For example,
      Difference after bounce 2 = Height of bounce 2 – Height of bounce 1
   c) Calculate the ratio of heights by dividing the height of successive bounces. For example,
      the ratio after bounce 2 = height after bounce 2 ÷ height of bounce 1

9. a) Does there seem to be a common difference in successive heights? If so, what is its value?
   b) Does there seem to be a common ratio in successive heights? If there is, what is this value?

c) Is the height of the ball better described by an arithmetic sequence or a geometric sequence? Predict the height of the ball after bounce 6. Check your prediction with the result recorded by the CBR.

UTILITY

If you need help to generate and graph a sequence using the TI-83, Utility 33 shows you how. If you use a different graphing calculator, consult the user's manual.

Example 1 Use technology to determine a term of a sequence.

A lottery has a grand prize paid as follows:

\$5 in week 1, \$10 in week 2, \$20 in week 3, and so on for 15 weeks

a) Describe the pattern of payments.

b) How much money was paid in the 12th week?

SOLUTION

a) The set of payments forms a sequence. To find each successive payment, multiply the previous amount by 2. The sequence is geometric, with a common ratio of 2.

b) Make sure your graphing calculator is in Sequence mode. Press [Y=] and enter nMin = 1, u(n) = u(n − 1)*2, and u(nMin) = 5 (shown below left). Use the TABLE command to obtain the table of values of the sequence. Scroll down the table to row 12 (shown below right).

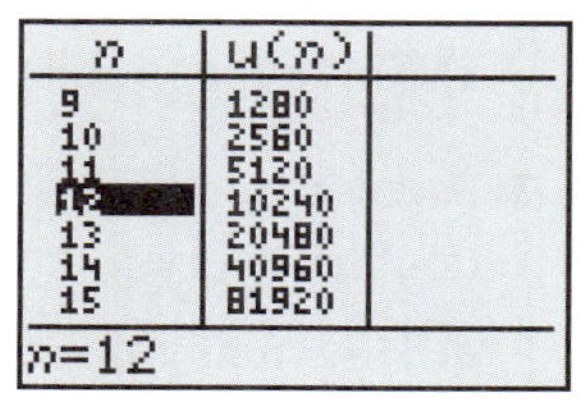

The amount paid in the 12th week is \$10 240.

DISCUSSING THE IDEAS

1. How can you test a sequence of numbers to determine if it is arithmetic or geometric?

2. Refer to *Investigation 1*. Describe another method of generating the sequence.

3. What happens to the terms of a geometric sequence if the common ratio is less than one?

4. Arithmetic sequences are often used to model data when growth is linear. What kind of growth is likely to generate data that form a geometric sequence? Explain.

EXERCISES

Checking Your Skills

1. Use a calculator to generate the following sequences and find the indicated term.
   a) 5, 7, 9, 11, …; the 14th term
   b) 5, 15, 45, 135, …; the 12th term
   c) 3.7, 2.3, 0.9, –0.5, …; the 15th term
   d) 2.4, 3.6, 5.4, 8.1, …; the 9th term
   e) 12, 3, 0.75, 0.1875, …; the 10th term

2. A salesperson at a car dealership receives a base salary of $275 per week, plus $50 for every car sold. Each week, the salesperson sells one more car than the previous week. In his first week, he sold one car.
   a) Create a graph to show the salesperson's earnings each week.
   b) Use your graph to determine the week in which earnings will exceed $1000, if the pattern continues.
   c) Solve the problem using another method.

3. A colony contains 100 insects at the start of May. Suppose the population doubles every month and no insects die.
   a) How many insects are there at the beginning of July?
   b) How many insects are there at the beginning of September?

4. A ball drops from a height of 2 m. After each bounce, it rises to 75% of its previous height.
   a) Use technology to generate a sequence of bounce heights.
   b) What height does the ball reach after each of the first 5 bounces?
   c) After how many bounces does it reach a height of approximately 20 cm?
   d) When the height is 0.1 cm, it may be considered to have stopped bouncing. How many times does it bounce?

5. A store sells fresh bread for $2 a loaf. Each subsequent day the price is reduced by 20%. How old is the bread when it sells for less than 90¢ a loaf?

6. A store is conducting a Dutch auction. It will take 10% off the previous day's cost of an item for each day it remains unsold. On the first day an item is priced at $250. How many days must it remain unsold before the sale price is under $25?

7. Jamie invested $10 000 in an account that paid interest compounded annually. There was $11 000 in the account after 1 year, $12 100 after 2 years, $13 310 after 3 years, and so on.

a) What was the annual interest rate?

b) What kind of sequence do the amounts form?

c) Use a graphing calculator to generate the first 20 terms of the sequence. Suppose the money accumulates at the same rate for 20 years. How much will Jamie have in the account after 20 years?

8. The pay of a systems analyst is $40 000 for the first year, with an annual increase of 5% at the beginning of subsequent years. Create a table to show the pay during each of the first 10 years of employment.

9. If a basketball is properly inflated, the height of each bounce is about one-half the height of the previous bounce. A basketball was dropped from a height of 3.2 m. After how many bounces does the ball rebound to a height of less than 0.5 m?

Extending Your Thinking

10. A rubber ball is dropped on a hard surface. It makes a sequence of bounces, each one four-fifths as high as the preceding one. Suppose the ball was dropped from a height of 20 m. What is the total distance the ball has travelled when it hits the ground the fifth time?

PROJECT

In this Tutorial, you have used technology to generate sequences. You will use this skill in the Chapter Project and in the following projects in the Project Book.

- The Ponzi Scheme
- Threatened and Endangered Species
- Building the Best Soft-Drink Container

COMMUNICATING THE IDEAS

Provide examples of arithmetic sequences, geometric sequences, and sequences that are neither arithmetic nor geometric. Explain why each sequence is arithmetic, geometric, or neither.

TUTORIAL 6.2 Modelling Using Sequences

In *Tutorial 6.1* you investigated arithmetic and geometric sequences. In this Tutorial you will use sequences that can be generated recursively to model real-life phenomena.

Some sequences can be generated with a spreadsheet program. Many such programs have graphing capabilities that allow you to view a sequence both numerically and graphically.

For *Investigation 1*, you need access to a computer with a spreadsheet program. Work with a partner.

Investigation 1 Graph a Sequence Using a Spreadsheet Program.

A forest of 7000 trees is being placed under a new forestry management plan by the local pulp mill. Each year 25% of the trees will be harvested, and 1500 new trees will be planted.

1. Create a spreadsheet to display these data.

The number of trees at the beginning of a year

The number of trees cut that year

The number of trees planted that year

The number of years since the plan began

The number of trees at the end of the year

- Enter the initial data and create formulas as required.
- There cannot be fractions of trees, so format the columns appropriately.
- Use the **Fill Down** command to extend the data in all columns to row 37.

2. How many trees are there in the forest at the end of the thirteenth year in this management plan?

3. Graph the data.

- Highlight the Number of Years column and the Number of Trees at the End of the Year column.
- Open the **Insert** menu and select **Chart**.
- Select **X, Y Scatter** as the type of graph to create.
- Follow the on-screen instructions to obtain a graph.

Adjust the size, if necessary, then print the graph.

4. Describe the growth of the forest over time. Will the forest size ever stabilize? If so, in how many years will it stabilize? How many trees will the forest contain?

5. Experiment with your input values to investigate the rate at which trees should be harvested if the forest manager wishes to reach a stable forest population of 5000 trees over a period of 20 years.

When you take a medicine, its effects diminish with time. Your body uses some, and some are eliminated. You may have to take doses of the medicine at regular intervals to maintain its effectiveness.

Many diabetics take insulin every day so they can digest their food. After an injection, the level of insulin in the bloodstream is boosted to an acceptable level, but then declines by approximately 5% every minute. In the following Investigation, you will use a graphing calculator to explore the rate at which insulin declines.

Investigation 2 The Metabolism of Insulin

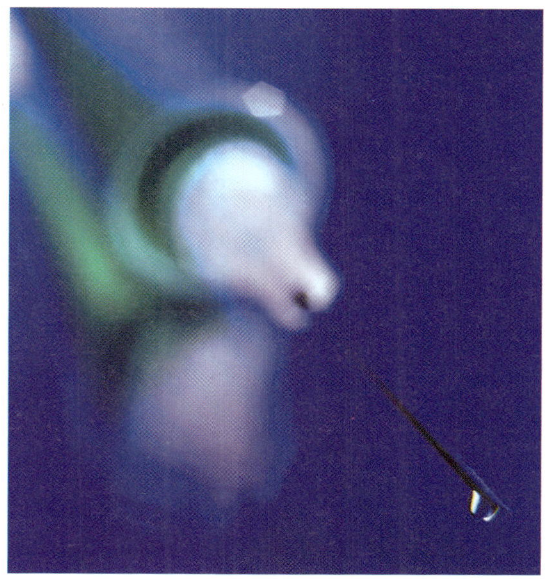

Assume that 10 units of insulin are injected into a person's bloodstream. Also assume that of the insulin present at the beginning of the minute only 95% of that amount remains at the end of each minute.

1. Write a recursive rule to express the amount of insulin that remains at the end of a minute, $u(n)$, in terms of the insulin present at the start of the minute, $u(n-1)$.
2. Use your graphing calculator to enter the sequence. The term number will represent time in minutes, so set $n\text{Min} = 0$.

UTILITY

If you need help to generate and graph a sequence using a TI-83, refer to Utility 33. If you use a different graphing calculator, refer to the user's manual.

3. Change the window settings to those shown below.

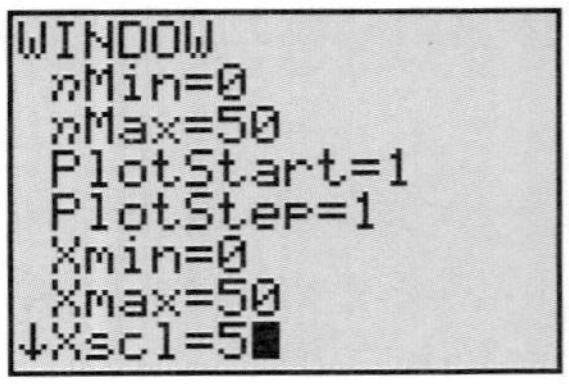

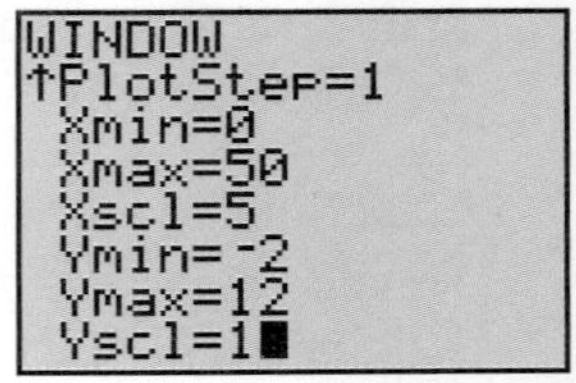

4. Graph the sequence. Your graph should look like the one to the right.

What do the values on the horizontal axis represent? What do the values on the vertical axis represent?

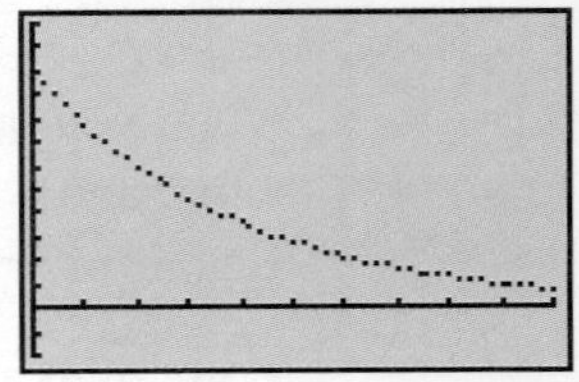

5. Use your graph to estimate how many units of insulin remain in the bloodstream after 27 min.

6. Doctors and pharmacists are usually interested in the *half-life* of a drug. This is the time taken for the drug to be reduced to one-half of its original amount. What is the approximate half-life of insulin?

7. A rule that pharmacists may use is that a drug has effectively disappeared from a person's system after 5 half-lives. Approximately how many minutes is it before this dose of insulin has effectively disappeared from the system?

8. Investigate a change in the initial dose of 10 units to 15 units, then to 5 units of insulin. Sketch the graphs.

9. Suppose a biochemist discovers a new drug that has the same effect as insulin but is eliminated more slowly, at a rate of 2% per minute. Sketch the graph for this drug.

10. Explain why some insulin-dependent diabetics must eat soon after they take insulin.

When a person ingests a substance at regular time intervals, the amount of the substance in the blood eventually levels off, or stabilizes. This is called the *maintenance level* for the substance. It is the maximum level the substance will reach in the body.

Example **Modelling the effect of repeated intake of a substance**

A veterinarian prescribes a 25-mg pill to be given to a dog every morning. Twenty percent of the medication is eliminated from the dog's body every day.

a) Graph the level of medication in an average dog's bloodstream for 40 doses of this medication.

b) Assume the dog continues to receive one pill every day. How many days does it take before the level of medication in its bloodstream reaches its maintenance level?

SOLUTION

a) Use a graphing calculator in sequence mode. The sequence of levels of medication in the dog's bloodstream begins at time 0 h. The dog is given one pill every 24 h. The medication level at the time of each dose is 80% of the previous day's level. The starting value of the sequence is 25. Enter these values in the Y= list, as shown below left. Use the window settings shown below middle and right.

```
Plot1 Plot2 Plot3
 nMin=0
·.u(n)=.8*u(n-1)+
25
 u(nMin)=(25)
·.v(n)=
 v(nMin)=
·.w(n)=
```

```
WINDOW
 nMin=0
 nMax=40
 PlotStart=1
 PlotStep=1
 Xmin=0
 Xmax=40
↓Xscl=5
```

```
WINDOW
↑PlotStep=1
 Xmin=0
 Xmax=40
 Xscl=5
 Ymin=-2
 Ymax=135
 Yscl=10
```

Press GRAPH to obtain the graph shown below.

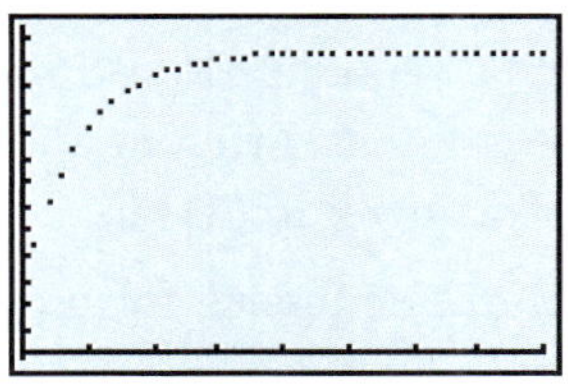

Each division on the x-axis corresponds to 1 dose, or 1 day.

Each division on the y-axis corresponds to 5 mg of the medication.

b) Use the TRACE feature. The amount of medication in the dog's bloodstream stabilizes at close to 124.9 mg when x is 30. This means that it takes about 30 days for the level of medication in the dog's bloodstream to stabilize at about 125 mg.

DISCUSSING THE IDEAS

1. What is the difference between a sequence and a function?

2. Why are geometric sequences often used to model biological activity?

3. What are the advantages of using a spreadsheet rather than a calculator to analyze a sequence?

4. In *Investigation 2*, if insulin were removed from the body at a rate of 13% per minute, what equation would you enter into your graphing calculator for u(n)?

EXERCISES

Checking Your Skills

1. Atmospheric pressure is 101.3 kPa at sea level. As the altitude increases, atmospheric pressure decreases at an approximate rate of 12% per kilometre above sea level. As air pressure decreases, the amount of oxygen decreases. Most humans have difficulty carrying out normal activities over 3 km above sea level.
   - **a)** Create a sequence to estimate the air pressure at different altitudes, from sea level to 50 km.
   - **b)** Graph the sequence from part a.
   - **c)** At what altitude will atmospheric pressure be one-half what it is at sea level?

2. The Foothills Nursery has 6000 cherry tree seedlings that grow in a greenhouse. Each year it plans to sell 15% of the older seedlings and plant 800 more.
   - **a)** Create a spreadsheet to show the number of seedlings at the end of each year.
   - **b)** How many seedlings will there be in the greenhouse at the end of 11 years?
   - **c)** If 20% of the seedlings were sold each year, how many would there be at the end of 20 years?

3. You have a cold and a local pharmacy recommends a non-prescription cough medicine to help relieve the symptoms. You take one 20-mL dose right away, then another 20-mL dose 30 min later. Just after you take the second dose, you read the directions and discover that the correct dosage is 20-mL once every 4 h. Your kidneys remove about 40% of the cough medicine from your blood every 4 h. You continue to take the cough medicine every 4 h, for the next 2 days.

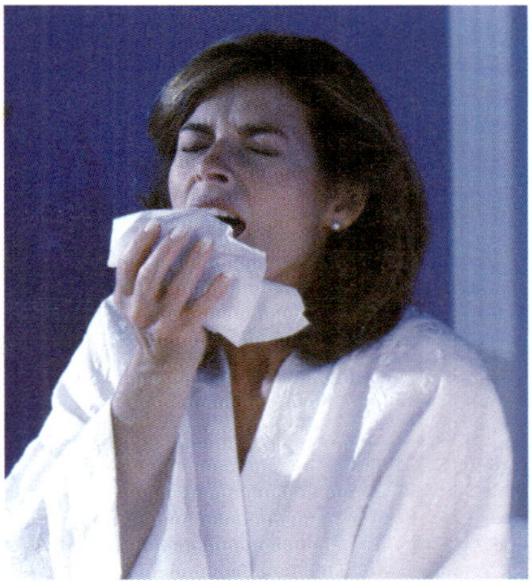

a) Use a spreadsheet to determine how many millilitres of cough medicine are in your bloodstream after the tenth dose. Assume you started with 40 mL of medicine in your bloodstream.

b) How many doses must you take to achieve the maintenance level? How many hours does that take?

c) What is the maintenance level?

4. A certain painkiller has a half-life of 0.5 h.

a) What fraction of the original drug remains in a person's body 4 h later?

b) Suppose Sarah takes two 300-mg tablets of painkiller every 4 h, for 24 h. How much painkiller remains in her bloodstream just before she takes her third dose?

c) What is the maintenance level of this painkiller?

5. Chlorine is often added to swimming pools to help purify the water. In a particular pool, 9 kg of chlorine is added each week. The pool filters absorb thirty percent of the chlorine each week.

a) How much chlorine is in the pool after 7 weeks?

b) Determine the maintenance level of chlorine in the pool.

6. Each year, 9% of the water in lake A flows into lake B. There is an oil-sands plant near lake A that pumps 45 t of pollutant into that lake each year. Presently, there are 250 t of pollutant in lake A and 75 t of pollutant in lake B. This pollutant is dissolved uniformly in the water of both lakes.

a) Calculate the levels of pollutant in each lake at the end of 5 years and at the end of 20 years.

b) Does the amount of pollutant in each of the lakes ever stabilize? If so, after how many years does it stabilize?

7. A mild infection spreads through a school. On any given day, 10% of the uninfected population become infected and 20% of the infected population are cured. Once cured, there is 0% chance of further infection. On the first day of school, 100 students are infected, 900 students are uninfected, and 0 students are cured.

a) Track the epidemic for 30 days.

b) Graph numbers of infected, uninfected, and cured as a function of time for 30 days on the same grid.

Extending Your Thinking

8. A lake in northeastern Manitoba has an initial population of 150 000 rainbow trout. The local fish and wildlife officer indicates that they stock 200 000 fingerling rainbow trout in this lake each year. She also indicates that anglers catch and remove 35% of the fish in the lake each year and that there is a 27% mortality rate from other causes for the fish in the lake each year.

a) Use the data to determine how many rainbow trout are in the lake after 10 years.

b) In what year does the population of the fish stabilize, and how many fish are in the lake in that year?

c) Suppose that after 5 years there is a very severe winter and 60% of the fish die. How many fish are in the lake after 10 years?

d) Suppose the number of fish stocked each year decreases to 100 000. How does this affect the stable population and the year when the population stabilizes?

e) Catch-and-release is an alternate form of sport fishing. If 60% of released fish live, describe how your answers to parts a through d will change if anglers only catch-and-release.

9. One of the most common antibiotics is penicillin. One hour after Jamie takes a dose, only 60% of the drug remain in his system. Suppose Jamie took 150 mg of penicillin at 8 A.M.

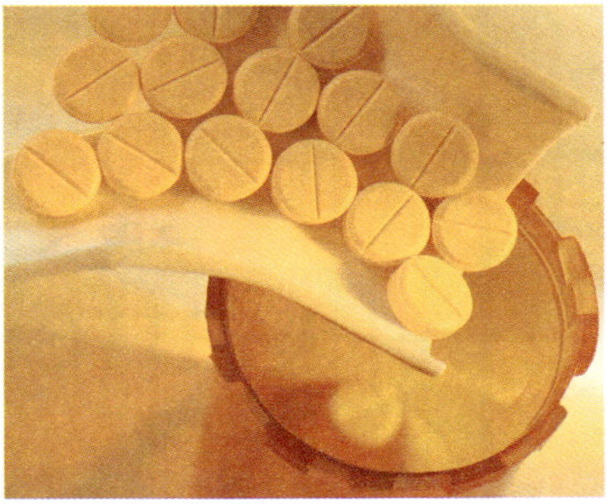

a) Create a table that shows the amount of penicillin that remains in Jamie's body every hour from 8 A.M. to 2 P.M.

b) Graph these data.

c) Determine a regression equation for these data.

d) Explain why an exponential regression equation was used.

e) Use the regression equation to produce a table of values that shows the amount of penicillin that remains after every quarter hour. Use the table to determine the half-life of penicillin.

PROJECT

In this Tutorial you have learned how to graph a sequence on a graphing calculator and how to analyze a sequence with a spreadsheet. You will use these skills in the Chapter Project and the following projects in the Project Book.

- The Ponzi Scheme
- Threatened and Endangered Species
- Building the Best Soft-Drink Container

COMMUNICATING THE IDEAS

Explain why the graphs in this Tutorial are exponential in nature.

How Long Does Caffeine Remain in Your Body?

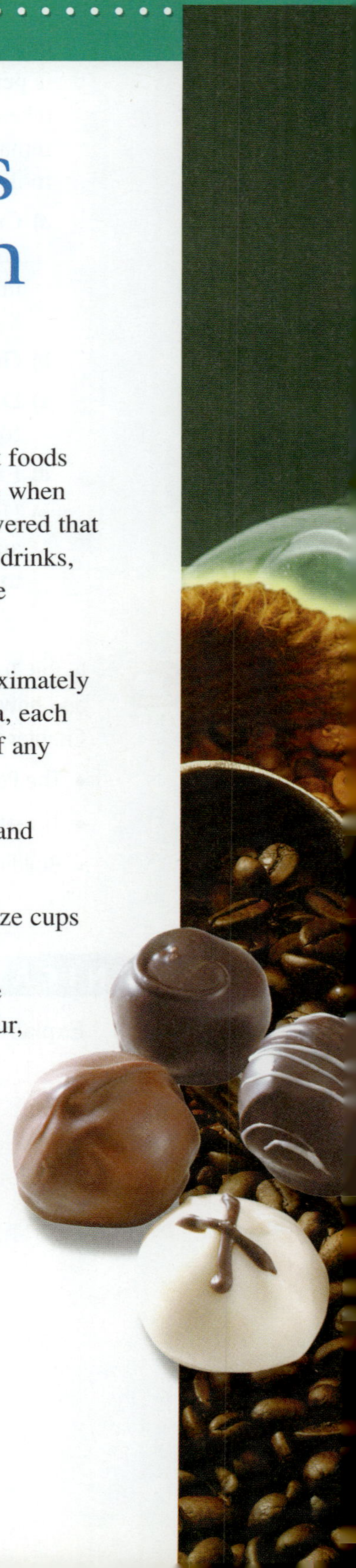

At the start of this project, you researched different foods that contain caffeine. Most people think of coffee when they hear "caffeine." However, you should have discovered that caffeine is found in many other foods such as tea, soft drinks, and chocolate, and in many pain relievers. Most people consume some caffeine on a daily basis.

A regular-size (8 oz) cup of drip coffee contains approximately 165 mg of caffeine. According to pharmacological data, each hour the human body eliminates approximately 13% of any caffeine present in the bloodstream.

Now take a more detailed look at caffeine breakdown and elimination.

1. At 8 A.M. you start the day by drinking 2 regular-size cups of coffee.

a) Create a table to illustrate the amount of caffeine that remains in your body at the end of every hour, for 12 h.

b) Graph the data.

c) Determine a regression equation for the data. Graph this equation on the same grid as the scatterplot. Sketch the graph.

d) Use your graph.

i) How much caffeine remains in your body after 1.5 h?

ii) What is the half-life of caffeine?

CHAPTER PROJECT

To complete this Chapter Project, you need to learn some graphing skills for recursively generated patterns. Complete *Tutorials 6.3* to *6.4* to develop an understanding of such patterns. You will return to this Chapter Project on page 292.

e) Caffeine is a stimulant, so it could affect your ability to sleep. Determine how much caffeine remains in your bloodstream when you go to bed at 10 P.M. Could this amount affect your sleep?

f) When is the caffeine level less than 10 mg?

2. Most people do not just drink coffee early in the morning. They often have additional cups during the day. Assume that in addition to 2 cups of coffee at 8 A.M., you consume 1 cup at 10 A.M., 1 cup at 1 P.M., and 1 cup at 7 P.M.

 a) Create a table to show the amount of caffeine in your bloodstream, at the end of each hour for 12 h.

 b) How much caffeine remains in your system at a bedtime of 10 P.M.?

Write a report to describe the effects of caffeine on the human body. Include supporting graphs and a recommendation about the daily consumption of caffeine.

Career Skills

Drugs are developed, tested, and manufactured to achieve specific results. The ability to analyze the effect of different drugs and the rate at which these drugs are metabolized is an important skill in the pharmaceutical industry. Many pharmaceutical workers perform calculations and analyses similar to those you have completed in this project.

TUTORIAL 6.3

Introduction to Fractals

Fractals are geometric figures that were first devised about 100 years ago. An interest in them has been revived by modern computer graphics. Fractals can be generated by *iterative* procedures. This means that a sequence of steps is repeated.

In this Tutorial, you will create and explore fractals that possess the quality of *self-similarity*, where a part of a fractal is geometrically similar to the whole fractal. There is a pattern that repeats as you zoom in, or zoom out. The design shown to the right is self-similar. It is known as a Koch snowflake, for the Swedish mathematican Helge von Koch (1870–1924) who devised it in 1904.

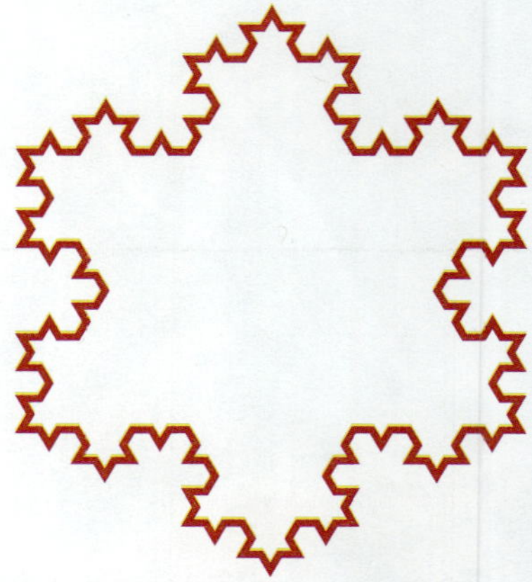

In the following Investigation, you will use an iterative procedure to create a 3-dimensional fractal that is self-similar.

Investigation 1

Create a 3-Dimensional Fractal.

You need an 8.5 by 11 inch sheet of stiff paper such as construction paper, scissors, and a ruler.

1. Fold the paper in half to form a 8.5 by 5.5 inch shape.
2. Use a ruler. Mark points on the fold one-quarter of the way in from each short edge of the folded sheet. Draw a line from each point, perpendicular to the fold, that extends one-half of the way to the top edge of the paper (Figure 1).

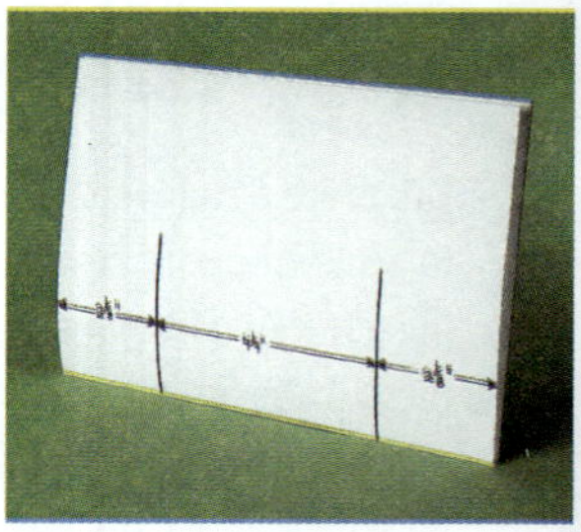

Figure 1

Figure 2

Figure 3

3. Cut along these lines. Fold along the line segment produced by the joining the ends of the cuts (Figure 2). Open the paper and fold the cut portion to the inside (Figure 3).

4. Repeat the drawing, cutting, and folding procedure described in exercises 2 and 3 until the paper becomes too thick to continue. The next iteration is shown below left.

5. Carefully open out the folds in reverse order. Gently push out the fractal (above right).

6. Copy the table below. Complete the entries in the table to summarize some of the properties of this 3-dimensional fractal.

	Number of Additional Boxes	Length (cm)	Width (cm)	Height (cm)	Volume of One New Box (cm^3)	Volume of All New Boxes (cm^3)	Total Volume of All Boxes (cm^3)
Original	1	10.8	7.0	7.0	529.2	529.2	529.2
Iteration 1	2	5.4	3.5	3.5			
Iteration 2	4	2.7					
Iteration 3							
Iteration 4							

7. It may be difficult to construct stage 5. Use any patterns you can see in the table to complete the entries for this stage.

8. Write the sequence that describes the number of new boxes at each stage. How many new boxes would there be at stage 10?

9. Write the sequence that describes the volume of 1 new box at each stage. What is the volume of a new box for stage 12?

Many fractals are named after the mathematicians who first created and analyzed them. The Sierpinski gasket is a fractal pattern named after the Polish mathematician Waclaw Sierpinski (1882–1969). One way to create this fractal begins with one triangle, and systematically removes smaller triangular regions from it. This way, the outer dimensions of the fractal remain constant. Another way to create a Sierpinski gasket is to start with one triangle and systematically add more triangular regions. This way, the outer dimensions of the fractal grow with each stage of complexity.

Investigation 2 **Use Technology to Create a Sierpinski Gasket.**

You need isometric paper, access to a computer with a draw or paint program. These instructions are for Microsoft Paint and a TI-83 graphing calculator. If you are using a different program, consult the user's manual or the Help menu. Work with a partner.

Part A: Use a Computer to Construct a Sierpinski Gasket.

1. Start the paint program. Draw a triangle with sides approximately 10 cm. Fill the triangle with a colour. Print or sketch this triangle. It represents the original.

2. Select the triangle and shrink it to 50% of its current size. This takes two steps.

- **i)** From the **Image** menu, select **Stretch/Skew**. Set the Horizontal stretch to 50%. Click OK.
- **ii)** From the **Image** menu, select **Stretch/Skew**. Set the Vertical stretch to 50%. Click OK.

3. Select the reduced triangle.

- **i)** Move the triangle to position 1 on the screen.
- **ii)** From the **Edit** menu, select **Copy**, then **Paste**, to make a copy of the reduced figure. Move the copy to position 2.
- **iii)** Make a second copy of the reduced figure. Move it to position 3.

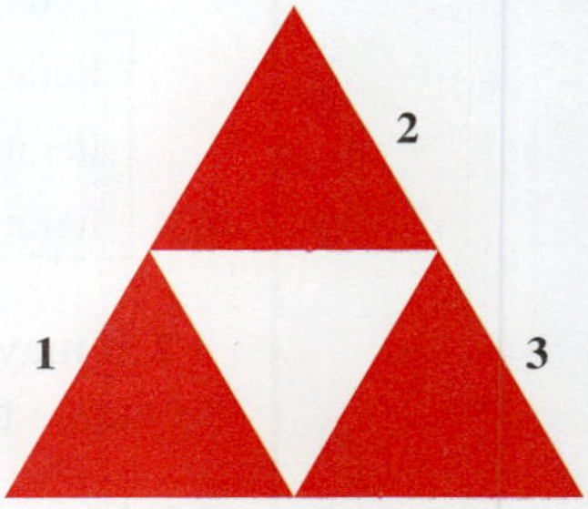

1st Iteration

4. Print or sketch the figure. It represents the 1st iteration of the Sierpinski gasket.

5. Take this new figure and repeat exercises 2 to 4.

6. Repeat exercises 2 to 4 as many times as you can to produce a more complex fractal.

Part B: Create a Sierpinski Gasket Using a TI-83 Calculator.

The Sierpinski gasket is so well known that many programs are available that generate the pattern.

UTILITY

If you need help to load and run a program on a TI-83, Utilities 5 and 13 show you how. If you use a different graphing calculator, consult the user's manual.

1. Enter the SIERPINS program into the memory of your graphing calculator. This is listed in the User's guide to the TI-83 and TI-83 Plus calculators. You may also obtain it from another calculator that already has the program stored in memory by linking the calculators.
2. Run SIERPINS. Recall and display the picture with the instruction RecallPic 6.
3. Make a copy of the picture. You may find isometric paper helpful.

Dynamic geometry software is ideally suited to creating self-similar figures. In *Investigation 3*, you will create a version of the Koch snowflake curve on page 276.

Investigation 3 Create a Koch Snowflake Curve Using Dynamic Geometry Software.

This Investigation is written for *The Geometer's Sketchpad*. If you use some other dynamic software, consult the guide that came with it.

UTILITY

If you need help to use *The Geometer's Sketchpad*, Utility 23 shows you how.

1. From the **Display** menu, choose **Preferences**. Make sure that **Autoshow** labels for points is checked.
2. From the **File** menu, select **New Script**. A Script window will appear on the right of the screen. Click on **REC**.
3. Construct a line segment AB. Select point A.
4. From the **Transform** menu, select **Mark Center**.
5. Select point B. From the **Transform** menu, select **Dilate**. Enter a scale factor of 0.333. The label of the new point should be B′.

6. Select point B. From the **Transform** menu, select **Dilate**. Enter a scale factor of 0.666. The label of the new point will also be B′.

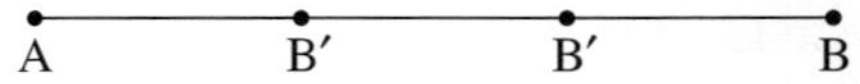

7. Select the first point B′, closer to A. From the **Transform** menu, select **Mark Center**.

8. Select the second point B′, closer to B. From the **Transform** menu, select **Rotate**. Enter 60°, and click OK.

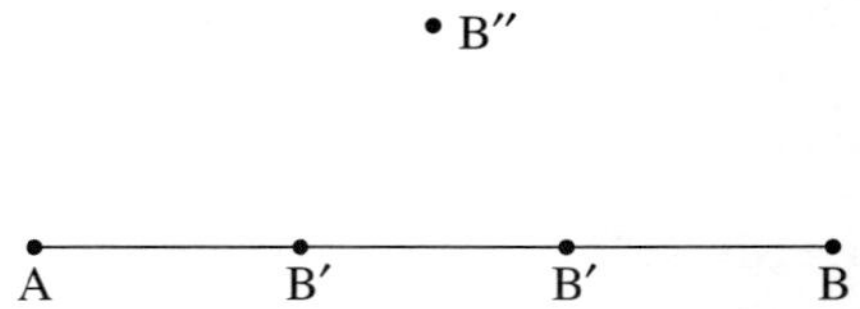

9. Select segment AB. From the **Display** menu, select **Hide Segment**.

10. Shift-click each pair of points in order from left to right and construct each segment.

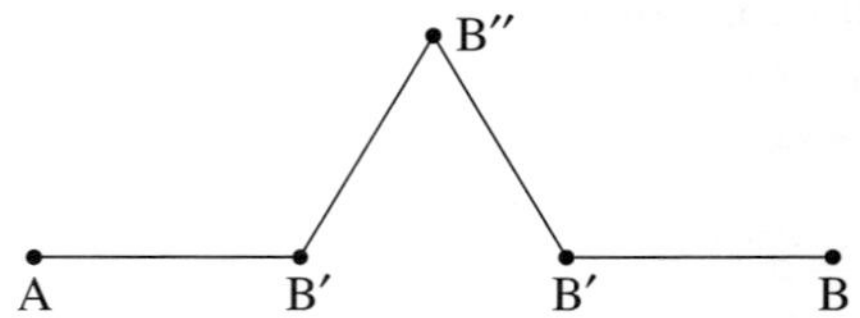

Next, loop the script so that it will build a hat on each of the four segments in the above hat. Exercises 11 to 13 explain how to do this.

11. Shift click on the points A and the first point B′ to select them in that order. Click on **LOOP** in your script. Nothing appears on screen, but this adds a step to the script.

12. Continue around the curve, looping on the end points of each segment. Be sure to select the points in order each time.

13. Once you have four recursive steps in your script, hide all points except points A and B.

14. Click on **STOP** in your script.

15. Open a new sketch. Construct points A and B; then shift-click to select the points in that order.

16. Run your script. Click **PLAY**. Enter 0 for the depth of recursion. The script should reproduce the figure you constructed in exercise 10.

17. Print or sketch this curve and label it Koch Curve — Depth 0.

18. Run the script again. Enter 1 for the depth of recursion. Print or sketch this curve and label it Koch Curve — Depth 1.

19. Run the script with recursive depths of 2 and 3.

DISCUSSING THE IDEAS

1. In *Investigation 1*, you created a 3-dimensional fractal. Can you create other 3-dimensional fractals by cutting and folding? Explain.

2. Refer to *Investigation 2*. What sequences can you discover in the Sierpinski triangle?

3. In *Investigation 3*, what happens to the length of the Koch curve at greater recursive depths?

4. Why are self-similar patterns particularly suited to calculator or computer programs?

EXERCISES

Checking Your Skills

1. Use *The Geometer's Sketchpad* to create a fractal called the Hat Curve. Repeat all the steps in *Investigation 3*, with these changes.

a) Add these steps between exercises 8 and 9.

i) Select point B″. From the **Transform** menu, select **Mark Center**.

ii) Select the first point B′. From the **Transform** menu, select **Rotate**. Enter 90°. Label the new point B″.

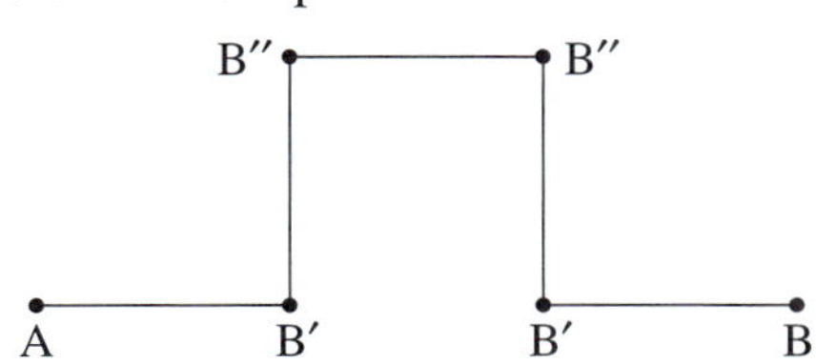

b) Run the script for 3 different recursive depths. Print or make sketches of the patterns.

2. Use *The Geometer's Sketchpad* to create a Koch snowflake design. Start with an equilateral triangle.

3. Create a Sierpinski carpet. Use grid paper or technology.

The general method is: start with a square; then remove a smaller square. Use shading to indicate that a square has been removed.

a) Draw an outline of a square on grid paper, with side length 27 units.

b) Draw an outline of a square in the centre of the original square with side length one-third that of the original. Shade this square.

c) Divide the region between the squares into 8 squares, each the same size as the central shaded square.

d) Repeat parts b and c for each of these 8 squares.

e) Complete the third iteration.

Extending Your Thinking

4. Create your own fractal. Use technology if it is available.

a) Start with some basic shape and use the process of self-similarity and recursion to develop the fractal. Use colour to enhance the appearance of your fractal.

b) Make a list of step-by-step instructions that someone else could follow to replicate your fractal.

c) Check your instructions. Exchange instructions with a partner and each follow the other's instructions. Correct any errors.

PROJECT

In this Tutorial, you have learned about fractals that possess the property of self-similarity. You will use these ideas when you return to the Chapter Project and in these projects in the Project Book:

- The Ponzi Scheme
- Building the Best Soft-Drink Container

COMMUNICATING THE IDEAS

In your own words, explain the meaning of "self-similarity" and "iteration."

Some Properties of Fractals

In *Tutorial 6.3*, you investigated some well-known fractals. In this Tutorial, you will explore how the perimeter and area of a fractal change at each iterative stage of its creation.

Suppose you had to estimate the length of the Athabasca River from Fort McMurray to Lake Athabasca. You could use a length of thread and a map with a scale of 1:14 500 000 to trace the river, then measure the thread. You would probably get a length of approximately 16 mm, which converts to a distance of about 232 km. Now, if we zoom in and repeat the procedure on a map with a scale of 1:6 000 000, the length of the thread is 41 mm. This produces a distance of about 246 km. Suppose we zoom in again, and use a map with a scale of 1:2 000 000. The length of the thread is 130 mm, which corresponds to a distance of 260 km. Each time we zoom in on the Athabaska River, more of its twists and turns are apparent and the estimate of its length increases.

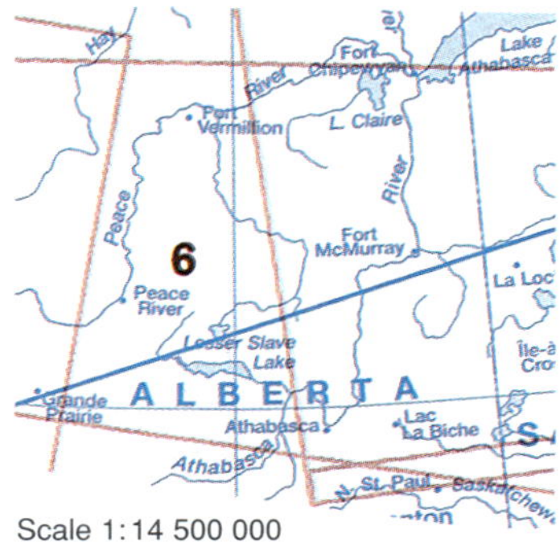

Scale 1:14 500 000

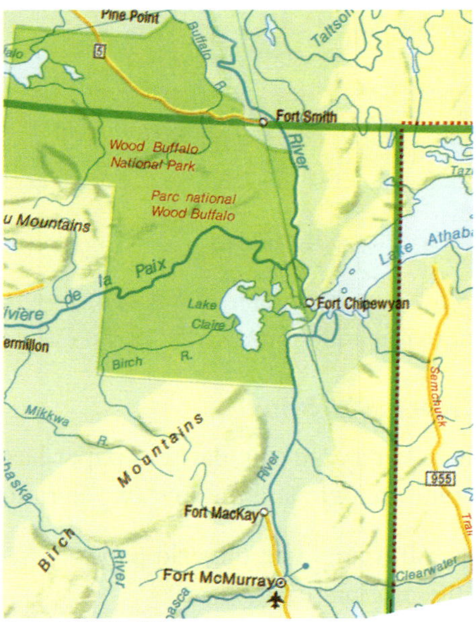

Scale 1:6 000 000

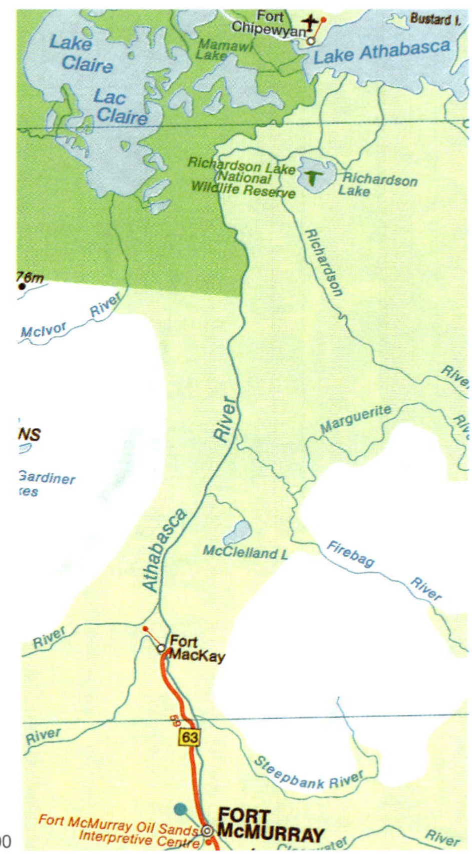

A similar effect occurs when we examine the properties of fractals in more detail.

Scale 1:2 000 000

Investigation 1 **Properties of a Koch Snowflake**

Part A: The Perimeter of a Koch Snowflake

1. The figures below were generated using *The Geometer's Sketchpad*. Assume that the original triangle has sides of length 9 cm. Visualize the procedure continuing.

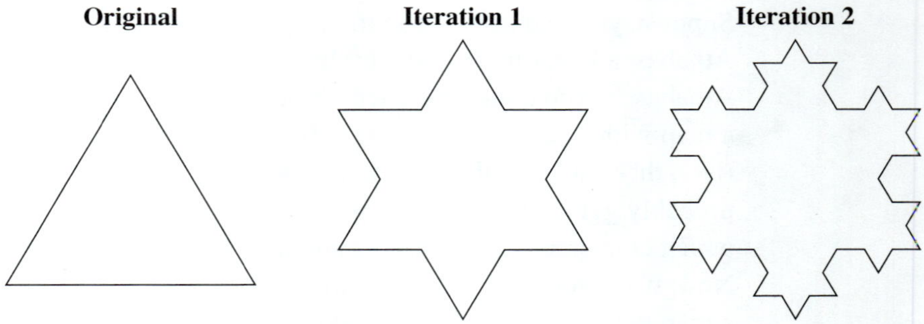

2. Copy this table. Complete it for iterations 1 and 2. Use any patterns you see to complete iterations 3 and 4.

Iteration	Number of Sides	Length of Each Side (cm)	Total Perimeter (cm)
Original	3	9	27
1			
2			
3			
4			

3. Predict the total perimeter after 6 iterations.

4. Predict what will happen to the perimeter of a Koch snowflake as the number of iterations increases indefinitely.

Part B: The Area of a Koch Snowflake

To find the area of a Koch snowflake, triangular grid paper has been added to the background of the original triangle and the first two iterations.

1. Suppose the area of a small triangle on the grid paper is 1 square unit. Determine the area of the original triangle.

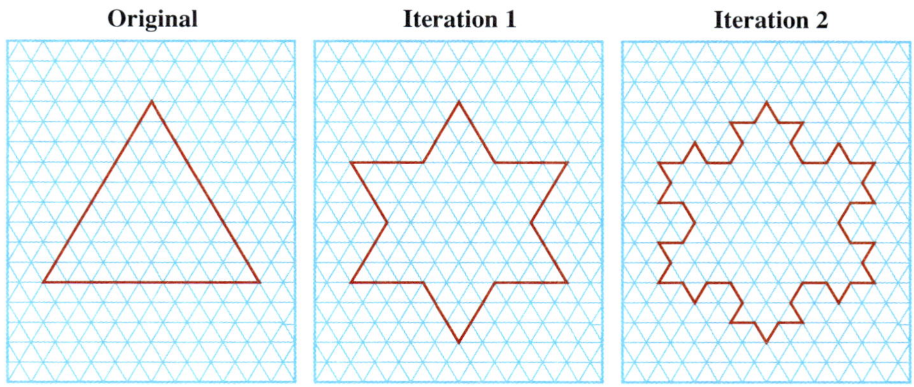

2. Copy this table. Use the diagrams above to complete it for iterations 1 and 2.

Iteration	Area of One New Triangle	Number of New Triangles Added	Total Area Added	Total Area
Original				81
1	9			
2	1			
3				
4				

3. How does the area of a new triangle change from iteration to iteration?

4. Use any patterns you see to complete the table.

5. What is the trend in the total area? What do you think the total area of the Koch snowflake will be as the number of iterations increases indefinitely?

Investigation 2 **Some Properties of a Sierpinski Gasket**

Use these figures in the Investigation that follows. The original figure and iterations 1 and 2 in the development of a Sierpinski gasket are shown. For simplicity, the triangles are equilateral.

Part A: The Perimeter of a Sierpinski Gasket

The perimeter of a Sierpinski gasket is defined as the total length of all the sides of the shaded triangles. Suppose the original triangle has side length 16 units.

1. Copy this table. Use the diagrams to complete the table for iterations 1 and 2.

Iteration	Number of Shaded Triangles	Length of Each Side	Total Perimeter
Original	1	16	48
1	3		
2			
3			
4			

2. The number of triangles forms a sequence. What kind of sequence is this? Write a recursive rule that describes how many shaded triangles occur at any stage. Use a calculator to generate the number of triangles needed for iterations 3 and 4.

3. Use any other patterns you can see to complete the table.

Part B: The Area of a Sierpinski Gasket

The area of a Sierpinski gasket is defined as the total area of the shaded triangles that make up the gasket. Use the diagrams in Part A. Suppose the original triangle has an area of 64 square units.

1. How is the area of a small shaded triangle in iteration 1 related to the area of the original triangle? How do you know this?

2. Copy this table.

Iteration	Number of Smallest Triangles	Area of Smallest Triangle	Total Area
Original	1	64	64
1			
2			
3			
4			

3. Use the diagrams and your knowledge of similar figures to complete the table for iterations 1 and 2.

4. Use any patterns you see to complete the table for iterations 3 and 4.

5. Write an expression for the area of the Sierpinski gasket after n iterations.

The Investigations have explored fractals that are 2-dimensional. In the previous Tutorial, you created a fractal that was 3-dimensional.

Example

A sequence of cylinders

Consider a cylinder with the base radius $r = 10$, and height $h = 20$. A second cylinder is similar to this but with base radius and height each one-half of the original. A third cylinder is similar with base radius and height one-half of the second cylinder. Suppose the sequence of similar cylinders continues.

a) Predict the surface area of the fifth cylinder.

b) Write an expression for the surface area of the *n*th cylinder in the sequence.

c) Predict the volume of the sixth cylinder.

d) Write an expression for the volume of the *n*th cylinder.

SOLUTION

a) Use the formula for the surface area, A, of a cylinder with base radius r and height h: $A = 2\pi r(r + h)$

The surface area of the original cylinder is:

$$A(1) = 2\pi(10)(10 + 20)$$
$$\doteq 1885$$

Recall that when the dimensions of an object are halved, the surface area is reduced by a factor of 4, or 0.25.

So the new surface area = 0.25 × (previous surface area).

Therefore, the sequence of surface area is generated by the recursive formula $A(n) = 0.25A(n - 1)$.

Use a graphing calculator in sequence mode to determine the fifth term of this sequence.

UTILITY

If you need help to use the sequence mode of a TI-83 calculator, refer to Utility 33.

Enter the data shown below left in the [Y=] screen. Press [2nd] [GRAPH] to obtain the table of values below right.

$A(5) \doteq 7.4$

The surface area of the fifth cylinder is approximately 7.4 square units.

b) From part a, $A(n) = 0.25A(n-1)$
This result is always true, so $A(n-1) = 0.25A(n-2)$.
Substitute this result in the first equation.

$$A(n) = 0.25A(n-1)$$
$$A(n) = 0.25[0.25A(n-2)]$$

Similarly, $A(n) = 0.25\{0.25[0.25A(n-3)]\}$, and so on
The sequence continues until $A(n)$ is reached in $n-1$ steps.

$$A(n) = (0.25)^{n-1}A(1)$$

c) The volume of a cylinder of base radius r and height h is:

$$V = \pi r2h$$

The volume of the original cylinder is: $V(1) = \pi(10^2)20$
$\doteq 6283 \text{ cm}^3$

Recall that when the dimensions of an object are halved, the volume is reduced by a factor of 8. So the recursive formula for the sequence of volumes is $V(n) = 0.125V(n-1)$.

Use a graphing calculator in sequence mode to generate the sequence of volumes (above). The volume of the sixth cylinder is approximately 0.19 cubic units.

d) Use a process similar to the one in part b. The volume of the nth cylinder is $V(n) = (0.128)^{n-1}V(1)$.

DISCUSSING THE IDEAS

1. Do you think that the area of a fractal will always approach 0 as in *Investigation 2*?

2. Can the perimeter of a fractal increase indefinitely? If so, give an example.

3. Can the area of a fractal ever increase indefinitely? Explain.

EXERCISES

Checking Your Skills

1. Consider the Hat Curve fractal.

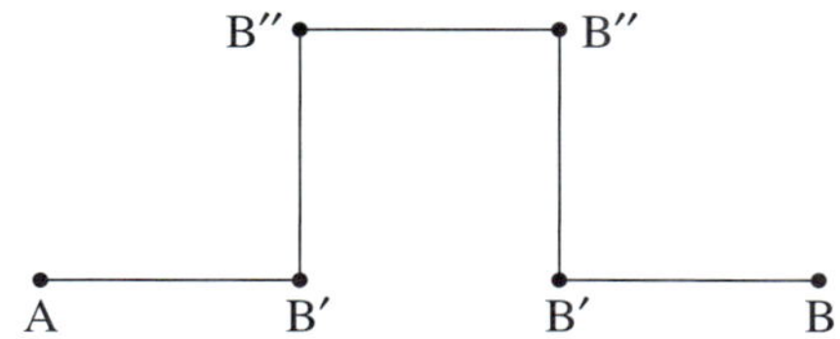

The Hat Curve fractal begins as a straight line.

a) Copy this table.

Iteration	Length of Each Segment	Number of Segments	Total Length
Original	243	1	243
1			
2			
3			
4			

b) Suppose the length of the original line is 243 units. In each successive iteration, the lengths of the segments are one-third the length of their parent segment. Use this information to complete the table.

c) What is the pattern in the number of segments at each successive iteration? Predict the number of segments in iteration 6.

d) What observations can you make about the total length of the Hat Curve fractal as the number of iterations increases?

2. In exercise 3 of *Tutorial 6.3* you created a fractal called the Sierpinski carpet. The original and the first two iterations are shown.

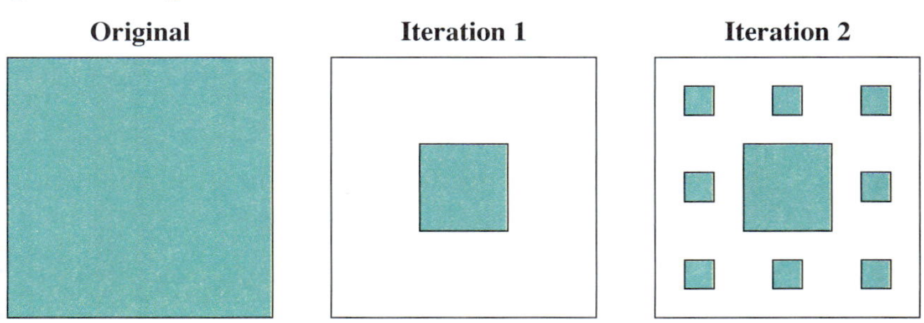

a) Assume the original square has a length of 27 units. Copy and complete this table.

Iteration	Side Length of New Square	Area of a New Square	Number of New Squares	Total Area of New Squares	Total Unshaded Area
1	9		1		
2					
3					
4					

b) What is the area of the figure as the number of iterations increases indefinitely?

3. Examine the following drawings created from circles.

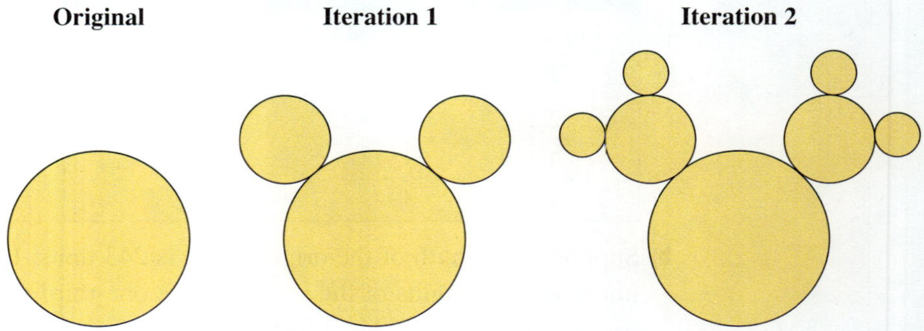

a) Assume that the radius of the original circle is 2 cm and that each successive circle has a radius that is one-half that of its parent. Copy and complete the following table up to iteration 4.

Iteration	Number of New Circles	Radius of One New Circle	Increase in Circumference	Total Circumference of All Circles
1				
2				
3				
4				

b) Use the patterns to predict the total circumference as the number of iterations increases indefinitely.

c) Explore how the area of this fractal pattern changes with each iteration. Copy and complete this table.

Iteration	Area of One New Circle	Number of New Circles Added	Increase in Area	Total Area
1		2		
2				
3				
4				

d) Use any patterns you see. Do you think there will be a maximum possible area? If so, what would that area be?

4. Consider this sequence. The first sphere has radius 64 cm, the second has radius 32 cm, the third has radius 16 cm, and so on.

a) Determine the surface area of the seventh sphere. Write an expression for the surface area of the nth sphere in the sequence.

b) Determine the volume of the sixth sphere. Write an expression for the volume of the nth sphere in the sequence.

5. Create your own fractal pattern. Use a simple basic shape. Use the methods of this Tutorial to predict the perimeter and area of your fractal for the fifth iteration, then the nth iteration.

Extending Your Thinking

6. Consider a sequence of cubes where the side length of the first cube is x cm. Each cube in the sequence has side length twice that of the previous cube. Determine an expression for the total length of the edges, the surface area, and the volume of the nth cube in the sequence.

PROJECT

In this Tutorial, you learned about the perimeter, area, and volume of a fractal pattern. You will use these skills in the Chapter Project and in these projects in the Project Book.

- The Ponzi Scheme
- Threatened and Endangered Species
- Building the Best Soft-Drink Container

COMMUNICATING THE IDEAS

Explain why the perimeter and the area of a fractal behave differently as you perform more and more iterations on that fractal.

How Does Caffeine Affect Your Body?

You have investigated the effect of caffeine on the human body. Now you will create a poster that illustrates your findings.

Create a pictorial representation of the metabolism of caffeine in the human body. You could use dynamic geometry software such as *The Geometer's Sketchpad* to create a pattern that illustrates the elimination of caffeine from the body, if this is available.

- Choose a suitable simple closed shape, such as a circle, to represent a cup of coffee. Determine its area.
- Use the area of the enclosed shape to represent the amount of caffeine in the body. Reproduce the shape, suitably decreased in area, at equal intervals on a line to represent the elimination of caffeine from the body each hour.
- Use your result to create a new diagram that shows the half-life of caffeine.
- Use your diagrams and your report to create a poster to explain the effect of caffeine on the human body.

Career Skills

The ability to communicate information truthfully and effectively is used by people in the training industry, and by people who work in the media. Communications can take many different forms: an oral presentation, a written report, or a pictorial representation. Employers value effective communication skills.

CONSOLIDATING YOUR SKILLS

What Do I Need To Know?

Tutorial	Skills and Concepts	Important Results and Formulas
6.1	• Determine the pattern in a sequence • Identify arithmetic and geometric sequences • Use technology and recursive formulas to generate a sequence	• Consecutive terms of an arithmetic sequence have a common difference. • Consecutive terms of a geometric sequence have a common ratio.
6.2	• Model problems using sequences • Graph, interpolate, and extrapolate from sequential data	
6.3	• Use self-similarity and iteration to construct fractals	• Fractals are generated by iterative procedures.
6.4	• Calculate the perimeter, area, or volume of a fractal	• Some fractals have a perimeter that increases without bound, but a finite area.

What Should I Be Able To Do?

After you have completed the work in Chapter 6, you should be able to solve the problems that follow. Part A is investigative in nature. As you work through the exercises, refer to the *Student Reference*, the *Utilities*, and the *Tutorials* as necessary.

Part A

1. The following instructions are for use with a draw or paint program on a computer. You could also adapt the instructions and use grid paper to create a fractal.

a) Draw a square.

b) Select and make a copy of the square. Then paste it onto the current screen.

c) Shrink the second square to 50% of its current size. Make 2 copies of this smaller square and attach the small squares in the middle of the top and bottom of the original square as shown. Keep one copy of the small square for the next step.

d) Select the remaining small square. Shrink it to 50% of its current size. Make 2 copies of this smaller square and attach them to the top and bottom of the fractal. Continue the procedure for 3 more steps.

e) Use colour to enhance the design. Print or sketch the fractal pattern.

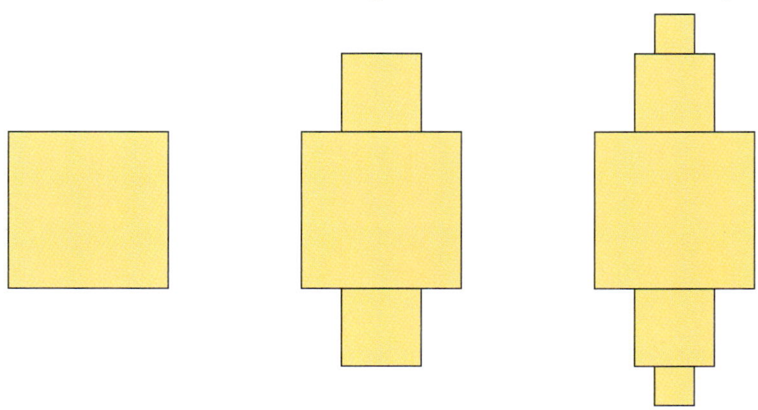

Original **1st iteration** **2nd iteration**

2. Assume the original square in exercise 1 had side length 64 cm.

a) Copy and complete the following table up to iteration 5.

Iteration	Length of Side of Newest Square (cm)	Increase in Area of Fractal (cm^2)	Total Area (cm^2)	Increase in Perimeter (cm)	Increase in Perimeter (cm)
Original	64				
1					
2					
3					
4					
5					

b) Predict the total perimeter and area after iteration 6.

c) Predict what will happen to the perimeter of this fractal as the number of iterations increases indefinitely. What will happen to the area?

Part B

3. A racquetball bounces to 75% of the height from which it is dropped. Use the sequence mode of your calculator to graph the first 10 successive rebound heights of a racquetball that has been dropped from an initial height of 3.4 m.

a) How high was the ball after the eighth bounce?

b) After how many bounces does the racquetball rebound to a height of less than 0.5 m?

4. Amoxicillin is an antibiotic often used to treat bacterial infections. It has a half-life of about 2 h. Suppose a child is given a 250-mg dose of Amoxicillin at 7 A.M.

a) Calculate the number of milligrams of Amoxicillin that remain in the child's body at the end of every two-hour period, up to 3 P.M.

b) Estimate the amount of Amoxicillin in the child's body at 11:45 A.M.

5. Ibuprofen is an anti-inflammatory drug to treat arthritis symptoms. It has a half-life of 2 h. Suppose Jeremy takes two 150-mg tablets of Ibuprofen every 4 h.

a) How much Ibuprofen will remain in his body after 12 h, if he has not yet taken his 4th dose?

b) What is the maintenance level of Ibuprofen?

6. Fish and wildlife officers study the deer population in northeastern Alberta. One summer, a population count estimated 2200 deer to be in the region. Later that year, hunting permits for 100 deer were issued and used. From calls they received from motorists, the officers know that at least 83 deer were killed by vehicles during the year. They also estimate the natural mortality rate at 5% per year. Combining deaths from all causes, wildlife officers estimate that the herd has been reduced by 293 animals. The next summer, another population count estimates 2290 deer in the region. This means that 383 fawns joined the herd in the spring.

a) If this same pattern continues every year, how many deer will there be in the region 10 years from the initial population count?

b) The officers estimate that the region can support a maximum of 3000 deer. In how many years will the officers have to either increase the number of hunting permits issued, or catch excess deer and ship them to other parts of the province?

7. The diagram below shows the first three stages in the generation of a fractal with a reduction factor of 0.5. The original figure is a horizontal line. In iteration 1, a line that is one-half the length of the original line bisects this horizontal line. In iteration 2, two horizontal lines that are one-half the length of the vertical bisector from iteration 1 bisect the vertical line.

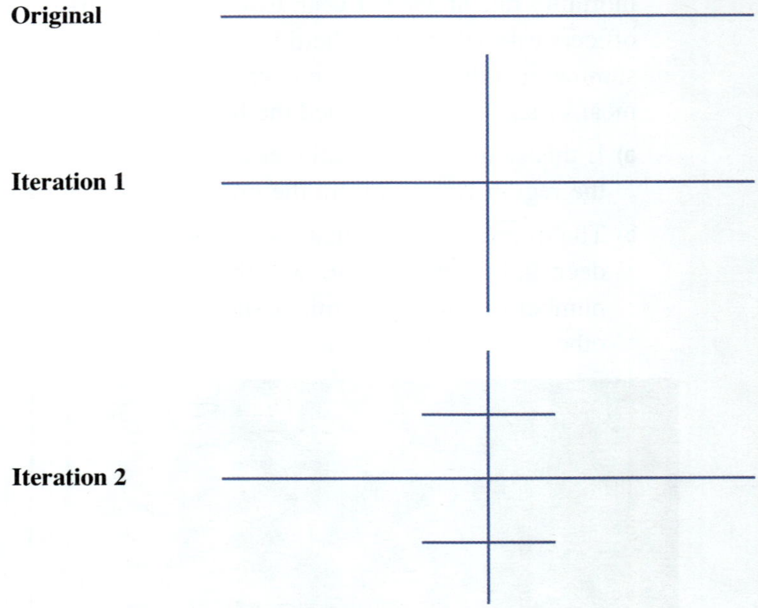

a) Draw iterations 3 and 4 of this fractal.

b) If the length of the original line is 128 mm, what is the total length of all the lines in iteration 5?

8. An artist wants to create a set of nested boxes. The boxes are open at one end, so that each smaller box fits inside a larger box. The largest box measures $10 \text{ cm} \times 10 \text{ cm} \times 16 \text{ cm}$, and one square end is open. Each successive box is 20% smaller.

a) If there are 6 boxes in the set, what is the volume of the sixth box? Ignore the thickness of the material.

b) If the boxes are made of thin sheets of copper and 10% of the material is wasted in cutting, what is the total surface area of copper sheet required to make all 6 boxes?

Part C

The Chapter Project combines many skills and concepts to solve a problem. Here is a sample of student work for the project. Review this sample to gain insight into your own project.

9. Here is a graph of the caffeine that remains in the body. Has the student entered the sequence correctly?

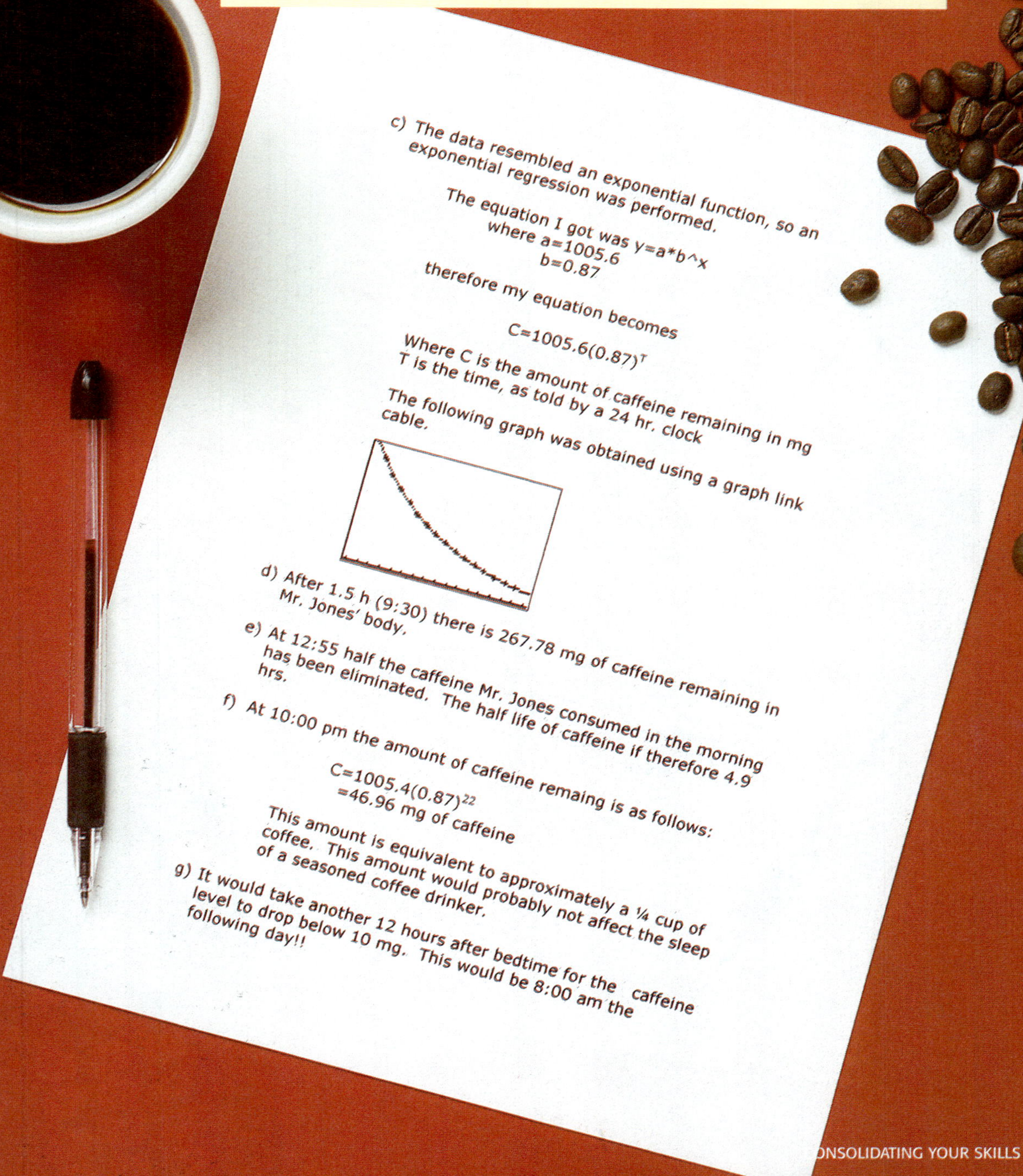

c) The data resembled an exponential function, so an exponential regression was performed.

The equation I got was y=a*b^x
where a=1005.6
b=0.87

therefore my equation becomes

$C=1005.6(0.87)^T$

Where C is the amount of caffeine remaining in mg
T is the time, as told by a 24 hr. clock

The following graph was obtained using a graph link cable.

d) After 1.5 h (9:30) there is 267.78 mg of caffeine remaining in Mr. Jones' body.

e) At 12:55 half the caffeine Mr. Jones consumed in the morning has been eliminated. The half life of caffeine if therefore 4.9 hrs.

f) At 10:00 pm the amount of caffeine remaing is as follows:

$C=1005.4(0.87)^{22}$
$=46.96$ mg of caffeine

This amount is equivalent to approximately a ¼ cup of coffee. This amount would probably not affect the sleep of a seasoned coffee drinker.

g) It would take another 12 hours after bedtime for the caffeine level to drop below 10 mg. This would be 8:00 am the following day!!

10. Here is part of that student's report on the elimination of caffeine. Are the findings justified?

7 Vectors

Overview

Chapter Section	Skills and Concepts
Chapter Project: Accident Reconstruction, page 302	Research the concept of momentum
Tutorial 7.1: Vectors and Scalars, page 304	Understand basic properties of geometric vectors
Chapter Project: Simulating an Accident, page 310	Simulate an automobile accident
Tutorial 7.2: Adding Vectors Using Scale Diagrams, page 312	Find the resultant of two vectors; find the equilibrant of two forces
Tutorial 7.3: Multiplying a Vector by a Scalar, page 320	Find the product of a scalar and a vector
Tutorial 7.4: Solving Vector Problems by Computation, page 324	Solve triangles to find vector magnitudes and directions
Tutorial 7.5: Vector Problems in Three Dimensions, page 330	Solve simple vector problems in three dimensions
Chapter Project: How Fast Was That Car Going? page 338	Find the velocity of a car before a crash
Consolidating Your Skills, page 340	Review the work of the chapter

PROJECT

You should do at least one project that uses the skills you learn in this chapter. The recommended projects in the Project Book are:

- Flight Simulator Development
- Tracking Pollution
- Training a Robot

CHAPTER PROJECT

Accident Reconstruction

One of the responsibilities of the RCMP is to gather evidence at the scene of a car accident, then use that evidence to determine what happened. In what direction were the cars travelling? How fast were the cars going? Which car caused the accident? One method used to reconstruct what happened during a car accident employs vectors.

In this project, you will learn some of the basic concepts used in accident reconstruction. Here are some initial questions to consider.

- What other occupations might need to have some knowledge of accident reconstruction? In what type of situations might they use this knowledge?
- When reconstructing a car accident, one factor that may need to be considered is the type of road surface on which the accident occurred. What other factors may have to be considered?
- Are there any businesses that deal primarily with accident reconstruction. What types of services do they offer? Is there computer software available to simplify the work?

One of the factors that should be considered in accident reconstruction is momentum. Research the topic of momentum and the Law of Conservation of Momentum. Potential sources of information include the Internet and the library. Physics textbooks might also be useful.

CHAPTER PROJECT

To complete this Chapter Project, you must develop an understanding of vectors. Complete *Tutorial 7.1* to develop the skills you need. You will return to the Chapter Project on page 310.

FYI Visit: www.awl.ca/school/connections

You can start your research from our Internet site: Click on MATHLINKS, followed by the *Applied Mathematics 12* logo. Then select a topic under Accident Reconstruction.

YIELD
SCHOOL BUS

TUTORIAL 7.1

Vectors and Scalars

Most of your experience in mathematics has been based on quantities, such as time intervals, or the length of an object, that can be described by specifying their magnitude. Such quantities are known as *scalar quantities*. Many other quantities, known as *vector quantities*, can only be fully described by specifying both a magnitude and a direction. An example of this is acceleration due to gravity, where we must give the magnitude (usually about 9.8 m/s^2) and direction (always vertically downward).

Suppose, one sunny day, you decide to go for a hike in Riding Mountain National Park. Unfortunately, you wander off the trail, trip over a tree root, and twist your ankle so badly that you cannot walk. Fortunately, you have a cellular phone and call the ranger station for assistance. Unfortunately, all you can tell the ranger is that you are about one kilometre from Turtle Pond. The ranger checks her map, but you could be one kilometre from Turtle Pond in any direction. She needs more information.

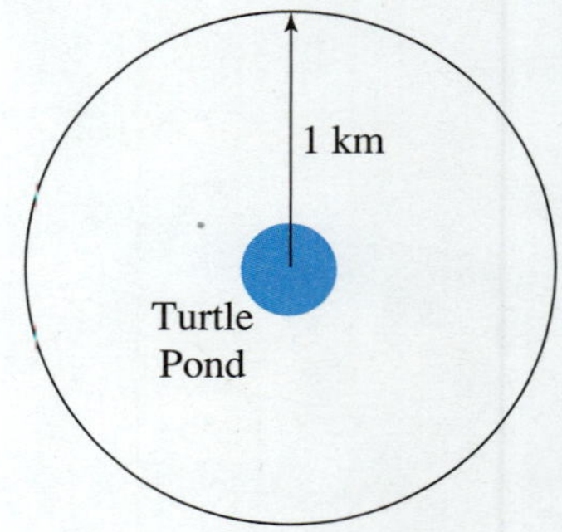

You realize that it is lunchtime and the sun is in the same direction as Turtle Pond, so you must be north of Turtle Pond. Now the ranger knows both your distance and your direction from Turtle Pond. She can draw a vector to your position on her map. You should be rescued soon.

In this chapter, you will use *geometric vectors* to solve practical problems. Geometric vectors are represented by directed line segments. The arrowhead indicates the direction of the vector, and the length of the segment indicates its magnitude.

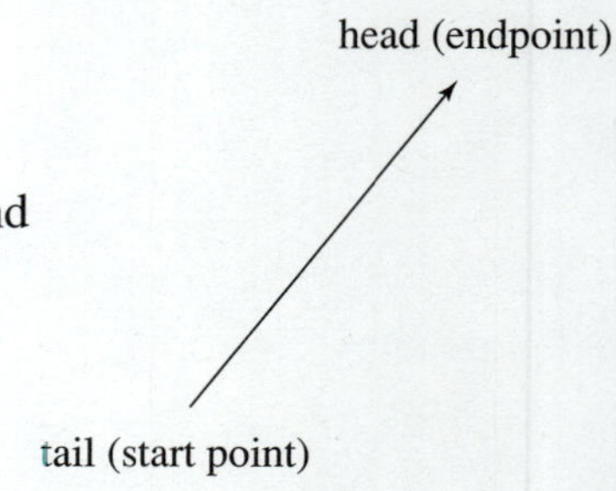

A geometric vector has two components, each of which is a scalar: the magnitude, or length, of the vector, and the direction in which the vector acts. To describe a geometric vector, both its magnitude and direction must be given.

A vector from point A to point B is written $\overrightarrow{AB}$. A vector can also be labelled by a single, lowercase letter such as $\vec{v}$. The magnitude of a vector is represented by $|\overrightarrow{AB}|$ or $|\vec{v}|$. The vertical bars tell us to use the absolute, or positive, value of the number contained within them. For example, $|-7| = 7$ and $|14| = 14$.

The direction of a vector can be described using the main compass bearings of north, east, south, or west; or a bearing such as 057°. Directions are sometimes written in square brackets, for example, [north] or [248°]. Unless stated otherwise, north is always at the top of the page.

STUDENT REFERENCE

To review bearings, refer to the Student Reference.

A *displacement* of 1 km north of Turtle Pond is a vector. *Velocity* is also a vector quantity describing the speed and direction of a moving object.

Some examples of vectors are:

- a velocity of 100 km/h [north]
- a force of 15 N vertically upward
- a displacement of 25 m [280°]

Some examples of scalars are:

- a speed of 80 km/h
- a mass of 96 kg
- a distance of 50 km from your school
- a time of 14 min

Example

Drawing vectors

Draw vectors to represent:

a) a displacement of 30 km to the northeast

b) a weight of 50 N acting vertically downward

c) a velocity of 230 km/h toward 310°

SOLUTION

a) Choose a convenient scale. Let 1 cm = 10 km. The segment will then be $30 \div 10 = 3$ cm long. Select a convenient starting point. Use a protractor to mark a direction 45° east of north. Construct a line segment 3 cm long. Add an arrowhead pointing toward the northeast. Label your vector.

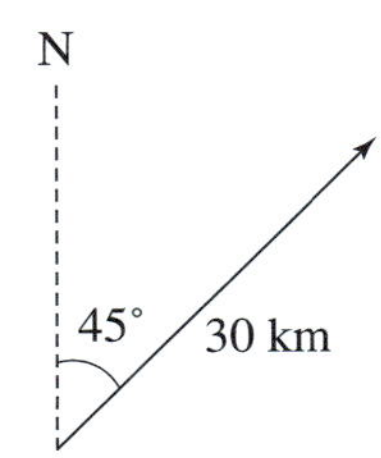

b) Let 1 cm = 20 N. A line segment of $50 \div 20 = 2.5$ cm is required. Let the top of the page represent the direction vertically upward.

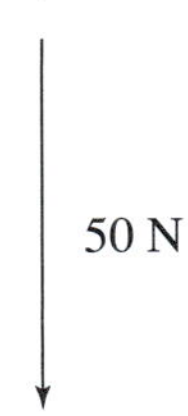

c) Use a scale of 1 cm = 50 km/h. Then, draw a line segment of $230 \div 50 = 4.6$ cm. Label the direction north. The compass bearing of 310° is found by measuring 310° clockwise from north, or by measuring 360° − 310° = 50° counterclockwise from north.

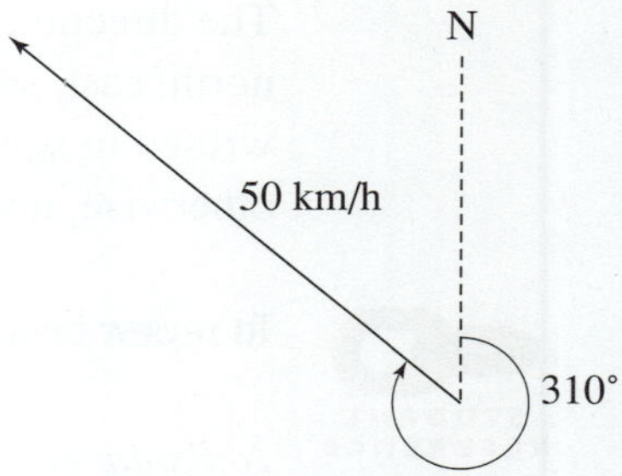

The Snowbirds are a precision-flying squadron of the Canadian Armed Forces. They are shown in the photograph flying in formation, that is, flying at exactly the same speed and in precisely the same direction. This means that, although the planes have different positions, their velocity vectors are equal. If their speeds are 350 km/h, their velocity vectors can be represented as follows (the direction of each vector is parallel to the direction of the aircraft that it represents):

Scale: 1 cm = 100 km/h

Equal vectors have the same magnitude and direction. The Snowbirds' velocity vectors are equal.

Opposite vectors have the same magnitude, but act in opposite directions.

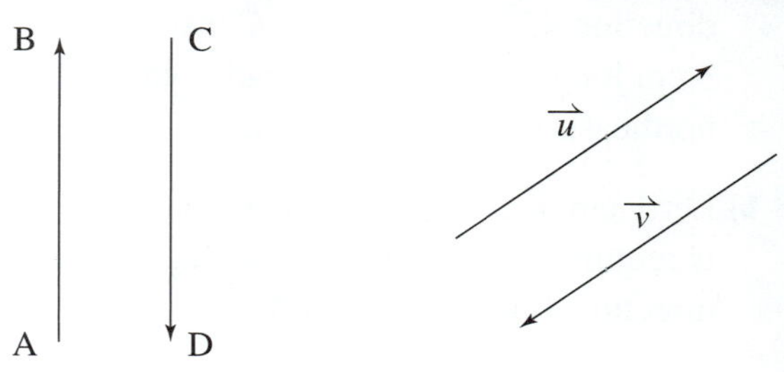

In the diagram, $|\overrightarrow{AB}| = |\overrightarrow{CD}|$, but $\overrightarrow{AB}$ points north while $\overrightarrow{CD}$ points south. Thus, $\overrightarrow{AB}$ and $\overrightarrow{CD}$ are opposites of each other. We write $\overrightarrow{AB} = -\overrightarrow{CD}$ or $-\overrightarrow{AB} = \overrightarrow{CD}$. Similarly, from the diagram, $\overrightarrow{v} = -\overrightarrow{u}$, or $\overrightarrow{u} = -\overrightarrow{v}$.

Two vectors $\overrightarrow{AB}$ and $\overrightarrow{CD}$ are **equal** if $|\overrightarrow{AB}| = |\overrightarrow{CD}|$ and the direction of $\overrightarrow{AB}$ is the same as the direction of $\overrightarrow{CD}$. Write $\overrightarrow{AB} = \overrightarrow{CD}$.

Two vectors $\overrightarrow{AB}$ and $\overrightarrow{CD}$ are **opposite** if $|\overrightarrow{AB}| = |\overrightarrow{CD}|$ and the direction of $\overrightarrow{AB}$ is opposite to the direction of $\overrightarrow{CD}$. Write $\overrightarrow{AB} = -\overrightarrow{CD}$.

DISCUSSING THE IDEAS

1. List 10 quantities that are scalars. List 10 quantities that are vectors.

2. Explain the difference between mass and weight. Give an example of when you would use mass and when you would use weight.

EXERCISES

Checking Your Skills

1. Which of the following can be described by a vector?

a) a wind of 35 km/h from the southwest

b) a distance of 14.7 km

c) a book containing 462 pages

d) a current of 6 km/h toward the east

e) a 72-L tank of fuel

f) a speed of 85 km/h

g) a weight on Mars of 300 N

h) an advance of 15 km due east

i) a 40-kg bag of fertilizer

2. Which of these quantities is a scalar?

a) age	**b)** volume	**c)** displacement
d) mass	**e)** density	**f)** area
g) temperature	**h)** weight	**i)** force of friction

3. Given the vectors below, use a ruler to measure them. Use the given scale to find the magnitudes. Use north, south, east, or west to describe the directions.

a)

Scale: 1 cm = 10 m/s

b)

Scale: 1 cm = 5 m

c)

Scale: 1 cm = 100 km/h

d)

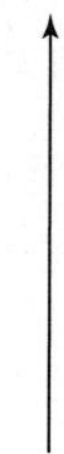

Scale: 1 mm = 10 m/s

e)

Scale: 1 cm = 50 N

f)

Scale: 1 mm = 5 m/s^2

4. Find the magnitude and direction of each vector below. Use the directions north, south, east, west, northwest, northeast, southwest, or southeast to describe the directions.

a)

Scale: 1 cm = 20 m/s

b)

Scale: 1 cm = 10 m

c) 

Scale: 1 cm = 10 km/h

d) 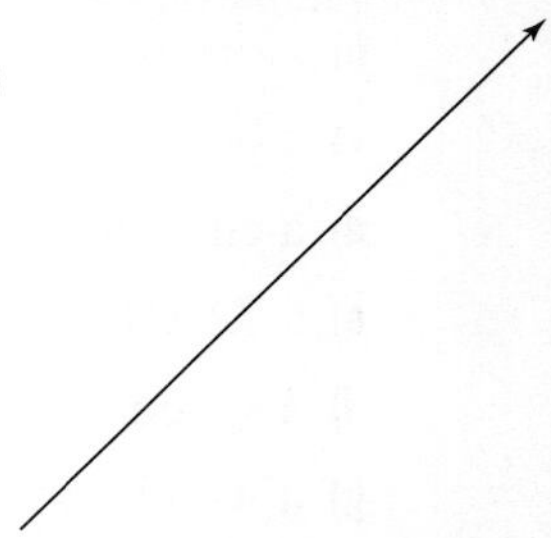

Scale: 1 mm = 1 m/s^2

e)

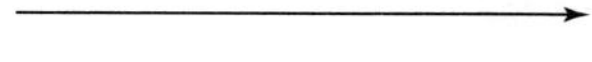

Scale: 1 cm = 25 knots

f) 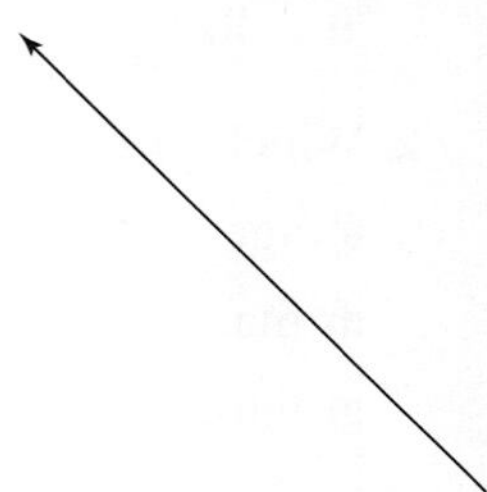

Scale: 1 mm = 5 m

5. Construct a scale drawing of the following vectors. Supply all necessary labels.

a) 50 km/h [north]

b) 12 m/s [095°]

c) 500 N [east]

d) acceleration of 2.5m/s^2 [southeast]

e) 7 paces [west]

6. For each of these vectors, describe the opposite vector then draw and label a scale drawing of the vector and its opposite.

a) 18 m/s [east]

b) 14 km [038°]

7. Hurricane Marge is situated 90 km southeast off Florida's coast. It is moving at a rate of 30 km/h in a northwest direction. Construct a scale diagram of the velocity vector.

8. A ball of Silly Putty is thrown horizontally against a wall with a velocity of 10 m/s. The ball rebounds with essentially the same velocity. Construct a scale diagram to represent this situation.

Extending Your Thinking

9. A car is travelling east along a highway. It passes one signpost at 10:23 A.M., and then passes another signpost 45 km along the road at 10:53 A.M. Construct a scale diagram to represent the velocity of the car.

PROJECT

In this Tutorial, you developed an understanding of scalars and vectors, and developed skills in the construction of scale diagrams of vectors. You will need these skills to complete the Chapter Project, and the following projects in the Project Book.

- Flight Simulator Development
- Tracking Pollution
- Training a Robot

COMMUNICATING THE IDEAS

Write a note to a friend to explain the difference between a scalar and a vector. Provide examples in your explanation.

Simulating an Accident

In the first part of the Chapter Project, you investigated the topic of momentum. You should have found that momentum is a vector quantity found by multiplying an object's mass by its velocity. You also found that according to the Law of Conservation of Momentum, the total momentum before a collision equals the total momentum after the collision.

Work with a partner to investigate a simple collision. To conduct this investigation you will need a level table, a large sheet of paper, and two objects — such as slotted masses or checkers — that you can slide over the paper. Object 1 should be double the mass of Object 2. If you are using checkers, glue two checkers together.

As an object slides across the paper, friction slows it down, until it stops. The faster an object is moving, the further it will slide. Use the square root of the length of a slide, L, to represent the initial velocity, v, of an object.

$$v = \sqrt{L}$$

1. Fasten the paper so that it lies flat in the middle of the table.

2. Place Object 1, the heavier object, in the middle of the paper and mark its position by tracing lightly around the base of the object with a pencil. Label the centre of that spot A.

3. Practise flicking Object 2 so that it slides along the paper and bounces off Object 1 at an angle. After the collision, both objects should remain on the paper. After each practice run, return Object 1 to spot A.

Before the collision:

CHAPTER PROJECT

To complete this Chapter Project, you must solve porblems with vectors. Complete *Tutorials 7.2* to *7.5* to develop the skills you need. You will return to the Chapter Project on page 338.

4. When you obtain a result that seems satisfactory, estimate the location of Object 2 at the moment of collision, and label the centre of that position D.

   At the moment of the collision:

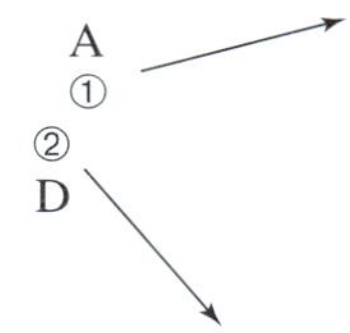

5. Mark the centre of final position of Object 1 as B, and the centre of the final position of Object 2 as C.

   After the collision:

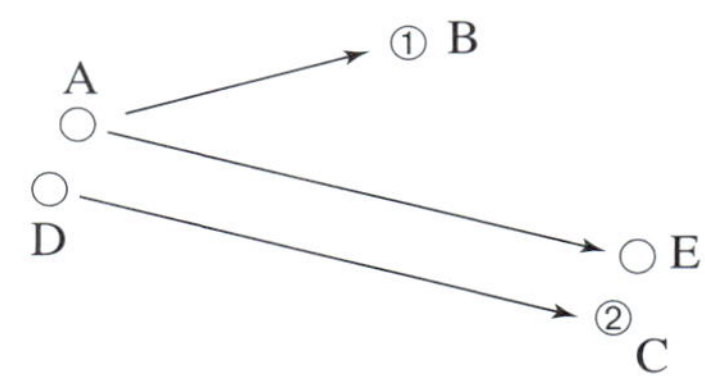

6. Suppose you have been called to investigate the collision between these objects. An eyewitness tells you that Object 1 was stationary at A when Object 2 ran into it from behind. Your job is to gather the available evidence and determine the initial velocity of Object 2. Save all your work. You will use it to complete the next part of the Chapter Project.

Career Skills

People who analyze collisions take careful measurements and interpret results. They must present reasoned, unbiased reports.

TUTORIAL 7.2

Adding Vectors Using Scale Diagrams

We must be able to combine vectors in some way that takes their direction into account. In this Tutorial, you will explore ways to add vectors.

Investigation 1

Adding Vectors That Represent Changes in Position

One summer's day, Jerri and Jamie were having a picnic on a small island in the upper reaches of the North Saskatchewan River. Jerri noticed that something sparkled in the sun and found an old milk bottle with a note inside.

> "Follow these directions to find one of the greatest treasures of all the ages. Only someone worthy of the treasure will find it. Start at the middle of the island. Walk 100 paces due north. Then turn toward 135° and walk 127 paces. Then turn toward 249° and walk another 95 paces. Finally, walk 28 paces on a bearing of 315°. There you will find a flat rock with the initials EM carved on it. Under the rock is your treasure."

Jamie exclaimed: "We can't follow those instructions! The island is only 50 paces by 50 paces. We will end up in the middle of the river!"

1. Construct a scale diagram of the island.
2. Using the same scale, construct a vector for each displacement. Draw the vectors *head-to-tail* (where one vector ends, the other vector begins). You may construct your scale drawing using a ruler and a protractor on paper or you may use dynamic geometry software such as *The Geometer's Sketchpad*.

UTILITY

If you need help to draw vectors using *The Geometer's Sketchpad*, Utility 34 shows you how.

3. Draw a vector to represent the displacement of the treasure from the centre of the island. Describe its magnitude and direction.
4. Suppose Jeri and Jamie are standing in the middle of the island. How can they find the treasure without getting their feet wet?

Vectors $\vec{u}$ and $\vec{v}$ can be added by placing them head-to-tail. The sum, $\vec{u} + \vec{v}$, of the vectors is another vector, called the *resultant*, which has the same start point as the first vector $\vec{u}$, and the same endpoint as the second vector $\vec{v}$. Use the scale diagram to find the magnitude and direction of the resultant.

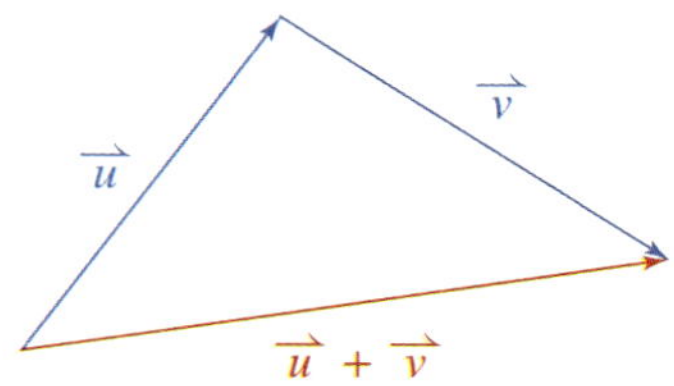

This is an example of the *triangle method* of adding vectors.

Consider an aircraft, driven forward by its engines, and blown sideways by a wind. The speed and direction of the aircraft relative to the ground can be found by adding two vectors arranged *tail-to-tail.*

In this Investigation, you will explore a method to add vectors when they have the same start point.

Investigation 2

The Parallelogram Method for Adding Two Vectors

An airplane is driven by its jet engines with a velocity of 600 km/h [090°]. A wind blows toward 015° at a speed of 100 km/h. Find the resultant.

1. Draw $\overrightarrow{AB}$ to represent the velocity of the plane toward 090°. Draw $\overrightarrow{AC}$ to represent the wind toward 015°. Use the scale 1 cm = 100 km/h.

2. Construct a line parallel to $\overrightarrow{AB}$ through point C. Construct another line, parallel to $\overrightarrow{AC}$, through point B to form a parallelogram ABDC.

3. Measure the magnitude and direction of $(\overrightarrow{AC} + \overrightarrow{CD})$ and $(\overrightarrow{AB} + \overrightarrow{BD})$.

4. How are $(\overrightarrow{AC} + \overrightarrow{CD})$ and $(\overrightarrow{AB} + \overrightarrow{BD})$ related to $\overrightarrow{AD}$?

   According to the *parallelogram method* of vector addition, the diagonal of the parallelogram, $\overrightarrow{AD}$, is equal to the sum of $\overrightarrow{AB}$ and $\overrightarrow{AC}$.

5. Describe the speed and direction of the aircraft relative to the ground.

- When two vectors are added, a third vector, called the resultant, is produced.
- When two vectors are arranged sequentially, they are head-to-tail. Head-to-tail vectors can be added using the triangle method of addition.
- When two vectors are acting on the same point, they are arranged tail-to-tail. Tail-to-tail vectors can be added using the parallelogram method of addition.
- Vectors can be added in any order:

$$\overrightarrow{u} + \overrightarrow{v} = \overrightarrow{v} + \overrightarrow{u}$$

You probably have an intuitive sense of a force as a push or a pull. A force has magnitude and direction, so it is a vector quantity. When more than one force acts on the same object, those forces can be added together. The resultant is the single force that could replace all the individual forces that act on that object. Forces tend to cause the object on which they are acting to move in the direction of the resultant force.

Investigation 3

A Study of Forces

For this Investigation you will need some heavy string, a set of lab masses up to 500 g, 2 force scales, and a protractor. You will also need two lab stands or some other supports to which the scales will be attached. Work in groups of 3.

1. Place each mass on a force scale and determine the downward force of gravity exerted on each mass. In your notebook, or on a spreadsheet, create and complete the following table.

Mass (g)	Force (N)
300	
400	
500	

2. Set up the equipment as shown.

You may need to place some heavy books on the bases or have 2 people hold down the lab stands.

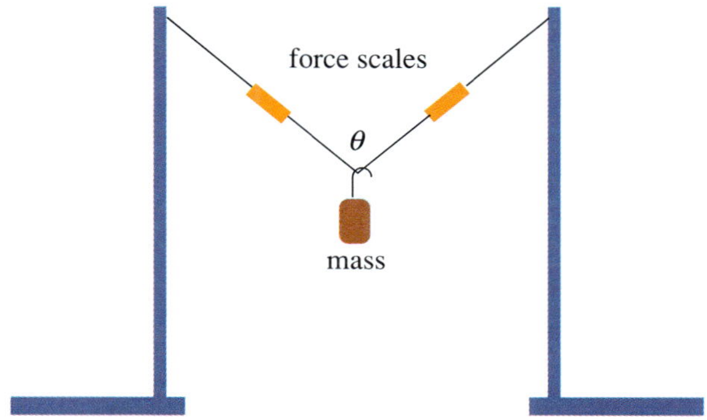

3. For each mass, record the force shown on each scale and measure the angle between the scales (θ). Record your results in a table like the one below.

Mass (g)	Force on First Scale (N)	Force on Second Scale (N)	Measure of θ (degrees)
300			
400			
500			

Draw and label a vector diagram for each situation.

4. Shorten the strings to change the size of the angle. Repeat exercise 3.

5. Lengthen the strings. Repeat exercise 3.

6. Use vector addition to find the resultant of each of the 9 pairs of vectors. Compare your resultants to your findings in exercise 1.

7. If you were to hang a heavy picture, which of the following methods would you use? Why?

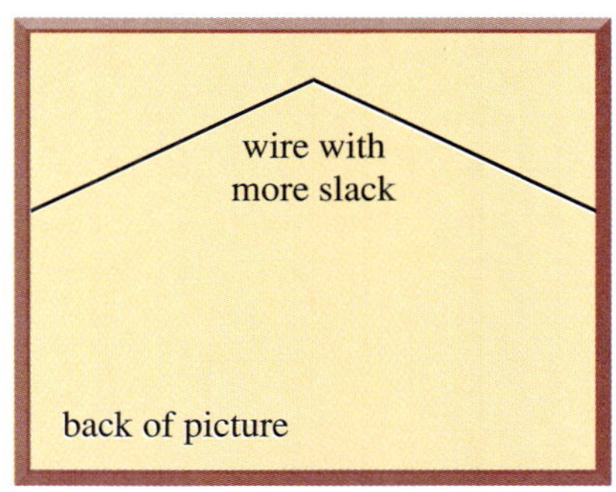

8. Summarize your findings.

If an object is acted on by forces, but does not move, then there must be another force present, equal to the resultant, but acting in the opposite direction. This force is called the *equilibrant* of the system of forces.

Example 1 **Determine a resultant force.**

Two tractors are being used to pull a tree stump out of the ground. The larger tractor exerts a force of 3000 N toward 108°. The smaller tractor pulls with a force of 2300 N toward 078°. It is a big stump and does not move.

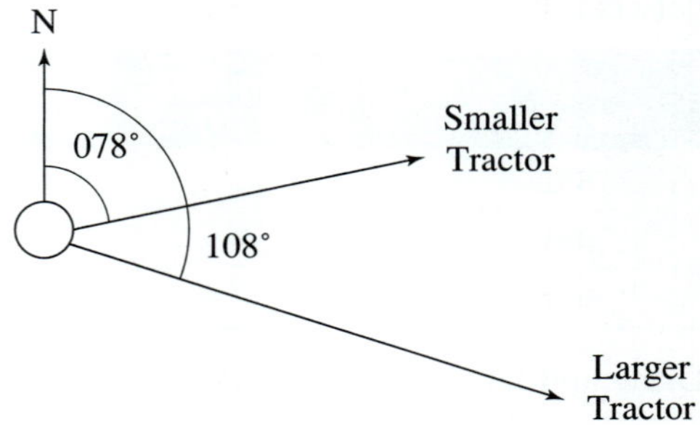

a) Determine the resultant force on the tree stump.

b) Determine the equilibrant force.

SOLUTION

a) The forces have the same start point. Use the parallelogram method.

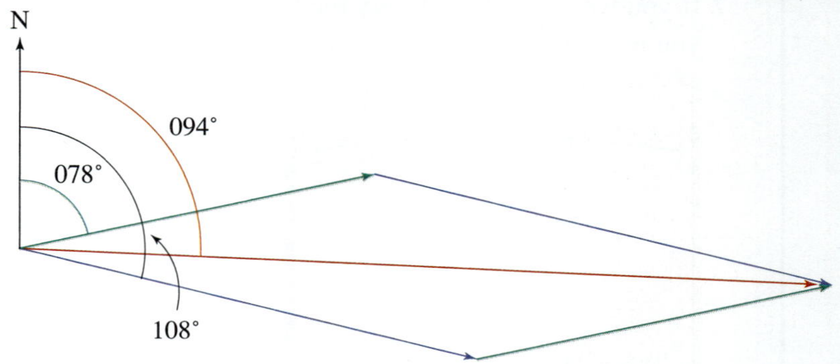

Scale: 1 cm = 500 N

The length of the resultant vector on the scale diagram is 10.1 cm. Use the scale to convert this measurement. The resultant force is 5050 N [094°].

b) The equilibrant force is equal in magnitude, but opposite in direction to the resultant force. The equilibrant force is 5050 N [274°].

Example 2 Adding vectors in a geometric figure

Determine each vector sum.

a) $\overrightarrow{AH} + \overrightarrow{HG}$ **b)** $\overrightarrow{HB} + \overrightarrow{BD}$

c) $\overrightarrow{GF} + \overrightarrow{BC}$ **d)** $\overrightarrow{GF} + \overrightarrow{CB}$

e) $\overrightarrow{FD} + \overrightarrow{DE}$

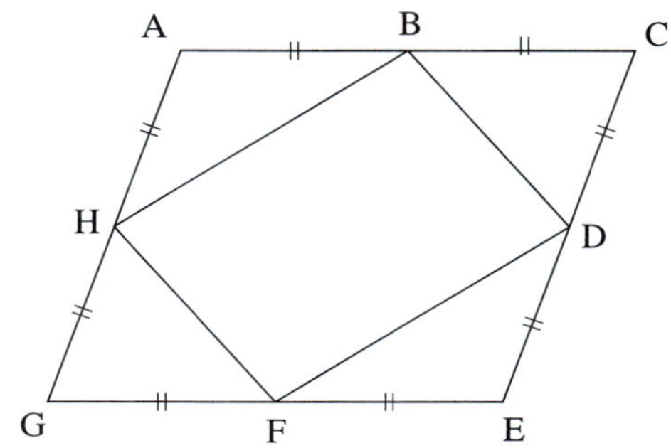

SOLUTION

a) $\overrightarrow{AH}$ and $\overrightarrow{HG}$ are pointing in the same direction and along the same straight line.
Thus, $\overrightarrow{AH} + \overrightarrow{HG} = \overrightarrow{AG}$.

b) $\overrightarrow{HB} + \overrightarrow{BD} = \overrightarrow{HD}$

When the endpoint of one vector is the same as the start point of another, the sum of the vectors can be found directly using the vector notation. Start from point H to point B, then from point B to point D. The net result is starting at H and ending up at D. The resultant displacement is $\overrightarrow{HD}$.

c) $\overrightarrow{GF} + \overrightarrow{BC} = \overrightarrow{GF} + \overrightarrow{FE}$ (Since $\overrightarrow{BC} = \overrightarrow{FE}$)
$= \overrightarrow{GE}$

d) $\overrightarrow{GF} + \overrightarrow{CB} = \overrightarrow{GF} + \overrightarrow{FG}$ (Since $\overrightarrow{CB} = \overrightarrow{FG}$)
$= \overrightarrow{0}$

e) $\overrightarrow{FD} + \overrightarrow{DE} = \overrightarrow{FE}$

DISCUSSING THE IDEAS

1. How can two different vectors in a diagram be equal?

2. What are some advantages and disadvantages of using scale diagrams to add vectors?

3. Describe how you could add three, or more, vectors.

4. If you were designing a machine, which forces would be more important: the applied forces, the resultant, or the equilibrant?

EXERCISES

Checking Your Skills

1. Jack jogs north at 15 km/h for 30 min, then turns east and jogs at 12 km/h for 20 min.

a) How far has Jack jogged in total?

b) How far is Jack from his starting point?

c) In what direction does Jack need to go to return to the start by the shortest path?

2. A jet travels due east from Edmonton for 1000 km, then turns toward [120°] and travels another 1200 km toward Winnipeg. Find the displacement of the jet from Edmonton.

3. A deep-sea diver is able to walk on the ocean floor at a speed of 0.2 m/s. He walks on a bearing of 120° for 90 s, then turns to a bearing of 205° and walks for another 50 s. How far from his start point, and in what direction, is the diver?

4. While taking part in an orienteering competition, Myra made a wrong turn. Then she discovered that her compass was not working and she was lost. The original directions were to walk 1.6 km on a bearing of 185°, then 0.9 km toward [200°]. When she radioed for help, she realized that she had walked the 0.9-km segment toward [020°]. How far, and in what direction, must a rescuer walk from the start point to find her?

5. A ski jumper encounters a horizontal friction of 85 N backward, a vertical weight of 750 N downward, and an air resistance of 340 N upward.

a) Draw the vector addition of these forces.

b) Use your drawing from part a to find the magnitude and direction of the resultant force.

6. Juan and Marie use ropes held horizontally to try to pull a rock to the side of a field. Juan pulls with a force of 400 N toward [115°], and Marie pulls with a force of 500 N toward [055°].

a) Sketch a vector diagram and estimate the magnitude and direction of the resultant force.

b) Check your estimate by calculating the resultant force.

7. Two ropes suspend a crate with a weight of 810 N. The force exerted by one rope is 610 N, directed to the left at an angle of 55° above the horizontal. What force is exerted by the other rope to keep the crate from moving?

8. A spaceship is equipped with thrusters on all 4 sides. If the south-facing thruster produces 200 N of thrust and the west-facing thruster produces 175 N of thrust, find the resultant thrust on the spaceship.

9. In unloading a heavy piece of machinery from a railway flatcar, a crane exerts a vertical force of 7500 N and a tow truck exerts a force of 3200 N at 12° above the horizontal. What is the resultant force on the machinery, just before it starts to move?

Extending Your Thinking

10. A Canadian Forces minesweeper leaves Sea Harbour on a practice patrol. It has orders to first proceed 40 km on a bearing of 135° to Bullion Island, then 60 km on a bearing of 225° to Port Karat, and finally 25 km on a bearing of 270° to Doubloon. What are the distance and bearing of Doubloon from Sea Harbour?

11. Ryley and his little sister, Shelby, are pulling a loaded raft along a stream. Ryley pulls with 20 N toward [055°]. Shelby pulls with 18 N toward [145°].

a) Draw a scale diagram.

b) Find the resultant force exerted on the raft.

c) In what direction should Ryley pull so that the raft travels straight down the stream?

PROJECT

In this Tutorial, you used scale diagrams to add vectors. You will need this skill to complete the Chapter Project, and the following projects in the Project Book.

- Flight Simulator Development
- Tracking Pollution
- Training a Robot

COMMUNICATING THE IDEAS

Describe the similarities and differences between the triangle method of vector addition and the parallelogram method.

TUTORIAL 7.3

Multiplying a Vector by a Scalar

Multiplication can be thought of as a shorthand way to write a repeated addition:

$$2 + 2 + 2 = 3 \times 2$$

In this Tutorial, you will explore a similar concept with vectors. Recall that the magnitude of a vector is always a positive scalar.

Investigation 1 Scalar Multiples of a Vector

Explore the result of multiplying a vector by a scalar.

1. Draw a vector $\overrightarrow{v}$, with $|\overrightarrow{v}| = 3$ cm [090°].
2. Draw the vector $\overrightarrow{z} = \overrightarrow{v} + \overrightarrow{v} + \overrightarrow{v}$; this can be represented by $\overrightarrow{z} = 3\overrightarrow{v}$.
3. What is the direction of $\overrightarrow{z}$? What is the magnitude of $\overrightarrow{z}$?
4. Draw the vector $\overrightarrow{x} = \overrightarrow{v} + \overrightarrow{v} + \overrightarrow{v} + \overrightarrow{v} + \overrightarrow{v}$; this can be represented by $\overrightarrow{x} = 5\overrightarrow{v}$.
5. What is the direction of $\overrightarrow{x}$? What is the magnitude of $\overrightarrow{x}$?
6. Draw the vector $\overrightarrow{u} = -\overrightarrow{v}$.
7. What is the direction of $\overrightarrow{u}$? What is the magnitude of $\overrightarrow{u}$?
8. Draw the vector $\overrightarrow{w} = -\overrightarrow{v} + (-\overrightarrow{v})$; this can be represented by $\overrightarrow{w} = -2\overrightarrow{v}$.
9. What is the direction of $\overrightarrow{w}$? What is the magnitude of $\overrightarrow{w}$?

The multiplication of a vector $\overrightarrow{u}$ by a scalar, k, creates a new vector with magnitude $|k\overrightarrow{u}|$.

If $k > 0$, the direction of $k\overrightarrow{u}$ is the same as that of $\overrightarrow{u}$. If $k < 0$, the direction of $k\overrightarrow{u}$ is opposite to that of $\overrightarrow{u}$.

If $k = 0$, then $k\overrightarrow{u}$ has no magnitude; it is the *zero vector*.

The enlargement and reduction of figures in computer graphics are common applications of scalar multiples of vectors.

Investigation 2 Enlarging Figures

1. In the middle of a sheet of grid paper, draw a square ABCD with sides 3 cm long.

2. Select a point P at a convenient location inside ABCD. Point P will act as the centre of expansion.

3. Draw the scalar multiples: $\overrightarrow{PE} = 2\overrightarrow{PA}$, $\overrightarrow{PF} = 2\overrightarrow{PB}$, $\overrightarrow{PG} = 2\overrightarrow{PC}$, and $\overrightarrow{PH} = 2\overrightarrow{PD}$. Connect points E, F, G and H.

4. What type of figure is EFGH? How long are the sides? What is the position of EFGH relative to ABCD.

5. Repeat exercises 1 to 4. Select a point P outside the square as the centre of expansion.

6. Repeat exercises 1 to 4. Select a point P on a side of the square as the centre of expansion.

7. Draw a triangle on grid paper. Now, use scalar multiplication of vectors, to create a version of your triangle that is three times as large, and positioned above the original triangle.

8. Describe how you could create a triangle that is one-third the size of the original triangle described in exercise 7.

Example Scale drawings of a scalar multiple of a vector

The vector $\vec{a}$ represents a velocity of 40 km/h [east]. Make a scale drawing of each vector. What velocity does each vector represent?

a) $3\vec{a}$ **b)** $7\vec{a}$ **c)** $-3\vec{a}$ **d)** $1.6\vec{a} + 4\vec{a}$

SOLUTION

Let 1 cm = 40 km/h.

a) $3\vec{a}$ represents a velocity of 3(40) = 120 km/h [east].

Scale: 1 cm = 40 km/h

b) $7\vec{a}$ represents a velocity of 7(40) = 280 km/h [east].

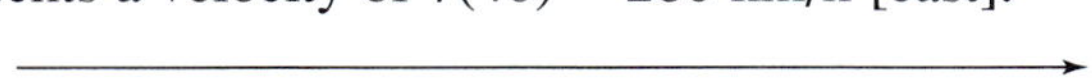

Scale: 1 cm = 40 km/h

c) $-3\overrightarrow{a}$ represents a velocity of $3(40) = 120$ km/h [west].

Scale: 1 cm = 40 km/h

d) $1.6\overrightarrow{a} + 4\overrightarrow{a} = (1.6 + 4)\overrightarrow{a}$
$= 5.6\overrightarrow{a}$

$5.6\overrightarrow{a}$ represents a velocity of $5.6(40) = 224$ km/h [east].

Scale: 1 cm = 40 km/h

DISCUSSING THE IDEAS

1. A vector $\overrightarrow{v}$ is multiplied by a scalar k. Determine k in each situation.
   a) $k\overrightarrow{v}$ is longer than $\overrightarrow{v}$, but is in the same direction.
   b) $k\overrightarrow{v}$ is shorter than $\overrightarrow{v}$, but is in the same direction.
   c) $k\overrightarrow{v}$ is in the opposite direction to $\overrightarrow{v}$.

2. What would happen if you tried to enlarge a circle using the techniques of *Investigation 2*?

EXERCISES

Checking Your Skills

1. A car is travelling at 85 km/h toward the northeast. Draw a scale diagram of its velocity. The car increases its speed by a factor of 1.25. Draw the new velocity vector.

2. The tension in an elevator cable increases as people enter the elevator, and decreases as people leave the elevator. Assume that the average person weighs 750 N and that the cable tension of an empty elevator is 20 000 N. Draw scale diagrams to represent the tension in the cable after each event.
   a) On the first floor, 6 people enter the empty elevator.
   b) On the third floor, another 4 people enter the elevator.
   c) On the fourth floor, 8 people leave, and 1 person enters.
   d) On the fifth floor, everybody leaves, then 2 people enter.

3. Hurricane Albert is situated 90 km southeast off Florida's east coast. It is moving at a rate of 30 km/h toward the northwest. Construct a scale vector to represent the movement of the hurricane. If the hurricane picks up speed to 40 km/h, draw the new scale velocity vector.

4. Suppose $|\overrightarrow{v}| = 1$ cm [north], $|\overrightarrow{w}| = 1$ cm [east], and $\overrightarrow{u} = k\overrightarrow{v} + m\overrightarrow{w}$. What values of k and m would produce the following results?

   a) $\overrightarrow{u}$ pointing toward the northeast
   b) $\overrightarrow{u}$ pointing toward the northwest
   c) $\overrightarrow{u}$ pointing toward the southeast
   d) $\overrightarrow{u}$ pointing toward the southwest

Extending Your Thinking

5. Given $\overrightarrow{u} = 1$ cm [north], $\overrightarrow{v} = 1$ cm [east], scalars k and m, and $\overrightarrow{w} = k\overrightarrow{u} + m\overrightarrow{v}$. For each of the following, construct a scale diagram and measure to find the values of k and m.

   a) $\overrightarrow{w} = 3$ cm [northeast]
   b) $\overrightarrow{w} = 5$ cm [southwest]
   c) $\overrightarrow{w} = 6$ cm [075°]
   d) $\overrightarrow{w} = 7.5$ cm [230°]
   e) $\overrightarrow{w} = 4.9$ cm [133°]

PROJECT

In this Tutorial, you multiplied a vector by a scalar. You will need this skill when you return to the Chapter Project, and for the following project in the Project Book.

- Training a Robot

COMMUNICATING THE IDEAS

Compare and contrast multiplication of real numbers with multiplication of a vector by a scalar.

TUTORIAL

7.4 Solving Vector Problems by Computation

In the preceding Tutorials you solved vector problems by constructing a scale diagram, then measured to find any missing information.

Scale diagrams provide approximate solutions to vector problems. Each geometric construction or measurement can result in a measurement error.

Trigonometry can be used to solve vector problems with greater accuracy.

Practise Your Prior Skills

In previous grades you used trigonometry and the Pythagorean Theorem to determine the measure of missing sides and angles in triangles. Try these exercises for review.

1. In right △ABC, below left, calculate tan A, ∠A, and the length of side AC to the nearest 0.1 cm.

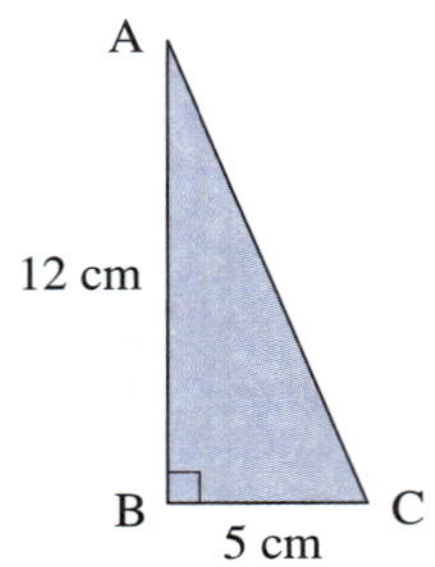

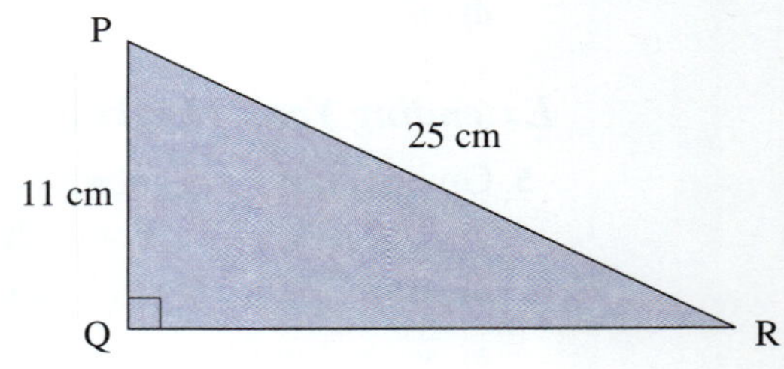

2. In right △PQR, above right, calculate sin R, ∠R, and the length of side QR to the nearest 0.1 cm.

3. In right △XYZ, below left, calculate the lengths of XZ and XY to the nearest 0.1 cm.

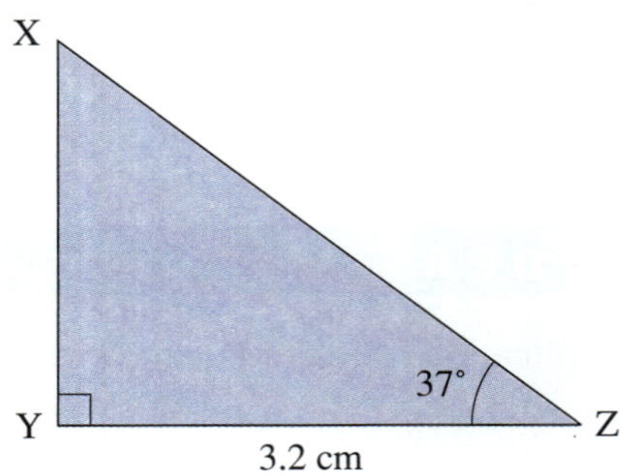

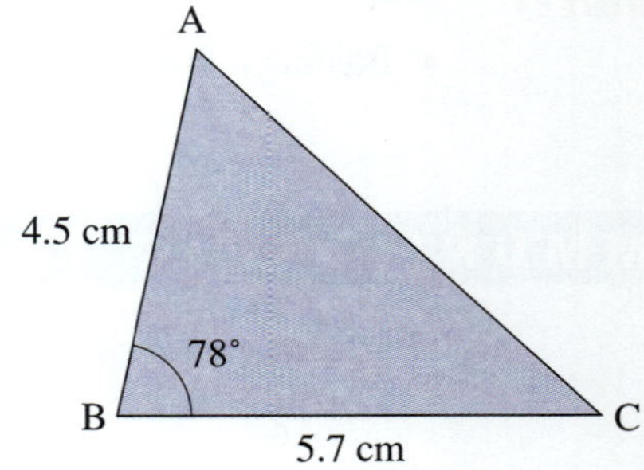

4. In △ABC, above right, calculate AC and ∠A.

5. In $\triangle ABC$, find side b and $\angle C$.

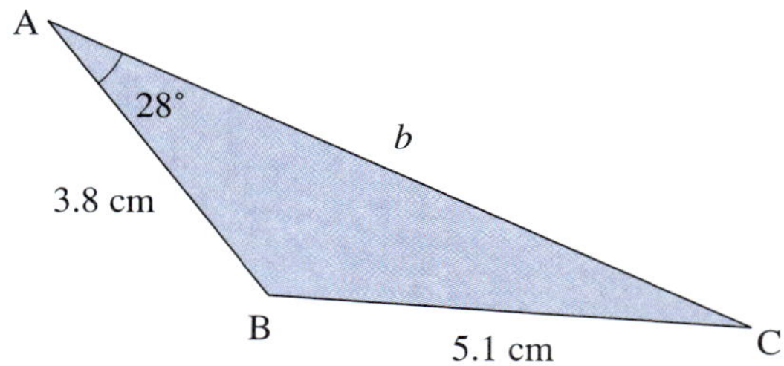

STUDENT REFERENCE

For worked examples that use the Pythagorean Theorem, Cosine Law, and the Sine Law, look in the Student Reference.

These are the steps to solve a vector problem using trigonometry:

Step 1: Draw a sketch of the physical situation and all the vectors involved.

Step 2: Sketch the corresponding vector diagram. Draw the resultant as the side of a triangle or as the diagonal of a parallelogram.

Step 3: Extract a simple triangle from the vector sketch. Indicate which sides and angles are to be calculated.

Step 4: Use trigonometry and the Pythagorean Theorem to solve the triangle.

Step 5: Interpret your solution. Explain what you have found, and include all appropriate units.

Example 1

Use right angle trigonometry and the Pythagorean Theorem.

A boat with a forward velocity of 14 m/s is travelling across a river, directly toward the opposite shore. At the same time, a current of 5 m/s carries the boat down the river.

a) How fast is the boat travelling relative to the shore?

b) Find the direction of the boat relative to the shore.

SOLUTION

Draw a situation sketch.

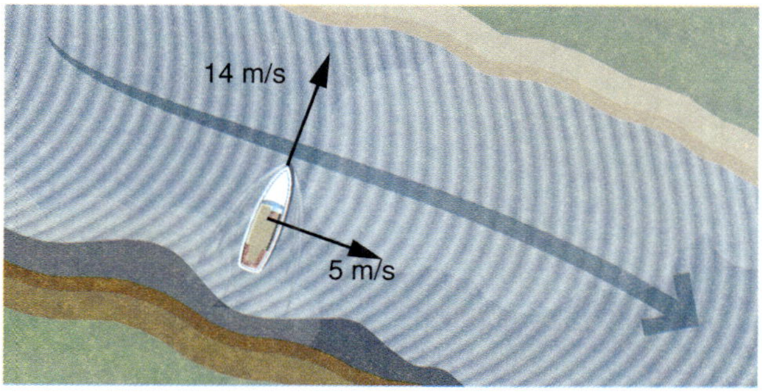

Draw the corresponding vector diagram and triangle diagram.

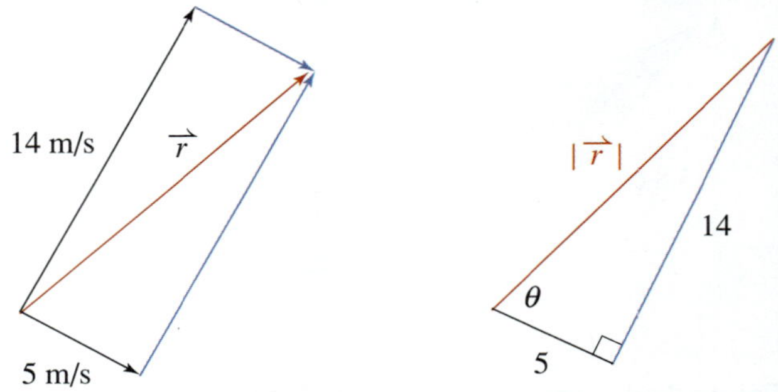

We must calculate the resultant obtained by adding the velocity of the boat as it heads across the river and the velocity of the boat as the current carries it down the river. The boat's heading is at right angles to the current, so the velocity parallelogram is a rectangle, and the triangles are right.

Use the Pythagorean Theorem to calculate the magnitude of the resultant, $\vec{r}$.

$|\vec{r}|^2 = 14^2 + 5^2$

$|\vec{r}| \doteq 14.9$

Use the tangent ratio to calculate the direction of the resultant $\vec{r}$.

$\tan\theta = \frac{14}{5}$

$\theta = \tan^{-1}\left(\frac{14}{5}\right)$

$\theta \doteq 70.3^\circ$

The boat is travelling at 14.9 m/s at an angle of 70.3°, relative to the shore.

When you are on a boat, or in a plane, and you steer the craft toward a particular direction, that direction is called your *heading*. A wind, or a current, can add another velocity component to your craft, so that your actual bearing relative to the ground is not the same as your heading.

Example 2 **Use the Sine Law and the Cosine Law.**

An aircraft is cruising at 150 km/h on a heading of 330°. The wind is blowing toward [085°] with a speed of 40 km/h. Determine the resultant velocity of the aircraft.

SOLUTION

Draw a situation sketch.

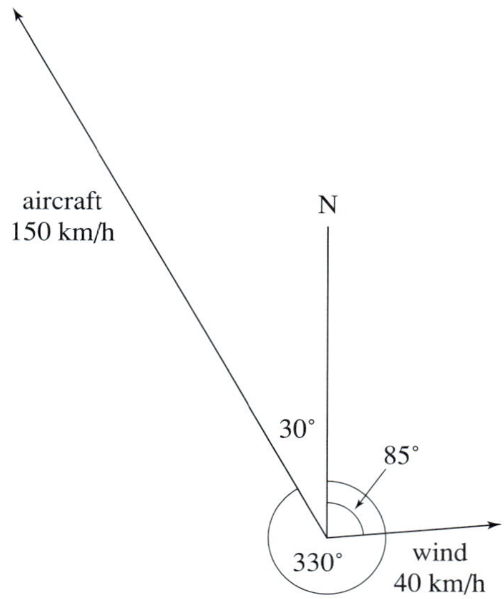

Draw the corresponding vector diagram and triangle diagram.

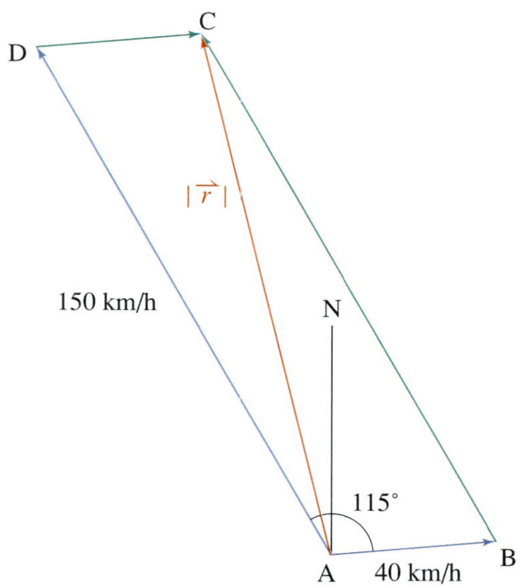

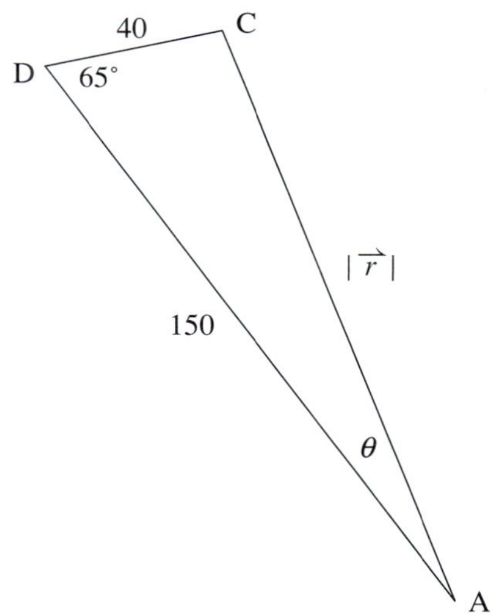

$\angle$BAD = 30° + 85° (from the given bearings)
= 115°

ABCD is a parallelogram, so:

$\angle$ADC = 180° − 115°
= 65°

Use the Cosine Law to calculate $|\overrightarrow{r}|$.

$$|\overrightarrow{r}|^2 = 40^2 + 150^2 - 2(40)(150)\cos(65^\circ)$$
$$|\overrightarrow{r}| \doteq 137.9$$

Use the Sine Law to determine θ.

$$\frac{\sin\theta}{40} = \frac{\sin 65^\circ}{137.9}$$
$$\theta \doteq 15.2^\circ$$

The bearing of $\overrightarrow{AD}$ is 330°, so the bearing of $\overrightarrow{r}$ is 330° + θ, or 345.2°.

The aircraft has a velocity of 137.9 km/h, on a bearing of 345.2°.

Discussing the Ideas

1. Why does solving a problem with trigonometry give a more accurate result than using a scale diagram?
2. In what situations would a scale diagram give a sufficiently accurate answer? When would it not?
3. Explain why sketches are helpful in solving vector problems.

Exercises

Checking Your Skills

1. You decide to swim across a river that flows at 2.2 m/s toward the south. If you swim at 1.5 m/s toward the west, what is your resultant velocity?
2. A pilot has set an aircraft's speed at 250 km/h toward the north. The wind velocity is 45 km/h toward the west. Determine the resultant velocity of the aircraft.
3. A golfer hits a golf ball with an initial velocity of 20 m/s, due north, toward the green. A crosswind blows at 7 m/s toward the east. Find the resultant velocity of the golf ball, immediately after it has been hit.
4. A small aircraft on a heading of [225°] is cruising at 150 km/h. It encounters a wind that blows toward [345°] at 35 km/h. Find the aircraft's resultant velocity.
5. Tiffany and Clinton push a crate across a smooth floor. If Tiffany pushes with a force of 50 N toward the west and Clinton pushes with a force of 55 N toward the south, find the resultant force exerted on the crate.

6. A rocket moves at 510 m/s horizontally, and 755 m/s vertically. What is the resultant velocity?

7. An aircraft is flying at 300 km/h toward [285°]. A constant wind of 90 km/h blows from [195°].

   a) Find the resultant velocity of the aircraft.

   b) Find the aircraft's change in location after a flight of 1 h and 15 min.

8. A kayak is paddled at 8 km/h toward [330°] while an ocean current carries the kayak at 3.2 km/h toward [170°]. What is the resultant velocity of the kayak?

9. Two astronauts use their jet packs to manoeuvre a satellite into position over Earth. The first astronaut applies a force of 75 N toward the west and the second astronaut applies a force of 80 N toward north. Determine the net force that acts on the satellite.

Extending Your Thinking

Navigation problems are more challenging when you are given the direction and ground speed that you want to achieve, and are asked to determine the speed and heading necessary to counter the effect of a wind or current.

10. A swimmer wishes to cross a river that flows at 0.75 m/s toward the west. She plans to swim at 0.5 m/s north, straight across the river. Determine the heading and speed needed to achieve this result.

11. A pilot wants to travel from Dauphin to Brandon (200 km due south) in 0.5 h. The wind is blowing from the west at 45 km/h. Determine the heading and speed needed to arrive in Brandon in the allotted time.

PROJECT

In this Tutorial, you learned to apply your prior knowledge to solve vector problems. You will use these skills in the Chapter Project, and in the following projects in the Project Book.

- Flight Simulator Development
- Tracking Pollution
- Training a Robot

Communicating the Ideas

Write a detailed set of notes you could use for review that describes how to solve a vector problem using trigonometry. Choose an exercise you have solved in this Tutorial to demonstrate your method.

TUTORIAL 7.5

Vector Problems in Three Dimensions

In a two-dimensional (2-D) vector problem, all the vectors can be drawn on a flat sheet of paper. All the problems you have worked on so far have been 2-D. If two vectors have a common start point, you can always position a flat sheet of glass so that both vectors lie along it. Furthermore, the resultant and equilibrant of two vectors always lie in the plane of the two vectors.

An air traffic controller must track the direction and distance of each aircraft that is approaching the airport. However, knowing that a plane is 2.5 km northwest of the airport is not enough. An air traffic controller must also know the plane's altitude. Air traffic controllers work with vectors in three dimensions (3-D).

In this Tutorial, you will examine some simple vector problems in 3-D.

Investigation 1 Displacements in Three Dimensions

You will need an open box such as a shoe box and a large piece of cardboard to complete this Investigation.

1. Measure and record all the dimensions of your box.

2. Place the box on a table, with a long side aligned north-south. Visualize an ant standing in the bottom, southwest corner of the box.

a) Suppose the ant walks east to the bottom, southeast corner of the box. In the box, draw the vector that represents the ant's path and label it $\overrightarrow{d_1}$.

b) Now, suppose the ant walks north to the bottom, northeast corner of the box. In the box, draw the vector that represents this path and label it $\overrightarrow{d_2}$.

c) The ant now walks up to the top, northeast corner of the box. Draw the vector that represents this and label it $\overrightarrow{d_3}$.

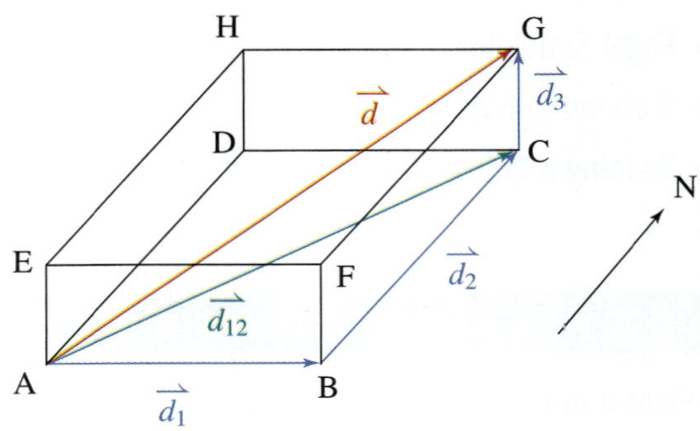

The path of the ant is illustrated in blue in the diagram. If each stage in the ant's journey is regarded as a displacement, then the resultant is $\vec{d} = \vec{d_1} + \vec{d_2} + \vec{d_3}$.

3. Consider the ant's travels along the bottom of the box from point A, to point B, to point C. In the box, draw a vector from A to C and label it $\vec{d_{12}}$ (read "d-one-two"). Determine $\vec{d_{12}}$ so that $\vec{d_{12}} = \vec{d_1} + \vec{d_2}$.

4. Use the cardboard to construct a rectangle ACGE that will fit diagonally inside the box. Label the corners of the rectangle so that when it is in the box, vertex A of the rectangle is in corner A of the box, and so on.

a) What is the measure of ∠ACG?

b) Draw the vectors $\vec{d_{12}}$ and $\vec{d_3}$ on the rectangle ACGE.

c) Determine $\vec{d}$ so that $\vec{d} = \vec{d_{12}} + \vec{d_3}$.

5. Summarize your results.

It is often useful to sketch 3-D vector problems as if they occur inside a transparent box frame. The box frame is drawn so that the start point of the first vector is in one corner of the box, and the endpoint of the final vector is in the diagonally opposite corner.

A box frame can be sketched in three simple steps:

Step 1: Draw a rectangle to represent the front of the box.

Step 2: Draw a second rectangle of the same size to represent the back of the box. Draw this second rectangle behind and to the right of the first rectangle.

Step 3: Connect the corresponding corners of the rectangles.

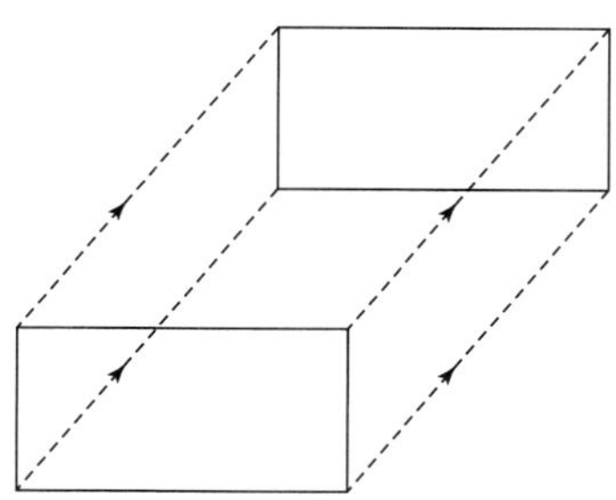

The length and width of the first rectangle and the distance between corresponding corners of the first and second rectangles determine the size of the box.

Example 1 Combine three forces, two of which are parallel.

A model rocket is launched vertically upward. The rocket has a weight of 40 N. The rocket motor generates a force of 85 N. The wind exerts a force of 17 N toward the east. Find the resultant force on the rocket.

SOLUTION

In this case, two of the forces are parallel to each other. The weight and the rocket thrust can be combined directly to find the resultant vertical force.

40 N [down] + 85 N [up] = 45 N [up]

There is a net vertical force of 45 N [up].

Now add the net vertical force and the wind force to find the resultant force $\vec{f}$.

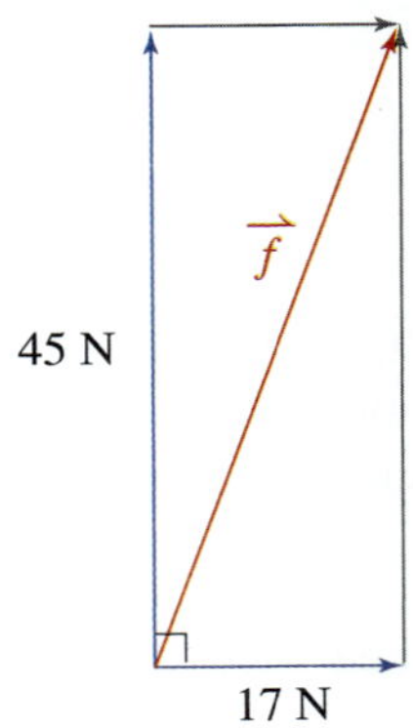

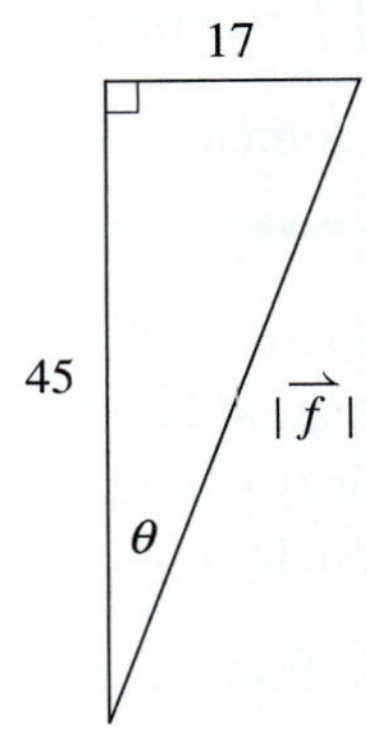

$|\vec{f}|^2 = 45^2 + 17^2$

$|\vec{f}| \doteq 48.1$

$\theta = \tan^{-1}\left(\frac{17}{45}\right)$

$\theta \doteq 20.7°$

The resultant force is 48.1 N, 20.7° east of vertical.

Example 2 Three perpendicular forces

A crate of delicate pottery is being unloaded, from a ship to a waiting truck. A crane is attached and lifts directly upward. Two other ropes are attached. One worker pulls horizontally toward the stern of the ship. Another worker pulls horizontally toward the side of the ship. Just before the crate starts to move, the crane is exerting a force of 1200 N, the worker in the stern is pulling with a force of 200 N, and the other worker is pulling with a force of 150 N. At that moment, what is the total force acting on the crate?

SOLUTION

There are 3 forces that are mutually perpendicular.

Let $\overrightarrow{a} = 1200$ N [up],
$\overrightarrow{b} = 200$ N [toward the stern], and
$\overrightarrow{c} = 150$ N [toward the side].

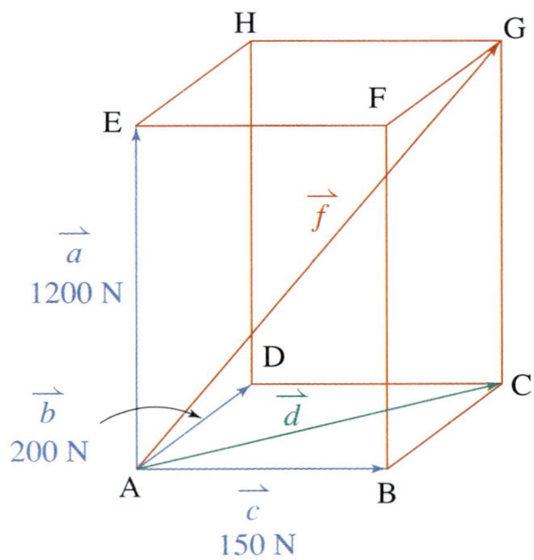

Top View

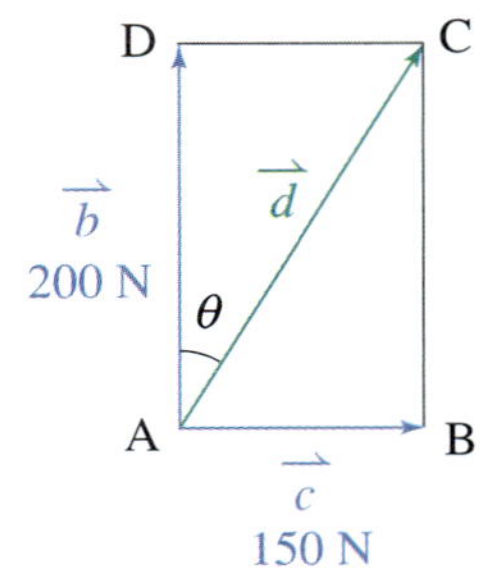

First, let $\overrightarrow{d} = \overrightarrow{b} + \overrightarrow{c}$.

$|\overrightarrow{d}|^2 = 200^2 + 150^2$ $\quad\theta = \tan^{-1}\left(\frac{150}{200}\right)$

$|\overrightarrow{d}| = 250$ $\quad\theta \doteq 36.9°$

Diagonal View

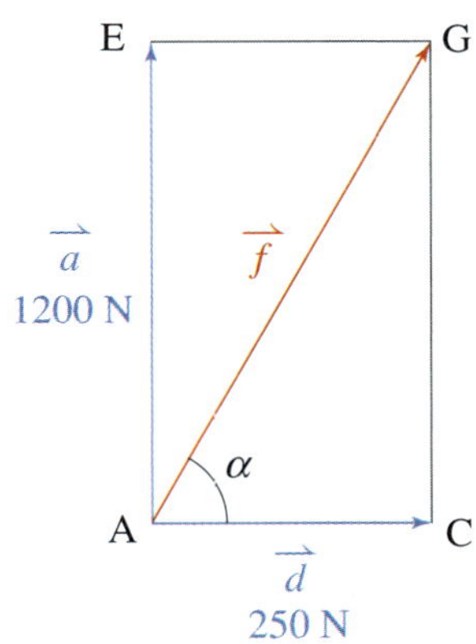

Now, let $\overrightarrow{f} = \overrightarrow{d} + \overrightarrow{a}$.

$|\overrightarrow{f}|^2 = 250^2 + 1200^2$ $\quad\alpha = \tan^{-1}\left(\frac{1200}{500}\right)$

$|\overrightarrow{f}| \doteq 1226$ $\quad\alpha \doteq 67.4°$

Just before the crate starts to move the total force is 1226 N acting about 67.4° above the horizontal, and about 37° from the stern, toward the side of the ship.

Example 3 Combine 3 three-dimensional vectors.

An aircraft flies with an airspeed of 80 m/s toward the north and climbs at an angle of 20° with the horizontal. A horizontal wind blows from west to east at 30 m/s. What is the magnitude of the velocity of the aircraft? What is the angle the aircraft's track makes with the horizontal?

SOLUTION

There are 3 forces:

Let $\overrightarrow{v_1} = 80$ m/s [north],
$\overrightarrow{v_2} = 30$ m/s [east], and
$\overrightarrow{v_3} =$ vertical speed [up]. (We need to calculate this.)

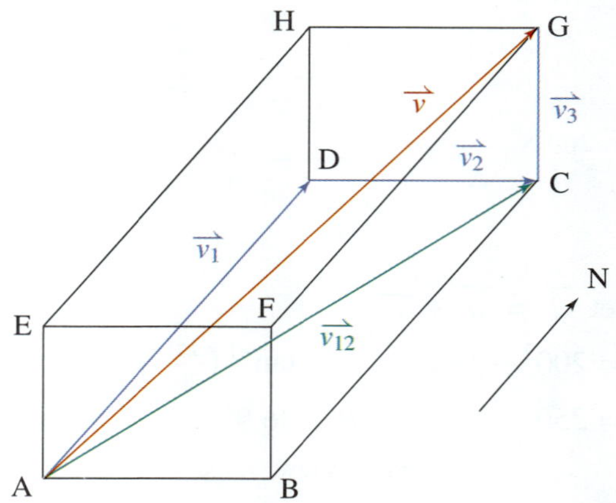

First calculate the vertical velocity of the aircraft, $\overrightarrow{v_3}$.

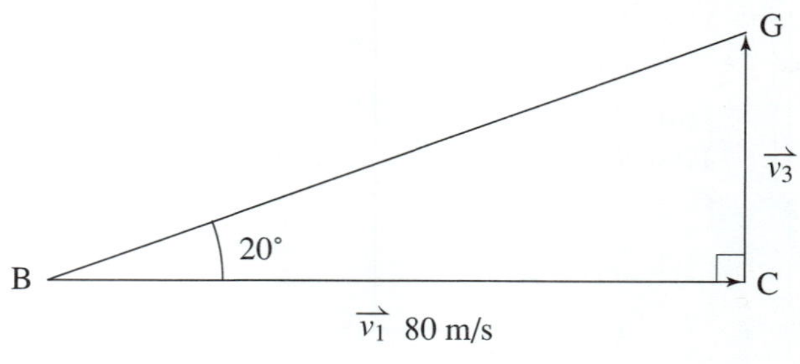

$$\overrightarrow{v_3} = 80 \tan(20°)$$
$$\overrightarrow{v_3} \doteq 29.1 \text{ m/s}$$

Determine the horizontal speed of the aircraft.

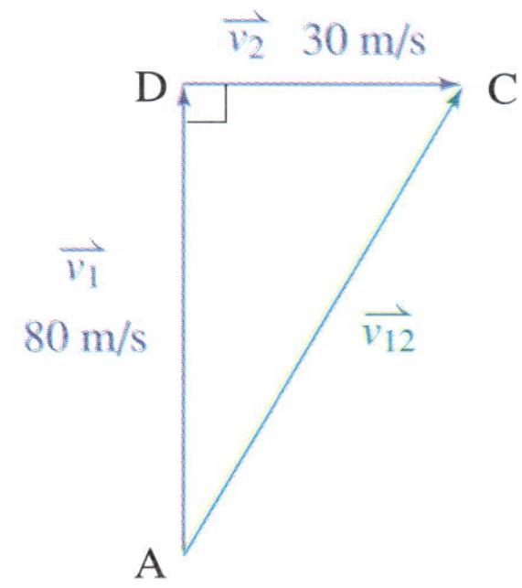

$$|\overrightarrow{v_{12}}|^2 = 80^2 + 30^2$$
$$|\overrightarrow{v_{12}}| \doteq 85.4 \text{ m/s}$$

Draw rectangle ACGE to represent the diagonal plane that includes the horizontal velocity, $\overrightarrow{v_{12}}$, the vertical velocity, $\overrightarrow{v_3}$, and the final velocity, $\overrightarrow{v}$. Find $\overrightarrow{v} = \overrightarrow{v_{12}} + \overrightarrow{v_3}$, then find $\angle\alpha$.

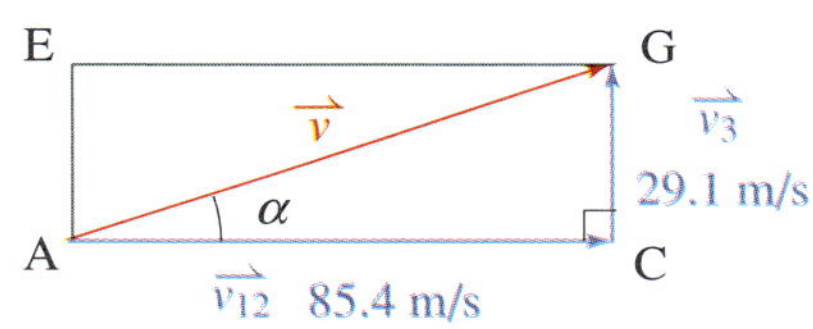

$$|\overrightarrow{v}|^2 = 85.4^2 + 29.1^2 \qquad \alpha = \tan^{-1}\left(\frac{29.1}{85.4}\right)$$
$$|\overrightarrow{v}| \doteq 90.2 \qquad \alpha \doteq 18.8^\circ$$

The aircraft has a speed of approximately 90.2 m/s relative to the ground, and its track makes an angle of approximately 18.8° with the horizontal.

DISCUSSING THE IDEAS

1. Why is the resultant of two vectors always in the same plane as the vectors?

2. Does order matter when you add vectors in 3 dimensions?

3. Can 3-D problems always be regarded as the combination of two 2-D problems?

EXERCISES

Checking Your Skills

1. A helicopter is experiencing the following forces during a flight.

gravity = 30 000 N [vertically down]
lift force created by rotors = 33 000 N [vertically up]
force of wind = 5000 N [west]

What is the net force on the helicopter?

2. Buster and Koko are two dogs playing with a stick. They pull sideways in opposite directions, Buster with 100 N and Koko with 80 N. To try to get the stick away from them, Harry lifts upward with 175 N. Find the net force that acts on the stick at that moment.

3. Two construction workers are preparing to move a beam into position. They have a crane to help them. The crane lifts up with a force of 2000 N. Each worker pulls with 750 N of force. One worker pulls east, the other south. Find the net force that acts on the beam.

4. A skydiver has a large fan attached to his back so he can steer towards a small target on the ground. After deploying his parachute, he reaches a terminal velocity of 5 m/s downward. The wind direction is toward the south at 8 m/s. He uses his fan to move west at 2 m/s. Find his resultant velocity.

5. An aircraft flies with an airspeed of 95 m/s toward the south and climbs at an angle of 23° with the horizontal. A horizontal wind blows from east to west at 21 m/s.

a) Find the speed of the aircraft relative to the ground.

b) Find the angle the aircraft's track makes with the horizontal.

Extending Your Thinking

6. A manned mission has been sent to Mars. The pilot is able to steer the descent to the surface with four thrusters, one on each side of the ship. The ship accelerates toward Mars at 3.8 m/s^2. The pilot fires the north thruster to achieve an acceleration of 1.5 m/s^2 toward the south. At the same time, the east thruster is fired to achieve an acceleration of 2.1 m/s^2 toward the west. Find the resultant acceleration of the spaceship.

7. A special fastener is used to anchor three cables to an east-facing wall in a factory. The fastener has a maximum strength of 1000 N. One cable applies a load of 300 N straight down. The second cable applies a load of 400 N, horizontally toward the south. The third cable applies a load of 500 N toward the north, but angled at 30° from the horizontal, down toward the floor. Will the fastener be strong enough to hold the total load?

8. A fighter pilot is flying at 300 km/h toward [025°]. Due to mechanical failure, his ejector seat rockets him upward at 45 km/h. If the wind blows at 60 km/h toward [345°], find the pilot's resultant velocity.

PROJECT

In this Tutorial, you developed an understanding of vectors in three dimensions. You will need this knowledge when you return to the Chapter Project, and for the following project in the Project Book.

- Training a Robot
- Flight Simulator Development
- Tracking Pollution

COMMUNICATING THE IDEAS

A classmate was sick today and asks you for help. Write a brief outline that explains how to solve a vector problem in 3-D.

CHAPTER PROJECT

How Fast Was That Car Going?

Use your work from the second part of the Chapter Project to complete your accident simulation.

1. The first step in your investigation is to take measurements at the accident scene. Draw $\overrightarrow{AE}$ equal to $\overrightarrow{DC}$. Measure $|\overrightarrow{AB}|$ and $|\overrightarrow{AE}|$. Measure $\angle BAE$.

2. Let
$$\begin{aligned} \vec{v} &= \text{velocity of Object 2 before the collision}, \\ \vec{v_1} &= \text{velocity of Object 1 after the collision} \\ &= \sqrt{|\overrightarrow{AB}|}, \text{ and} \\ \vec{v_2} &= \text{velocity of Object 2 after the collision} \\ &= \sqrt{|\overrightarrow{AE}|}. \end{aligned}$$
What was the velocity of Object 1 before the collision? Use the Law of Conservation of Momentum to write a vector equation that expresses $\vec{v}$ in terms of $\vec{v_1}$ and $\vec{v_2}$.

3. Use the results from exercises 1 and 2, to construct a vector diagram and find the resultant of $2\vec{v_1} + \vec{v_2}$.

4. Write a brief "accident report" for the objects. Describe what happened just before, and just after, the collision.

How fast was a car going just before an accident? According to the Law of Conservation of Momentum, the total momentum of the cars before a collision must equal the total momentum of the cars after a collision. When the masses and all but one velocity are known, the Law of Conservation of Momentum can be used to find an unknown velocity.

5. Suppose Car A, with a mass of 1500 kg, travels at 75 km/h toward the east, and Car B, with a mass of 1850 kg, moves at 90 km/h toward the north. Car A faces a stop sign at the intersection, but does not stop and a crash occurs.

   Draw vectors to represent each car's momentum before the crash. Perform vector addition to determine the total momentum of the system.

Career Skills

People who analyze car collisions collect data, draw accurate diagrams, and prepare concise reports for police and insurance purposes. Communication skills, computer skills, and an understanding of vectors are essential skills.

6. Suppose that after the crash, Car A had a velocity of 65.3 km/h toward [023°] and Car B had a velocity of 57.5 km/h toward [044°]. Determine the total momentum after the crash. Is the total momentum conserved?

7. The direction of the cars after the crash can be determined by the location of debris. The after-crash velocities can be determined by a formula for constant deceleration. If you knew the velocity of Car B before the crash, but not the velocity of Car A, how could you find Car A's pre-crash velocity using the Law of Conservation of Momentum? Why would anyone want to know this?

8. Use the method you devised above to solve this problem.

A truck with a mass of 2000 kg, but with an unknown velocity, is heading east [090°] toward an intersection. A car with a mass of 2150 kg and a velocity of 50 km/h is heading north toward the same intersection. The truck is supposed to yield to the car, but a collision occurs.

Based on information collected at the scene, it is determined that after the crash, the truck had a velocity of 10 m/s toward [054°]. The car had a velocity of 8 m/s toward [086°]. Determine the truck's velocity before the crash.

9. Write a short report on accident reconstruction. Include any spreadsheets or *The Geometer's Sketchpad* files you created.

What Do I Need To Know?

Tutorial	Skills and Concepts	Important Results and Formulas
7.1	• Understand the basic properties of geometric vectors • Construct a scale diagram of a vector	• Vectors are equal if they have the same magnitude and direction • Vectors are opposite if they have the same magnitude but point in opposite directions
7.2	• Construct scale diagrams of vector addition in 2-D • Find the resultant of two vectors • Find the equilibrant	• Triangle law of addition • Parallelogram law of addition • Equilibrant = –Resultant
7.3	• Find the product of a scalar and a vector • Enlarge or reduce a figure using scalar multiples	• The magnitude of a scalar times a vector is a positive quantity equal to $\lvert k\vec{v}\rvert$ • If $k > 0$, $k\vec{v}$ is in the same direction as $\vec{v}$ • If $k < 0$, $k\vec{v}$ is in the direction opposite to $\vec{v}$ • If $k = 0$, $k\vec{v}$ is the zero vector
7.4	• Use trigonometry to solve vector problems	• Pythagorean Theorem • Cosine Law • Sine Law
7.5	• Solve simple vector problems in 3-D • Represent problems within a box frame	

What Should I Be Able To Do?

When you have completed the work of this Chapter, you should be able to solve the problems that follow. Part A is investigative in nature.

Part A

For this activity you will need a large piece of paper, two pieces of heavy string or cord, and a science lab mass. Work in a group of three.

1. Tie two strings to the mass at the same place. Lay the paper on a smooth table and place the mass in the centre of the paper. Two people will pull the mass across the table in perpendicular directions. The third person will record the positions of the mass on the paper.

a) Put the mass in the centre of the paper and pull gently on the strings with enough force only to make them taut, do not move the mass yet. Have the recorder trace the initial positions of the mass and the strings.

b) Pull the mass in perpendicular directions. Both people should try to pull with the same amount of force. The recorder will draw the final position of the mass. In what direction did the mass move?

c) Draw vectors to represent the two forces acting on the mass. Use the distance travelled as a measure of the force exerted on the mass. Determine the resultant force.

d) Repeat parts a to c using forces that are not perpendicular to one another.

e) Repeat parts a to c using unequal forces.

f) Summarize your findings.

Part B

2. You are given the following vectors.

$\vec{a} = 5$ N [south] $\quad$ $\vec{b} = 4$ N [north]

$\vec{c} = 2$ N [west] $\quad$ $\vec{d} = 4$ N [east]

Draw scale diagrams to find:

a) $\vec{a} + 2\vec{a}$ $\quad$ **b)** $\vec{c} - \vec{d}$

c) $\vec{a} - 3\vec{d}$ $\quad$ **d)** $1.2\vec{b} + 3\vec{c}$

3. An explorer finds a pirate's map that gives instructions to find some buried treasure. The instructions are: "Start at a large rock, take 60 paces straight north, then 38 paces northeast." How far away, and in what direction from the rock is the buried treasure?

4. A motorboat cruises at 5 m/s toward [012°] as it crosses a river that flows west. The river's current is 2.7 m/s. Construct a scale drawing to find the resultant velocity.

5. Sean is kayaking across Lake Dauphin. He is headed toward [275°] at 2.5 m/s. If a current is flowing at 0.6 m/s toward [205°], what is his resultant velocity?

6. The Luke family has rented a houseboat for their vacation. They cross Clear Lake at a speed of 3.8 km/h on a heading of [072°]. The wind blows at 6 km/h toward [350°] and pushes the houseboat at 2.3 km/h in that direction. Find their resultant velocity.

7. At one moment during a volleyball game, the ball is in the air directly over the net and two players from opposing teams hit the ball. Player A exerts a force of 80 N toward [175°] and Player B exerts a force of 72 N toward [356°].

   a) Find the resultant force exerted on the ball at that instant.

   b) The net runs east–west. If no other player touches the ball, on which side of the net will the ball land?

8. An aircraft travels at 400 km/h on a heading of [095°]. The wind blows at 85 km/h from [165°]. What is the resultant velocity of the aircraft?

9. Mary uses a rope to pull a canoe toward the shore of a stream. The current exerts a horizontal force of 12 N on the canoe and pushes it toward [020°]. She pulls toward [165°] with a horizontal force of 18 N, and at the same time pulls vertically with a force of 5 N. Find the resultant force on the canoe.

Part C

The Chapter Project combines many skills and concepts in a single problem. Here is a sample of a student's work from the project. Review this sample to gain insights into your own project.

10. Here is part of one student's response for the third part of the project. Are his calculations correct? Has he explained how the Conservation of Momentum is used?

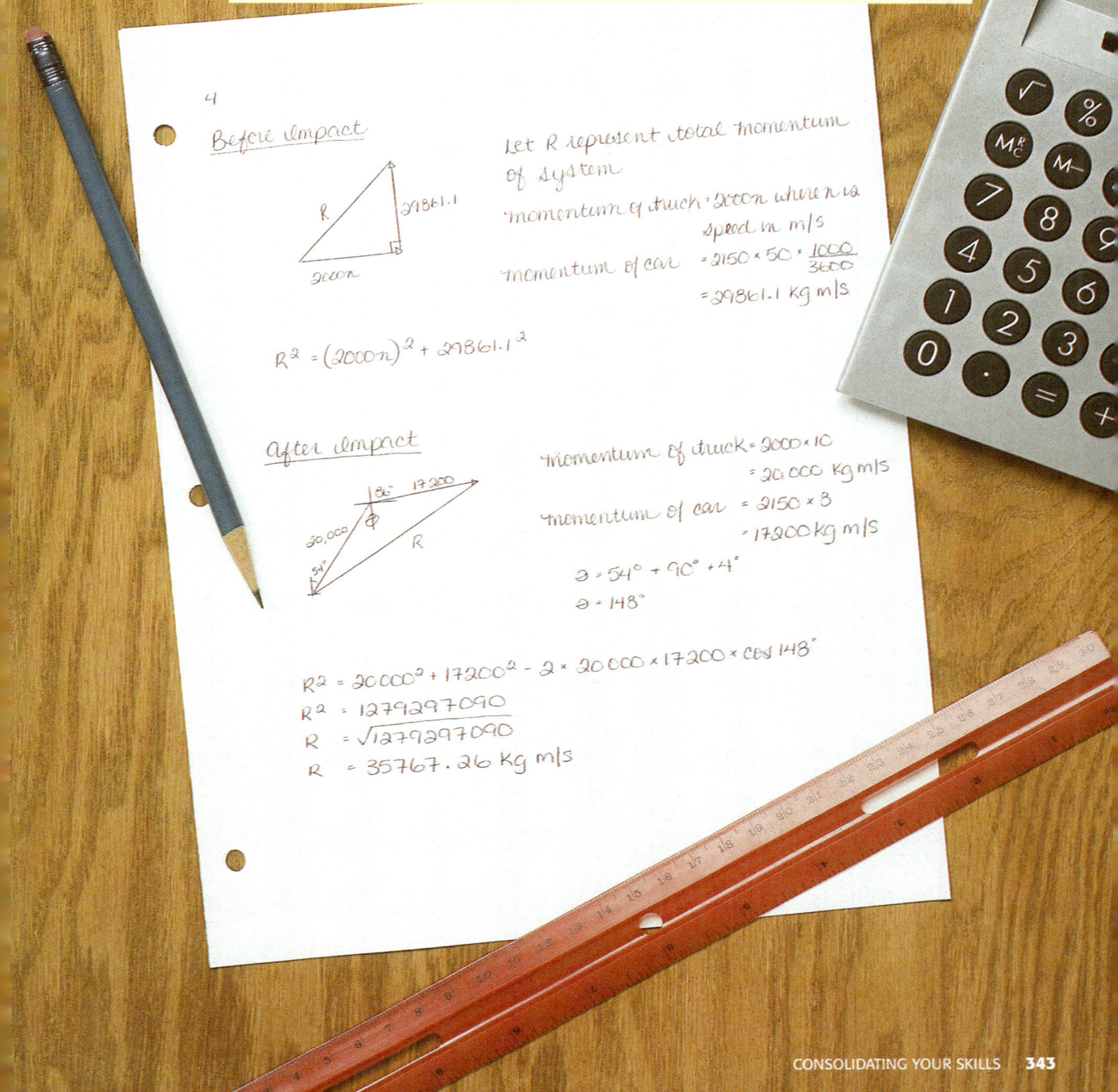

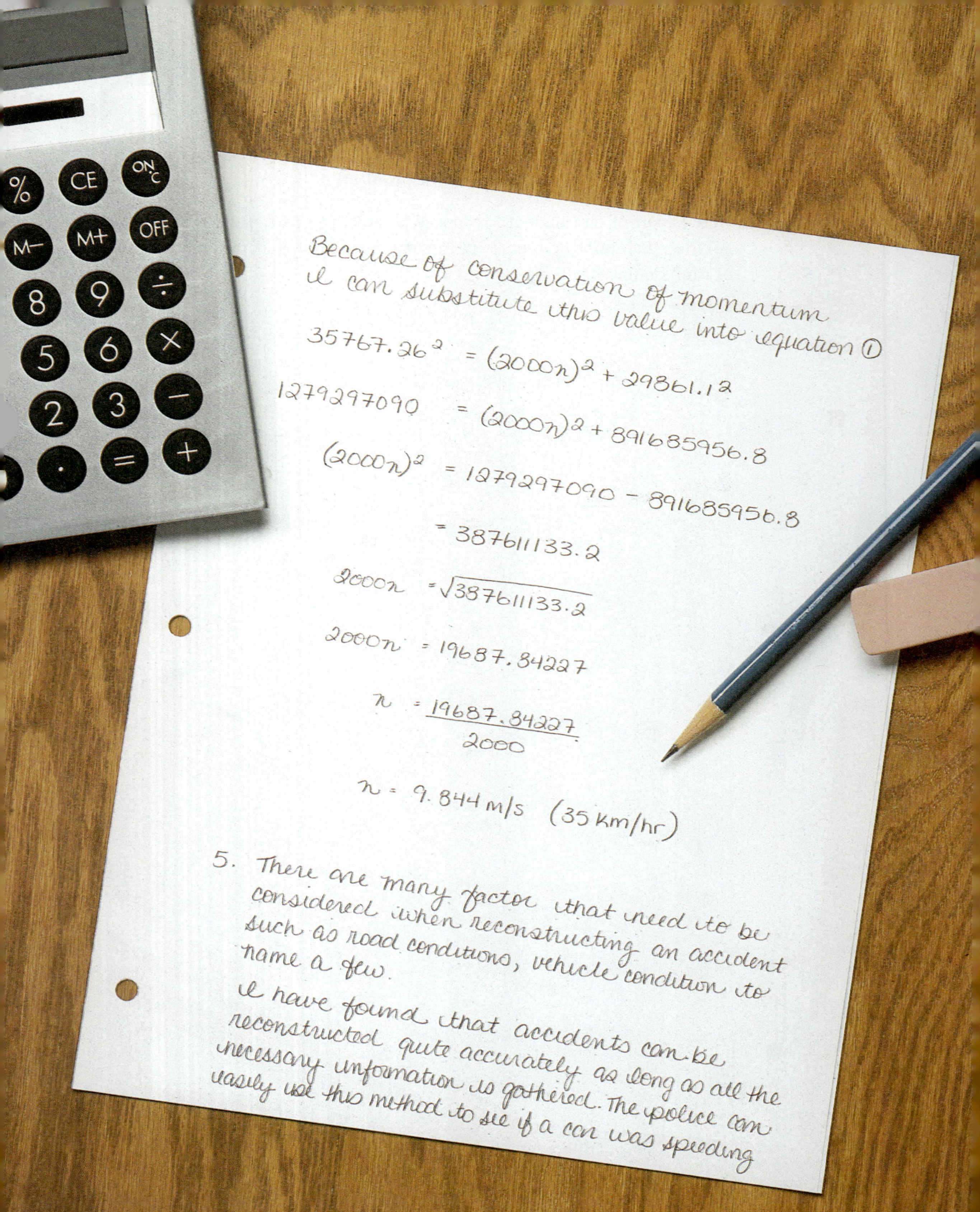
%
CE
ON/C
M–
M+
OFF
8
9
÷
5
6
×
2
3
–
=
+
Because of conservation of momentum
I can substitute this value into equation ①
35767.26² = (2000v)² + 29861.1²
1279297090 = (2000v)² + 891685956.8
(2000v)² = 1279297090 – 891685956.8
= 387611133.2
2000v = √387611133.2
2000v = 19687.84227
v = 19687.84227 / 2000
v = 9.844 m/s (35 km/hr)
5. There are many factor that need to be considered when reconstructing an accident such as road conditions, vehicle condition to name a few.
I have found that accidents can be reconstructed quite accurately as long as all the necessary information is gathered. The police can easily use this method to see if a car was speeding

CONTENTS

Utilities

Utilities 1–24 are on the CD-ROM that accompanies the *Teacher's Resource Book* for *Applied Mathematics 12*.

Using the TI-83 to Simulate a Binomial Experiment

You can simulate a binomial experiment using the randBin command on a TI-83 graphing calculator. The syntax of the command is randBin(number of trials, probability of success on one trial, number of simulations).

Simulate 200 tosses of a coin:

1. To access randBin, press [MATH] [◄] **7**. In this case, the number of trials is 1, the probability of success on any given trial is 0.5, and the number of simulations is 200.

2. Press **1** [,] **0.5** [,] **200** [ )] [ENTER]. A screen similar to the one below appears.

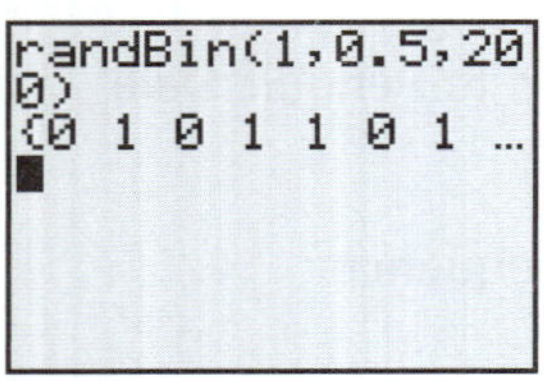

When there is a success (getting heads) the result is 1, otherwise the result is 0. In the screen above, heads appeared on the 2nd, 4th, 5th, and 7th trials. To determine the total number of heads that result, store the numbers in the memory and then add them.

3. Press [STO►] [2nd] **1** [ENTER] to store the numbers in list L1.

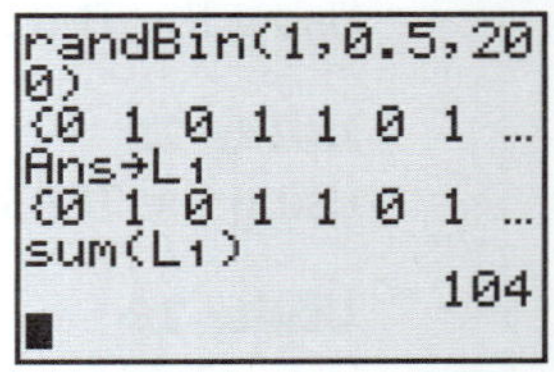

4. To add the numbers, press [2nd] [STAT] [►] [►] **5** [2nd] **1** [ )] [ENTER]. A screen similar to that above appears.

In this simulation, 1 occurred 104 times in 200 trials, so the experimental probability of heads is $\frac{104}{200}$, or 0.52.

Using the TI-83 to Perform Matrix Operations

The TI-83 can be used to perform matrix operations. The matrices A, B, C, and D will be used to illustrate these operations.

$$A = \begin{bmatrix} 1 & -4 \\ 3 & 7 \\ -2 & 5 \end{bmatrix} \qquad B = \begin{bmatrix} 5 & 0 \\ -1 & 2 \\ 4 & -3 \end{bmatrix}$$

$$C = \begin{bmatrix} 6 & 1 & 4 \\ -5 & 0 & 1 \end{bmatrix} \qquad D = \begin{bmatrix} 8 & 1 \\ 3 & -2 \end{bmatrix}$$

1. Define a matrix:
Follow these steps to define matrix A on a TI-83.

- Press [MATRX] [▶] [▶] **1** to select the edit mode for matrix A.
- Enter the dimensions of A, by pressing 3 [ENTER] 2 [ENTER].
- Enter the elements of A, pressing [ENTER] after each entry. Your screen should look like the one below.

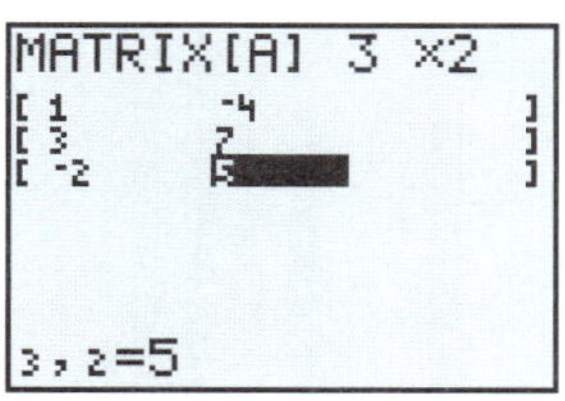

- When you are finished, press [2nd] [MODE] to QUIT.
- Repeat these steps to define matrices B, C, and D.
- Once you define a matrix, its dimensions will be displayed.

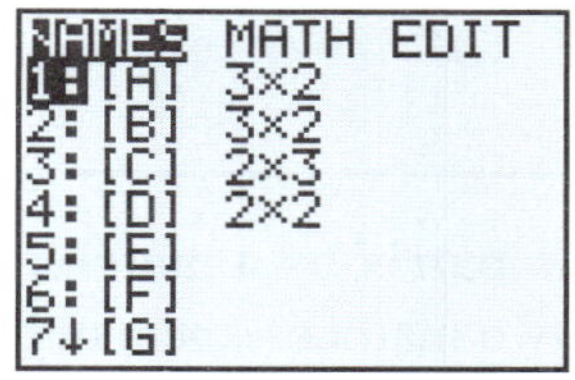

2. Display the contents of a matrix:
To display the contents of a matrix, select NAMES from the MATRIX menu and then enter the number for the appropriate matrix and press [ENTER].

3. Add and subtract matrices:
To add or subtract matrices, they must have the same dimensions.

- To evaluate $A + B$, press [MATRX] **1** [+] [MATRX] **2** [ENTER]. Your screen should look like the one below left.

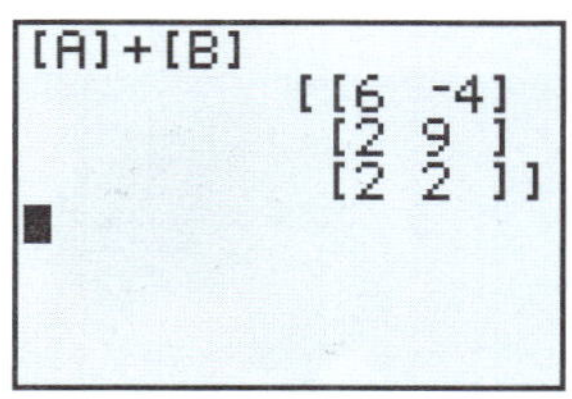

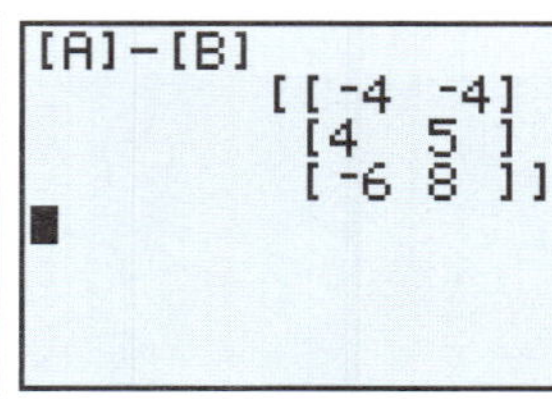

- To evaluate $A - B$, press [MATRX] **1** [−] [MATRX] **2** [ENTER]. Your screen should look like the one above right.

If you try to add or subtract matrices that do not have the same dimensions, the *Dimension Mismatch* error message will be displayed. To illustrate this error, try to calculate $A - D$.

4. Multiply a matrix by a scalar:

- To evaluate $2 \times A$, press **2** [×] [MATRX] **1** [ENTER].

  Your screen should look like this.

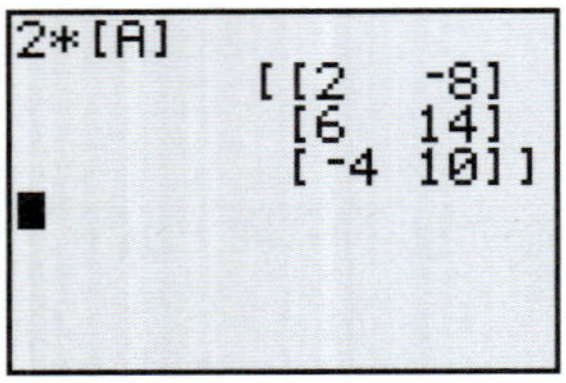

5. Multiply a matrix by a matrix:
To multiply two matrices, the number of columns in the first matrix must equal the number of rows in the second matrix. If A is an $m \times n$ matrix, and C is an $n \times p$ matrix, then $A \times C$ will be an $m \times p$ matrix.

- To evaluate $A \times C$, press [MATRX] **1** [×] [MATRX] **3** [ENTER]. Your screen should look like the one below.

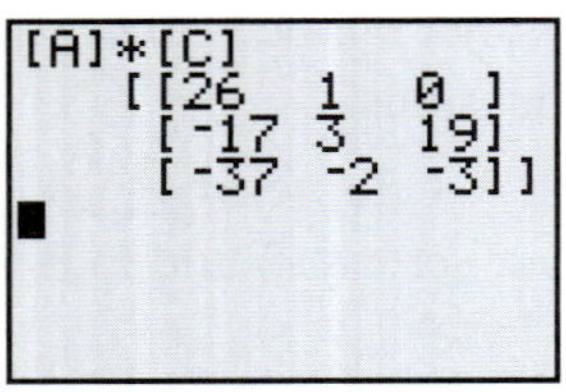

If you try to multiply matrices and the number of columns in the first matrix does not equal the number of rows in the second matrix, the *Dimension Mismatch* error message will be displayed. To illustrate this error, try to calculate $A \times B$.

If a matrix is square, you can multiply it by itself.

- To evaluate D^2, press [MATRX] **4** [^] **2** [ENTER]. Your screen should look like the one below.

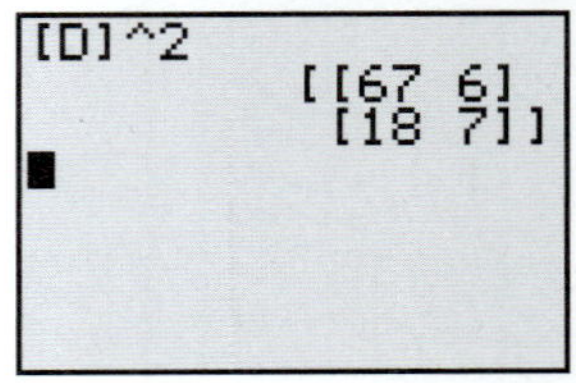

Using the TI-83 to Calculate and Graph a Binomial Distribution

The TI-83 can be used to calculate the probabilities for a binomial distribution. The results can then be displayed in table form using lists or graphed as a histogram. The syntax is **binompdf(**number of trials, probability of success on each trial).

To illustrate this feature, consider an experiment in which a coin is tossed 12 times. On each toss the probability of getting a head is 0.5. In such an experiment, what are the probabilities of tossing 0 heads, 1 head, 2 heads, 3 heads, …, 12 heads?

Calculate the probabilities:

1. Obtain the **binompdf(** command; press [2nd] [VARS] **0**.

2. Complete the binomial command; press **12** [,] **0.5** [)] [ENTER].

The command **binompdf(12, 0.5)** calculates the probabilities for all the possible outcomes in 12 trials, where the probability of success on each trial is 0.5. The possible outcomes are the whole numbers from 0 to 12, inclusive. Use the [▶] key to view all of the probabilities.

3. Store the results in list L2; press [STO▶] [2nd] **2** [ENTER]. Your screen should appear as the one shown below.

4. Clear list L1; press [STAT] **1** [▲] [CLEAR] [ENTER]. Then enter 0, 1, 2, 3, …, 12 in list L1. The contents of lists L1 and L2 represent the binomial distribution in table form as shown in the screen below.

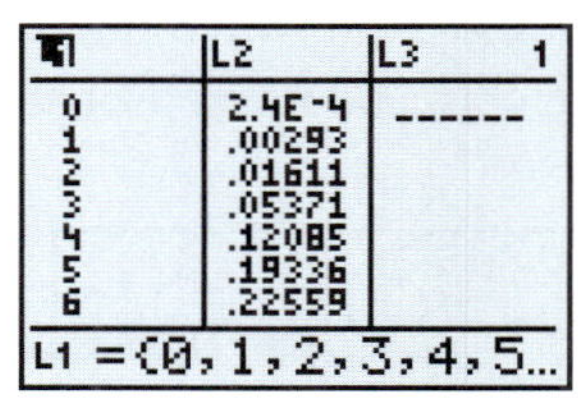

From this table we can see that the probability of getting 3 heads in 12 tosses of a coin is 0.053 71.

Graph the probabilities:

5. These window settings are appropriate for this distribution: Xmin = 0, Xmax = 13, Xscl = 1, Ymin = –0.1, Ymax = 0.3, Yscl = 0.1.

6. Press [2nd] [Y=] **1** to select Plot 1. On the first line, highlight **On**. On the second line, select the histogram icon. On the third line, enter L1 beside **Xlist**. On the fourth line, enter L2 beside **Freq**.

7. Press [GRAPH] to produce the graph of the binomial distribution shown below.

8. Press [TRACE] and move the cursor to the top of each bar to view the probabilities for each possible outcome. Note where the number of heads and the corresponding probability appear.

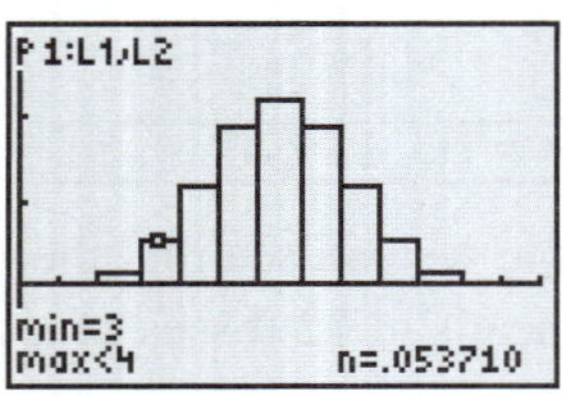

This screen shows that the probability of getting 3 heads in 12 tosses of a coin is 0.053710. Note that this is the same value we obtained from the table.

Using the TI-83 to Calculate Mean and Standard Deviation

The **1-Var Stats** command on the TI-83 can be used to calculate the mean and standard deviation of a data set or a probability distribution.

1. Find the mean and standard deviation of a data set:
Twenty-four households were surveyed and the number of children in each household was recorded. Table A shows the results as raw data.

Table A

2	0	3	2
1	3	1	3
4	1	4	2
2	3	2	3
2	1	0	0
1	2	3	2

To calculate the mean and standard deviation for raw data:

- Enter the data from Table A in list L1.
- Press [STAT] [►] 1 [2nd] 1 [ENTER] to get the results shown.

```
1-Var Stats
 x̄=1.958333333
 Σx=47
 Σx²=123
 Sx=1.160178647
 σx=1.135751098
↓n=24
```

- The TI-83 graphing calculator uses the symbols $\overline{x}$ for the mean and σx for the standard deviation.
- The mean is approximately 1.958 children per household and the standard deviation is approximately 1.136.

2. Find the mean and standard deviation of a frequency table:
Table B shows the results of the survey of twenty-four households as a frequency table.

Table B

Number of Children	Frequency
0	3
1	5
2	8
3	6
4	2

To calculate the mean and standard deviation for a frequency table:

- Enter the data for Number of Children in list L1 and the Frequency data in list L2.
- Press [STAT] [►] 1 [2nd] 1 [,] [2nd] 2 [ENTER]

```
1-Var Stats
 x̄=1.958333333
 Σx=47
 Σx²=123
 Sx=1.160178647
 σx=1.135751098
↓n=24
```

The results are the same as those from part 1.

3. Find the mean and standard deviation of a binomial distribution:

The table below shows the probability distribution for the experiment of tossing a coin 4 times and recording the number of heads.

Number of Heads	Probability
0	0.0625
1	0.25
2	0.375
3	0.25
4	0.0625

To calculate the mean and standard deviation for a binomial distribution:

- Enter the data for Number of Heads in list L1 and the Probability data in list L2.
- Press [STAT] [►] 1 [2nd] 1 [,] [2nd] 2 [ENTER]

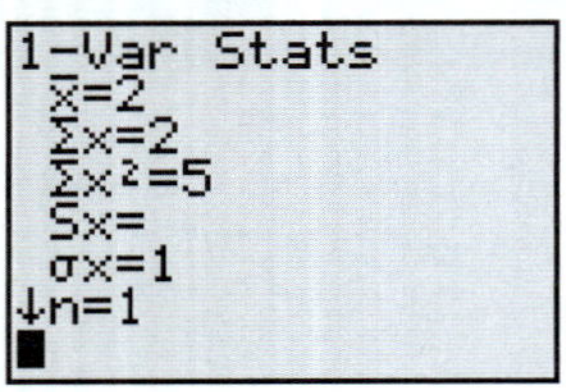

In this case, the mean is 2 heads and the standard deviation is 1.

Using Tables to Find Areas under the Standard Normal Curve

The tables on the following pages provide the area under the standard normal curve to the left of z, for values of z between –3.49 and 3.49.

1. Find the probability that z is less than a specific value:

Find the probability that z is less than –0.85.

To find the area to the left of –0.85, move down the left column of the table to –0.8, then along the line to the entry in the column headed 0.05 to obtain 0.1977.

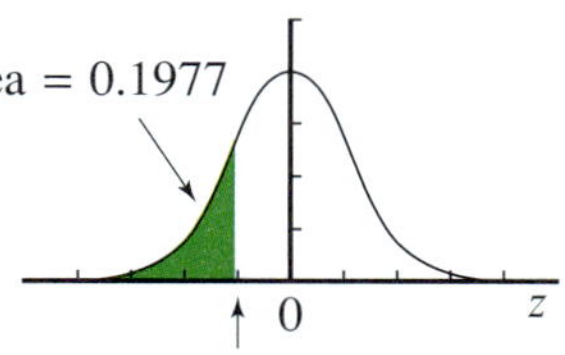

About 0.1977 of the area under a standard normal curve lies to the left of –0.85. Therefore, $P(z < -0.85) = 0.1977$.

2. Find the probability that z is greater than a specific value:

Find the probability that z is greater than –0.29.

The table provides the area *to the left* of the stated z value. We know that the total area under the standard normal curve is 1. To find the area to the right of z, we find the area to the left or z and subtract it from 1.

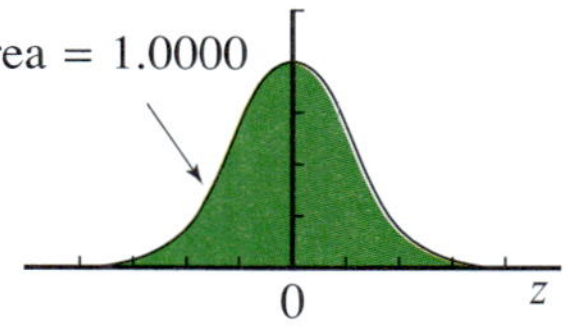

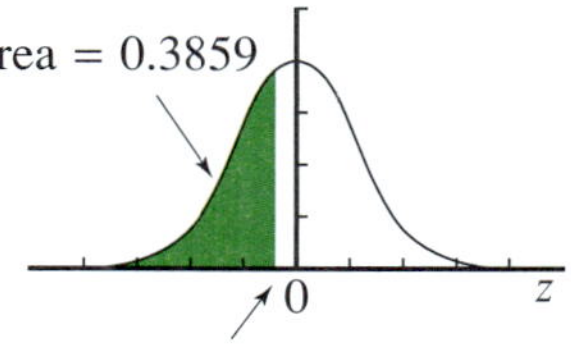

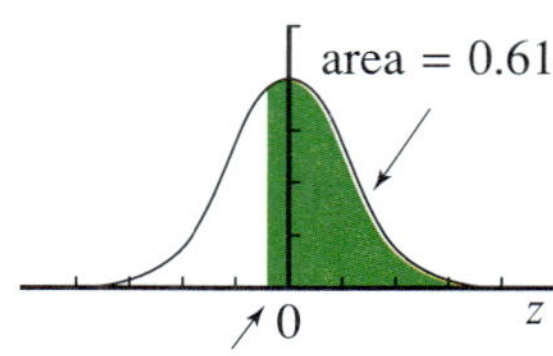

$$\begin{aligned} P(z > -0.29) &= 1 - P(z < -0.29) \\ &= 1 - 0.3859 \\ &= 0.6141 \end{aligned}$$

The area to the right of –0.29 is 0.6141. Therefore, $P(z > -0.29) = 0.6141$.

3. Find the probability that z is between two specified values:

Find the probability that z is between –0.67 and 1.52.

To find the probability that z is between –0.67 and 1.52, find the area to the left of 1.52, then subtract from it, the area to the left of –0.67.

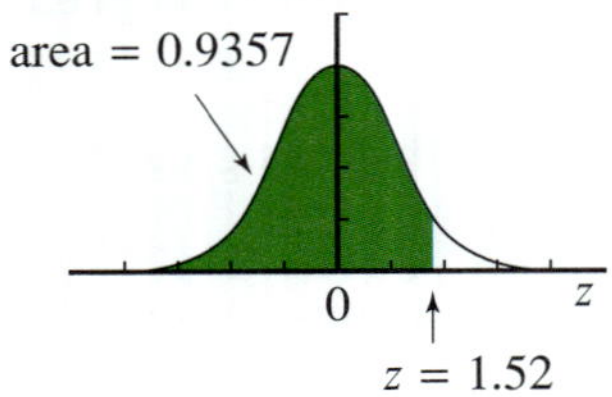

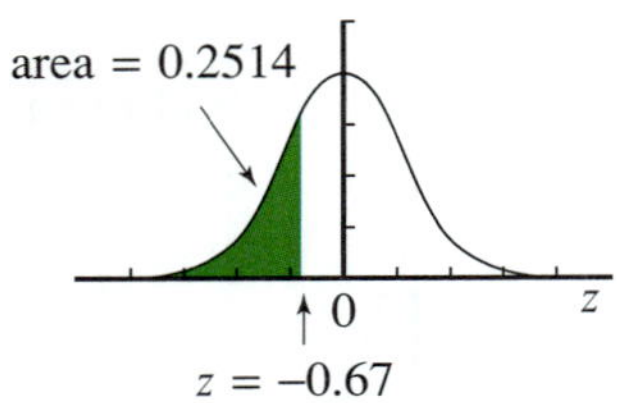

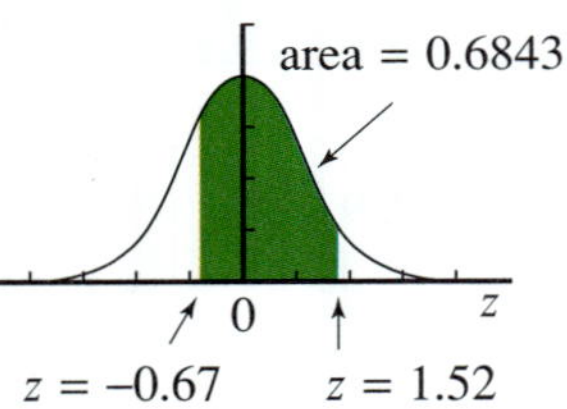

$$
\begin{aligned}
&P(-1.52 < z < 1.52) \\
&= P(z < 1.52) - P(z < -0.67) \\
&= 0.9357 - 0.2514 \\
&= 0.6843
\end{aligned}
$$

The area between –0.67 and 1.52 is about 0.6843.

Therefore, $P(-0.67 < z < 1.52) = 0.6843$.

Areas under a Standard Normal Curve to the left of *z*

z	0.09	0.08	0.07	0.06	0.05	0.04	0.03	0.02	0.01	0.00
−3.4	0.0002	0.0003	0.0003	0.0003	0.0003	0.0003	0.0003	0.0003	0.0003	0.0003
−3.3	0.0003	0.0004	0.0004	0.0004	0.0004	0.0004	0.0004	0.0005	0.0005	0.0005
−3.2	0.0005	0.0005	0.0005	0.0006	0.0006	0.0006	0.0006	0.0006	0.0007	0.0007
−3.1	0.0007	0.0007	0.0008	0.0008	0.0008	0.0008	0.0009	0.0009	0.0009	0.0010
−3.0	0.0010	0.0010	0.0011	0.0011	0.0011	0.0012	0.0012	0.0013	0.0013	0.0013
−2.9	0.0014	0.0014	0.0015	0.0015	0.0016	0.0016	0.0017	0.0018	0.0018	0.0019
−2.8	0.0019	0.0020	0.0021	0.0021	0.0022	0.0023	0.0023	0.0024	0.0025	0.0026
−2.7	0.0026	0.0027	0.0028	0.0029	0.0030	0.0031	0.0032	0.0033	0.0034	0.0035
−2.6	0.0036	0.0037	0.0038	0.0039	0.0040	0.0041	0.0043	0.0044	0.0045	0.0047
−2.5	0.0048	0.0049	0.0051	0.0052	0.0054	0.0055	0.0057	0.0059	0.0060	0.0062
−2.4	0.0064	0.0066	0.0068	0.0069	0.0071	0.0073	0.0075	0.0078	0.0080	0.0082
−2.3	0.0084	0.0087	0.0089	0.0091	0.0094	0.0096	0.0099	0.0102	0.0104	0.0107
−2.2	0.0110	0.0113	0.0116	0.0119	0.0122	0.0125	0.0129	0.0132	0.0136	0.0139
−2.1	0.0143	0.0146	0.0150	0.0154	0.0158	0.0162	0.0166	0.0170	0.0174	0.0179
−2.0	0.0183	0.0188	0.0192	0.0197	0.0202	0.0207	0.0212	0.0217	0.0222	0.0228
−1.9	0.0233	0.0239	0.0244	0.0250	0.0256	0.0262	0.0268	0.0274	0.0281	0.0287
−1.8	0.0294	0.0301	0.0307	0.0314	0.0322	0.0329	0.0336	0.0344	0.0351	0.0359
−1.7	0.0367	0.0375	0.0384	0.0392	0.0401	0.0409	0.0418	0.0427	0.0436	0.0446
−1.6	0.0455	0.0465	0.0475	0.0485	0.0495	0.0505	0.0516	0.0526	0.0537	0.0548
−1.5	0.0559	0.0571	0.0582	0.0594	0.0606	0.0618	0.0630	0.0643	0.0655	0.0668
−1.4	0.0681	0.0694	0.0708	0.0721	0.0735	0.0749	0.0764	0.0778	0.0793	0.0808
−1.3	0.0823	0.0838	0.0853	0.0869	0.0885	0.0901	0.0918	0.0934	0.0951	0.0968
−1.2	0.0985	0.1003	0.1020	0.1038	0.1056	0.1075	0.1093	0.1112	0.1131	0.1151
−1.1	0.1170	0.1190	0.1210	0.1230	0.1251	0.1271	0.1292	0.1314	0.1335	0.1357
−1.0	0.1379	0.1401	0.1423	0.1446	0.1469	0.1492	0.1515	0.1539	0.1562	0.1587
−0.9	0.1611	0.1635	0.1660	0.1685	0.1711	0.1736	0.1762	0.1788	0.1814	0.1841
−0.8	0.1867	0.1894	0.1922	0.1949	0.1977	0.2005	0.2033	0.2061	0.2090	0.2119
−0.7	0.2148	0.2177	0.2206	0.2236	0.2266	0.2296	0.2327	0.2358	0.2389	0.2420
−0.6	0.2451	0.2483	0.2514	0.2546	0.2578	0.2611	0.2643	0.2676	0.2709	0.2743
−0.5	0.2776	0.2810	0.2843	0.2877	0.2912	0.2946	0.2981	0.3015	0.3050	0.3085
−0.4	0.3121	0.3156	0.3192	0.3228	0.3264	0.3300	0.3336	0.3372	0.3409	0.3446
−0.3	0.3483	0.3520	0.3557	0.3594	0.3632	0.3669	0.3707	0.3745	0.3783	0.3821
−0.2	0.3859	0.3897	0.3936	0.3974	0.4013	0.4052	0.4090	0.4129	0.4168	0.4207
−0.1	0.4247	0.4286	0.4325	0.4364	0.4404	0.4443	0.4483	0.4522	0.4562	0.4602
−0.0	0.4641	0.4681	0.4721	0.4761	0.4801	0.4840	0.4880	0.4920	0.4960	0.5000

Areas under a Standard Normal Curve to the left of *z*

z	0.00	0.01	0.02	0.03	0.04	0.05	0.06	0.07	0.08	0.09
0.0	0.5000	0.5040	0.5080	0.5120	0.5160	0.5199	0.5239	0.5279	0.5319	0.5359
0.1	0.5398	0.5438	0.5478	0.5517	0.5557	0.5596	0.5636	0.5675	0.5714	0.5753
0.2	0.5793	0.5832	0.5871	0.5910	0.5948	0.5987	0.6026	0.6064	0.6103	0.6141
0.3	0.6179	0.6217	0.6255	0.6293	0.6331	0.6368	0.6406	0.6443	0.6480	0.6517
0.4	0.6554	0.6591	0.6628	0.6664	0.6700	0.6736	0.6772	0.6808	0.6844	0.6879
0.5	0.6915	0.6950	0.6985	0.7019	0.7054	0.7088	0.7123	0.7157	0.7190	0.7224
0.6	0.7257	0.7291	0.7324	0.7357	0.7389	0.7422	0.7454	0.7486	0.7517	0.7549
0.7	0.7580	0.7611	0.7642	0.7673	0.7704	0.7734	0.7764	0.7794	0.7823	0.7852
0.8	0.7881	0.7910	0.7939	0.7967	0.7995	0.8023	0.8051	0.8078	0.8106	0.8133
0.9	0.8159	0.8186	0.8212	0.8238	0.8264	0.8289	0.8315	0.8340	0.8365	0.8389
1.0	0.8413	0.8438	0.8461	0.8485	0.8508	0.8531	0.8554	0.8577	0.8599	0.8621
1.1	0.8643	0.8665	0.8686	0.8708	0.8729	0.8749	0.8770	0.8790	0.8810	0.8830
1.2	0.8849	0.8869	0.8888	0.8907	0.8925	0.8944	0.8962	0.8980	0.8997	0.9015
1.3	0.9032	0.9049	0.9066	0.9082	0.9099	0.9115	0.9131	0.9147	0.9162	0.9177
1.4	0.9192	0.9207	0.9222	0.9236	0.9251	0.9265	0.9279	0.9292	0.9306	0.9319
1.5	0.9332	0.9345	0.9357	0.9370	0.9382	0.9394	0.9406	0.9418	0.9429	0.9441
1.6	0.9452	0.9463	0.9474	0.9484	0.9495	0.9505	0.9515	0.9525	0.9535	0.9545
1.7	0.9554	0.9564	0.9573	0.9582	0.9591	0.9599	0.9608	0.9616	0.9625	0.9633
1.8	0.9641	0.9649	0.9656	0.9664	0.9671	0.9678	0.9686	0.9693	0.9699	0.9706
1.9	0.9713	0.9719	0.9726	0.9732	0.9738	0.9744	0.9750	0.9756	0.9761	0.9767
2.0	0.9772	0.9778	0.9783	0.9788	0.9793	0.9798	0.9803	0.9808	0.9812	0.9817
2.1	0.9821	0.9826	0.9830	0.9834	0.9838	0.9842	0.9846	0.9850	0.9854	0.9857
2.2	0.9861	0.9864	0.9868	0.9871	0.9875	0.9878	0.9881	0.9884	0.9887	0.9890
2.3	0.9893	0.9896	0.9898	0.9901	0.9904	0.9906	0.9909	0.9911	0.9913	0.9916
2.4	0.9918	0.9920	0.9922	0.9925	0.9927	0.9929	0.9931	0.9932	0.9934	0.9936
2.5	0.9938	0.9940	0.9941	0.9943	0.9945	0.9946	0.9948	0.9949	0.9951	0.9952
2.6	0.9953	0.9955	0.9956	0.9957	0.9959	0.9960	0.9961	0.9962	0.9963	0.9964
2.7	0.9965	0.9966	0.9967	0.9968	0.9969	0.9970	0.9971	0.9972	0.9973	0.9974
2.8	0.9974	0.9975	0.9976	0.9977	0.9977	0.9978	0.9979	0.9979	0.9980	0.9981
2.9	0.9981	0.9982	0.9982	0.9983	0.9984	0.9984	0.9985	0.9985	0.9986	0.9986
3.0	0.9987	0.9987	0.9987	0.9988	0.9988	0.9989	0.9989	0.9989	0.9990	0.9990
3.1	0.9990	0.9991	0.9991	0.9991	0.9992	0.9992	0.9992	0.9992	0.9993	0.9993
3.2	0.9993	0.9993	0.9994	0.9994	0.9994	0.9994	0.9994	0.9995	0.9995	0.9995
3.3	0.9995	0.9995	0.9995	0.9996	0.9996	0.9996	0.9996	0.9996	0.9996	0.9997
3.4	0.9997	0.9997	0.9997	0.9997	0.9997	0.9997	0.9997	0.9997	0.9997	0.9998

UTILITY 30

Using the TI-83 to Calculate Probabilities with the Standard Normal Distribution

The command **ShadeNorm(**lower bound, upper bound) shades the area under the standard normal curve between the lower bound and the upper bound, and gives the area of the shaded region. The area of the shaded region is the probability that a particular value of *c* falls within that region.

Before using the **ShadeNorm(** command:

- Clear all equations in the Y= list.
- Turn off all Plots. Press [2nd] [Y=] **1**, then select **Off**.
- Set the window to Xmin = –4.7, Xmax = 4.7, Xscl = 1, Ymin = –0.15, Ymax = 0.5, Yscl = 0.1.

1. Find the probability that *z* is less than a specific value:

Find the probability that *z* is less than –0.85.

- With the **ShadeNorm(** command, you must always enter two values, the lower bound (or the left boundary of the area) and the upper bound (or the right boundary of the area). Although the curve continues indefinitely to the left and right, the areas beyond –5 and +5 are so small that they can be ignored.
- Use –5 as the lower bound.
- Press [2nd] [VARS] [►] **1** [(–)] **5** [,] [(–)] **0.85** [)] [ENTER] to display this screen.

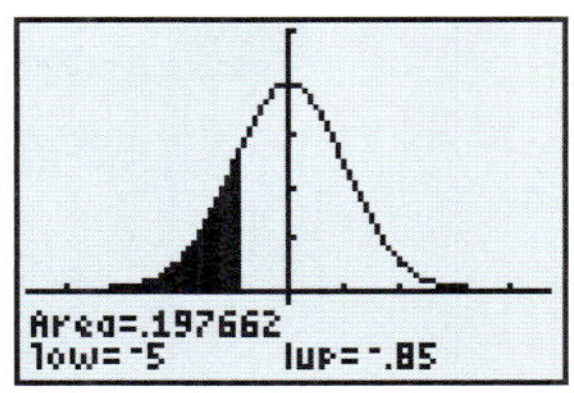

The area to the left of –0.85 is about 0.1977. Therefore, $P(z < -0.85) = 0.1977$.

- To clear the drawing on the screen, press [2nd] [PRGM] [ENTER].

2. Find the probability that *z* is greater than a specific value:

Find the probability that *z* is greater than –0.29.

- Use 5 as the upper bound.
- Press [2nd] [VARS] [▶] **1** [(–)] **0.29** [,] **5** [)] [ENTER] to display this screen.

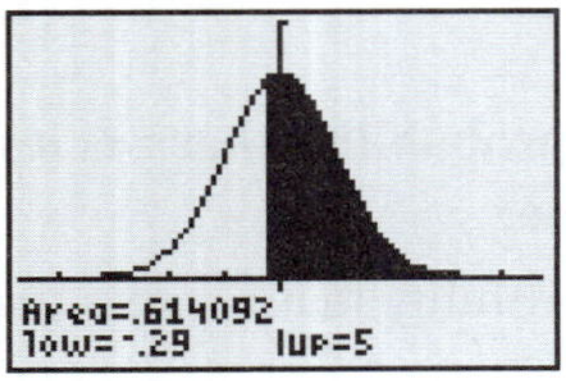

The area to the right of –0.29 is about 0.6141. Therefore, $P(z > -0.29) = 0.6141$.

3. Find the probability that *z* is between two specified values:

Find the probability that *z* is between –0.67 and 1.52.

- Press [2nd] [VARS] [▶] **1** [(–)] **0.67** [,] **1.52** [)] [ENTER] to display this screen.

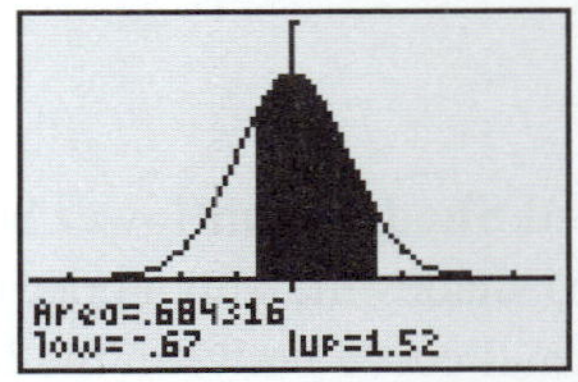

The area between –0.67 and 1.52 is about 0.6843.

Therefore, $P(-0.67 < z < 1.52) = 0.6843$.

Formatting Spreadsheets

1. Creating a title:

A spreadsheet should have a title. Any instructions should also be provided in the first few rows of a spreadsheet. For example, enter the title, *Investing a Fixed Sum of Money*, in cell A1.

2. Changing the width of a column:

Column widths should be adjusted to accommodate the data they contain. A column can be made wider, or narrower, by dragging on the column label at the top of the spreadsheet, or by selecting *Format*, *Column*, *Width*, and entering a numerical value for the width of the column. Changing the width of one cell will change the width of the whole column. If a cell is not wide enough for a particular numerical value, that cell will be filled with ##### signs.

3. Changing the height of a row:

The height of a row can be changed by selecting the row number on the left side of a spreadsheet, and then dragging the vertical-adjust cursor, up or down, or by selecting *Format*, *Row*, *Height*, and entering a numerical value for the height of the row.

4. Defining variables:

Variables should be defined near the top of a spreadsheet.

- Enter the names of the variables: *Principal*, *Annual Interest Rate*, and *Compounding Periods Per Year* in column B, as shown.

The text in column B has been right-justifed, which allows the text to "spill" back into column A. The width of column B still has to be increased so that each of the variable names will appear on just one line.

- Enter the initial values of these variables in column C, as shown.

	A	B	C
1	**Investing a Fixed Sum of Money**		
2			
3		Principal	$5,000
4		Annual Interest Rate	10%
5		Compounding Periods Per Year	2

5. Assigning names to variables:

Names are assigned to variables to help the user understand any formulas in the spreadsheet, and to reduce the possibility of cell location errors.

- Select cell C3. You should see the amount 5000 in the Formula Bar. In the Name Box at the left, you should see C3.

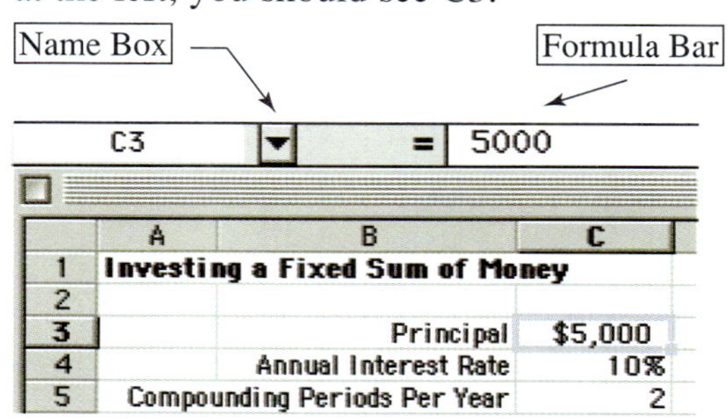

- Highlight C3 in the name box, enter the variable name *Principal*, and press Enter.
- Select cell C4, name this cell *Annual_Interest_Rate*, and press Enter.

- Select cell C5, name this cell *Compounding_Periods_Per_Year*, and press Enter.

Check your work. Select the drop-down arrow of the name box and you should find the three assigned names listed. Select any one of the names and the cursor will automatically move to the cell designated by that name.

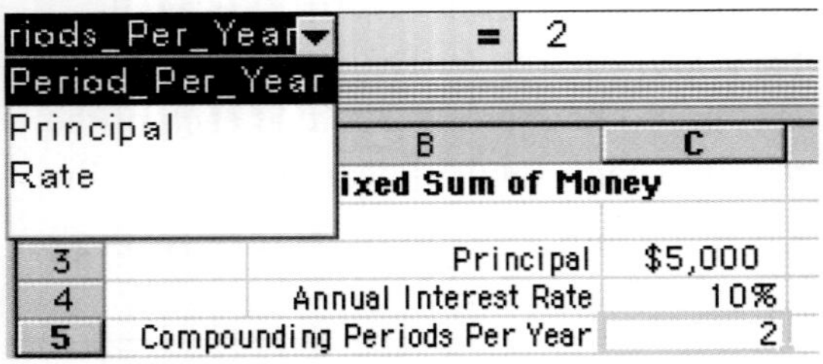

The contents of C3, C4, and C5 can now be referred to by the variable names *Principal*, *Annual_Interest_Rate*, and *Compounding_ Periods_Per_Year*, respectively.

6. Formating cells:

The appearance of a cell in a spreadsheet can be modified using the *Format* command. For example:

- To display a value as currency, highlight the cell then select *Format*, *Cells*, *Number*, and *Currency*.
- To display a value as a fraction, highlight the cell then select *Format*, *Cells*, *Number*, *Fraction*, and *Up to 2 Digits*.

7. Adding borders:

Single or double borders can be added to make your column headings stand out. Highlight the required cells and select *Format*, *Cells*, and *Border*.

8. Wrapping text:

Column headings such as **Opening Balance** and **Closing Balance** are frequently too wide to fit inside a column designed for numbers. To make titles fit within columns, select the entire row, then *Format*, *Cells*, *Alignment*, and *Wrap Text*.

9. Merging cells:

A group of cells can be merged into one cell to emphasize a title, or to create a large work space for writing comments.

- Highlight the cells E3 to F6.
- Select *Format*, *Cells*, *Alignment*, and *Merge cells*. This command will merge cells E3 to F6 into one large cell.
- To control the length of a line of text in a merged cell, use the combined keystrokes ALT-ENTER to end each line of text.

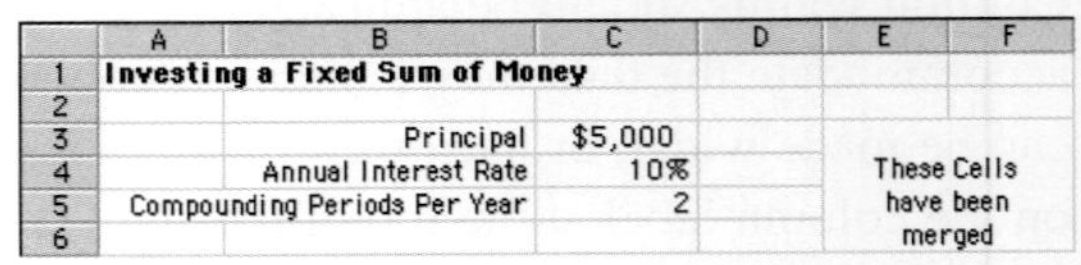

	A	B	C	D	E	F
1	**Investing a Fixed Sum of Money**					
2						
3		Principal	$5,000		These Cells have been merged	
4		Annual Interest Rate	10%			
5	Compounding Periods Per Year		2			
6						

10. Inserting a new column or row:

- Highlight the column, or row, where you want the new one to appear.
- Select *Insert*, *Columns* or *Insert*, *Rows* as required. The contents of the old column, or row, will shift. Any formulas that involve the contents of the shifted column, or row, will be automatically adjusted.

11. Saving a spreadsheet as a template:

- Delete the contents of variable cells, except for variables that are used as divisors. Enter 1 into these cells. If you delete the contents of a divisor and then run the spreadsheet, you will get a *Divide by 0* error.
- Select *File*, *Save As*, *Save As Type*, and *Template (*.XLT)*.
- Enter a name for your template. Excel will automatically save it on your C drive (on a PC), under Templates. The template MUST be saved in this location to allow it to work as a template.

Using the TI-83 to Fit a Regression Equation to Periodic Data

The statistical features of the TI-83 can be used to determine the equation of a curve that comes closest to passing through the points of a scatterplot of data. For linear regressions refer to Utility 14. For quadratic, exponential, and cubic regressions refer to Utility 18. You may also want to refer to Utility 7 to review entering data, and plotting data.

A sinusoidal regression, SinReg, is used to fit a curve of the form $y = a\sin(bx + c) + d$ to data that vary periodically. To perform a sinusoidal regression, the calculator must be in radian mode, and you must have at least five data points (better results are obtained with more points).

We will use this data set as an example.

x	0.0	0.5	1.1	1.5	2.0
y	−3.7	−1.7	−0.5	−0.3	−2.6

x	2.3	3.2	4.0	5.0
y	−3.8	−5.9	−3.8	−0.6

To carry out a sinusoidal regression, follow these steps:

1. Put the calculator into radian mode.

2. Clear the Y= list.

3. Enter the x-values in L1 and the y-values in L2.

4. Define an appropriate graphing window. A good window for this example is shown below.

```
WINDOW
 Xmin=-1
 Xmax=6
 Xscl=1
 Ymin=-6
 Ymax=1
 Yscl=1
 Xres=1
```

5. Turn on Plot1, and press GRAPH to obtain a scatterplot of the data, as shown below.

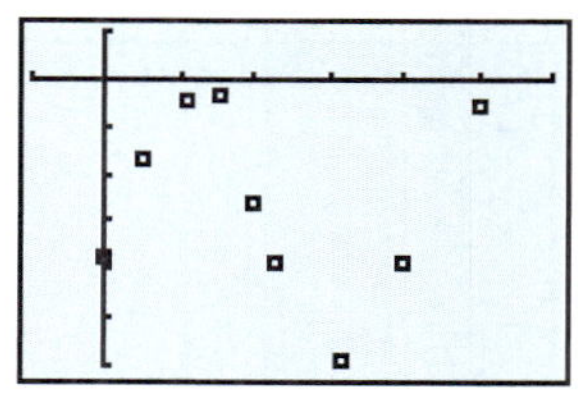

6. When we use the SinReg feature of the TI-83, the results are most reliable when we input an estimate for the period. Use the scatterplot to estimate the period of the data. In this case, the period is approximately 5 units.

7. Press [STAT] [▶] [ALPHA] [PRGM] to access the SinReg command.

- Enter five parameters, separated by commas, to prepare the SinReg command, and to automatically store the regression equation in Y1.
  - Press **3** [,]. This sets the calculator to run a medium level regression search.
  - Press [2nd] **1** [,] to indicate that the *x*-values are in L1.
  - Press [2nd] **2** [,] to indicate that the *y*-values are in L2.
  - Press **5** [,] to indicate that the period is about 5 units.
  - Press [VARS] [▶] [ENTER] [ENTER] to indicate that the regression equation is to be automatically loaded into Y1.

  The home screen should look like this:

- Press [ENTER] to calculate the regression equation. The results are shown below.

8. Press [GRAPH] to see a plot of your original data, and the regression curve below. The curve should be a close fit to the data.

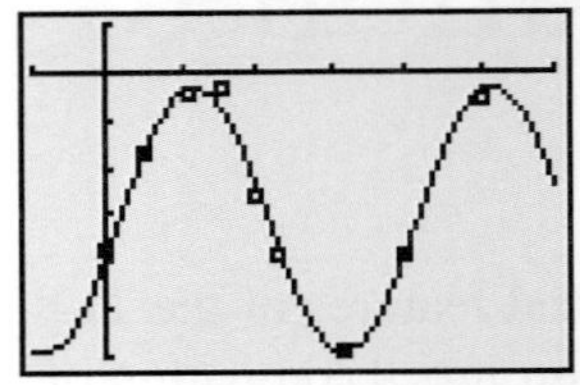

In practice, you would round the values of *a*, *b*, *c*, and *d* to one more significant figure than was contained in the original data.

Thus, the equation $y = 2.78 \sin(1.57x - 0.24) - 3.12$ would be used to represent the pattern in the original data.

Using the TI-83 to Generate and Graph a Sequence

The TI-83 can be used to generate and graph a sequence that is defined recursively. The notation u(n) refers to the value of the nth term in a sequence, and n refers the number of a term. Thus, u(3) = 24 indicates the value of the 3rd term is 24. We can refer to the terms before u(n) as u(n – 1), u(n – 2), and so on.

For example, in the sequence 5, 7, 9, 11, …, after the first term, each term is generated by adding 2 to the preceding term. Thus, u(1) = 5, and u(n) = u(n – 1) + 2.

A copying service charges 7¢ for the first sheet, and 4¢ for each subsequent sheet. The value of the nth term, u(n), is u(n – 1) + 4. We will use this example to illustrate the steps to generate and graph a sequence on the TI-83.

Part A: Generate a Sequence Recursively

1. Press **7** [ENTER] to enter the first term.

2. Press [+] **4** [ENTER] to produce the 2nd term. Repeat this 8 more times to generate 10 terms.

Part B: Generate a Sequence Using the Sequence Command

1. To select the Sequence Mode, press [MODE] [▼] [▼] [▼] [►] [►] [►] [ENTER] as shown.

2. Set the Format Menu to the default settings. Press [2nd] [ZOOM] and ensure that your calculator is set up as shown below.

3. To define the sequence, press [Y=] to obtain the screen below.

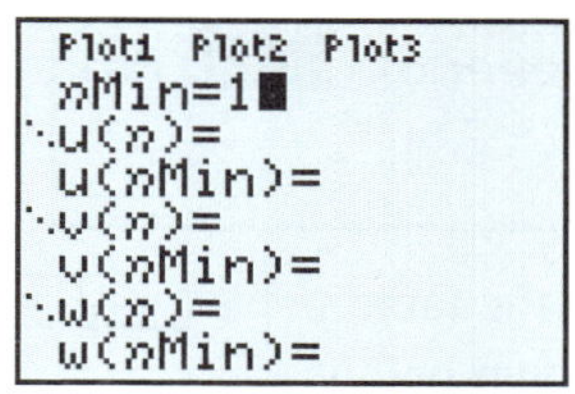

- nMin is the number of the first term in the sequence. Set nMin = 1.
- u(n) is the equation that defines the sequence. In this case, u(n) = u(n – 1) + 4. Press [2nd] **7** [(] [X,T,θ,n] [–] **1** [)] [+] **4** to define this sequence.
- u(nMin) is the value of the first term. Set u(nMin) = 7.
- Press [2nd] [MODE] to quit and return to the home screen.

4. To generate the first ten terms of the sequence (shown below), press [2nd] **7** [(] **1** [,] **10** [)] [ENTER]. The syntax for this command is u(*nstart*, *nstop*).

```
u(1,10)
{7 11 15 19 23 …
■
```

5. An alternative way to display the sequence is to use the table command.

- Press [2nd] [WINDOW] to access the TABLE SETUP menu (below).
  Set TblStart = 1 and △Tbl = 1. Press [2nd] [MODE] to quit.

```
TABLE SETUP
 TblStart=1
 △Tbl=1■
Indpnt: Auto Ask
Depend: Auto Ask
```

- To display the table press [2nd] [GRAPH]. The table is shown below.

n	u(*n*)
1	7
2	11
3	15
4	19
5	23
6	27
7	31

n=7

Part C: Graph a Sequence that was Generated Using the Sequence Command

1. We now wish to graph the sequence defined in Part B, step 3. Press [WINDOW] to set the window. The window settings in Sequence Mode contain four additional items before Xmin, Xmax, Xscl, and so on.

```
WINDOW
 nMin=1■
 nMax=10
 PlotStart=1
 PlotStep=1
 Xmin=0
 Xmax=10
↓Xscl=1
```

```
WINDOW
↑PlotStep=1
 Xmin=0
 Xmax=10
 Xscl=1
 Ymin=0
 Ymax=45
 Yscl=5■
```

- *n*Min is the first term to evaluate. Set *n*Min = 1.
- *n*Max is the last term to evaluate. Set *n*Max = 10.
- PlotStart indicates the first term to be graphed. Set PlotStart = 1.
- PlotStep is the incremental *n* value. Set PlotStep = 1.
- Set Xmin, Xmax, and so on as shown above.

When you graph a sequence, the horizontal axis indicates the term number, and the vertical axis indicates the value of the term.

2. Press [GRAPH] to obtain a plot of the sequence. Use [TRACE] to verify that u(1) = 7, u(2) = 11, u(3) = 15, and so on.

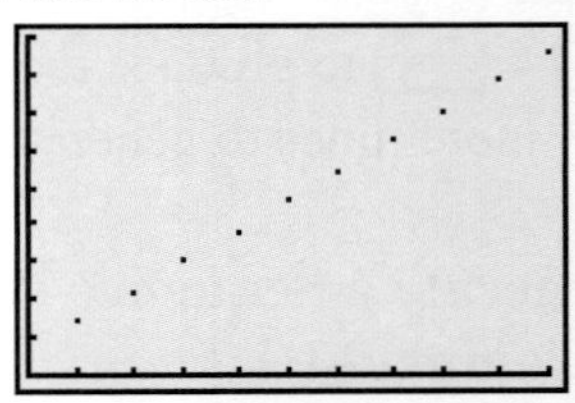

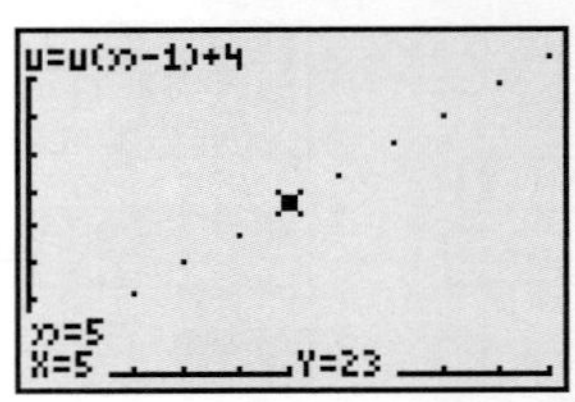

Using *The Geometer's Sketchpad* to Add Vectors

For basic instructions on using *The Geometer's Sketchpad,* refer to Utility 23.

We will use *The Geometer's Sketchpad* to solve this problem.

A boat travels 30 km toward [130°] then turns toward [285°] for another 40 km. What is the boat's displacement from its original position?

1. From the **File** menu, choose **New Sketch**.

2. Under the **Display** menu, select **Preferences**. Make sure the distance unit is centimetres with precision to the nearest tenth. The angle unit should be degrees with precision to the nearest unit.

3. Under the **Graph** menu, select **Show Grid** and turn off **Snap to Grid**.

4. Select the text tool and create a text box. In the text box, type in the scale 1 cm = 10 km.

5. From the **File** menu, select **Open**. The current directory should be c:\sketch. In the **Directories** column, double-click on [samples], then [scripts], and then [utility]. In the **Files** column, double-click on arrowclo.gss. This is a script file used to create vectors.

6. To draw the first vector, 30 km [130°]:

- Use the point tool to create a point on the origin.
- *The Geometer's Sketchpad* uses a convention in which the angle is measured from due east, and rotates counter-clockwise. To convert a bearing to an angle using the convention of *The Geometer's Sketchpad*, subtract the bearing from 450°.
- Select the point. Under the **Transform** menu, select **Translate**. Make sure **By Polar Vector** is selected. Enter 450° − 130° = 320° for **Direction**. Use the scale to convert 30 km to 3 cm and enter this value **Magnitude**. Click **OK**.
- Another point has been created. Select the two points using shift-click. Select the tail first, then the head. Click on the **Fast** button on the script arrowclo.gss. This creates a vector with an arrowhead.

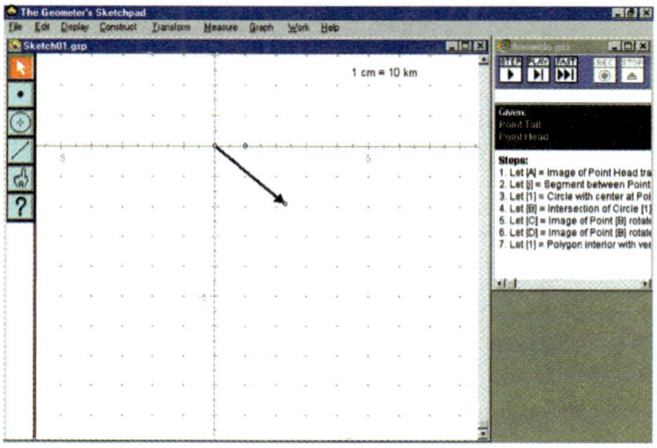

7. To draw the second vector, 40 km [285°], head-to-tail with the first vector:

- Select the point at the head of the first vector. Under the **Transform** menu, select **Translate**. Enter 450° – 285° = 165° for direction and 4 cm for **Magnitude**. Click **OK**.
- Another point has been created. Select the two points using shift-click. Select the tail first, then the head. Click on the **Fast** button on the script arrowclo.gss. This creates another vector head-to-tail with the first.

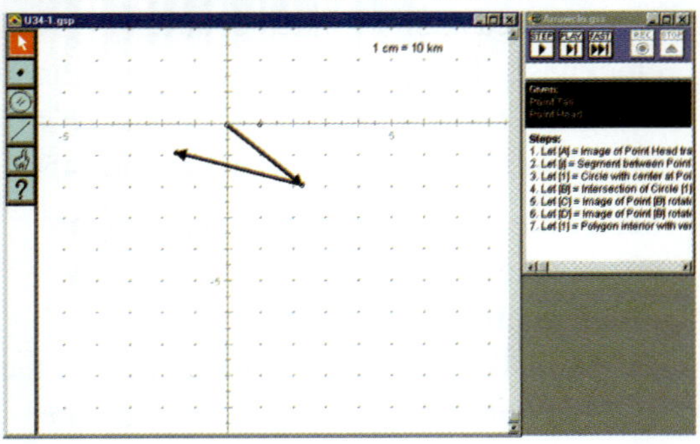

8. To draw the resultant vector:

- Shift-click to select the point at the tail of the first vector and the point at the head of the second vector in that order. Click on the **Fast** button on the script to create the resultant.
- Select the resultant vector. From the **Measure** menu, select **Length**. The magnitude of the resultant is 1.8 cm. Use the scale to convert this to 18 km.
- In this order, shift-click the point at the head of the resultant vector, the point at the tail of the resultant, and the point at the head of the first vector. From the **Measure** menu, select **Angle**. The angle between the resultant and the first vector is 110°. We can add this to the bearing of the first vector to find the bearing of the resultant.
  130° + 110° = 240°

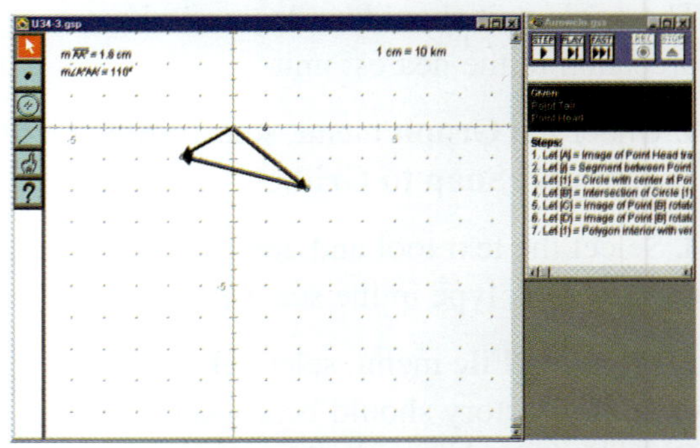

The boat's displacement from its original position is 18 km [240°].

absolute cell reference: in a spreadsheet formula, a reference to one particular cell of the spreadsheet that will not change when the formula is moved to a different cell

$B$3 indicates that the value or expression contained in cell B3 is to be used in the formula.

accuracy: when referring to a measurement, it indicates how close the measurement comes to its true value

active cell: the cell of a spreadsheet into which an item of data is placed when you start to type

acute angle: an angle whose measure is less than 90°

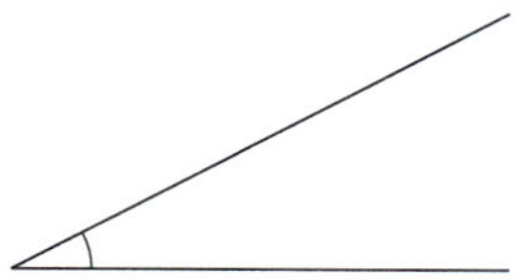

adjacent side: in a right triangle, the side next to the named angle that is not the hypotenuse

For ∠A

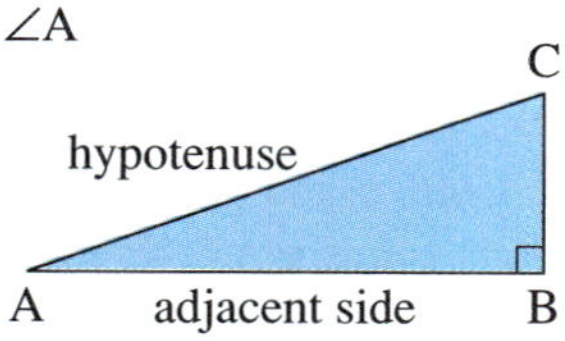

For ∠C

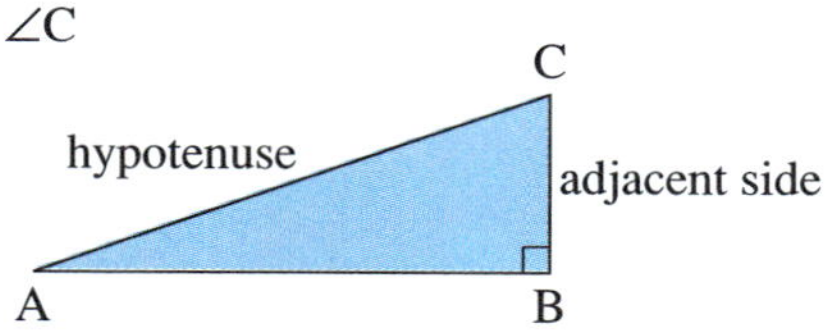

algebraic expression: a mathematical expression that contains at least one variable

$6x + 4$ is an algebraic expression.

alleles: alternative forms of a gene

alternate angles: see *parallel lines*

altitude: the perpendicular distance from the base of a figure to the opposite side or vertex; the height of an aircraft above the ground

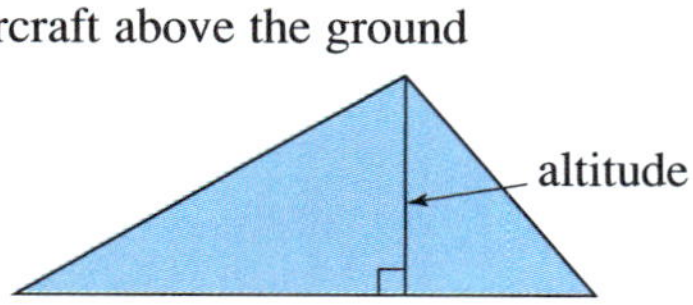

amortization period: the length of time over which a loan is paid off, usually used for a large amount of loan such as a house purchase

amount of an investment: the value of the principal plus interest

Calculate the amount of an investment of $200 after 3 years at 6.5%, compounded annually.

$P = 200 \quad i = 0.065 \quad n = 3 \quad A = P(1 + i)^n$

$A = 200(1 + 0.065)^3$

$= 200(1.207\ 949\ 625)$

$\doteq 241.59$

The investment was worth $241.59.

amplitude: half the distance between the maximum and minimum values of a periodic function;

$\text{amplitude} = \dfrac{\text{maximum value} - \text{minimum value}}{2}$;

see *Tutorials 5.2*, *5.3*, and *5.4*

angle of declination of the sun: the angle of the sun above or below the equator as measured on the equator at noon; see *Tutorial 5.6*

angle of depression: the angle between the horizontal and the oblique line joining the observer's eye to a point lower than eye level

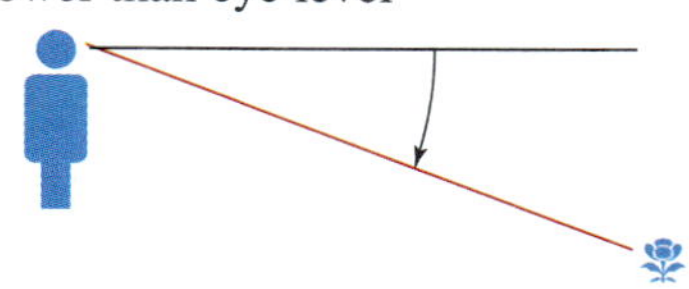

angle of elevation: the angle between the horizontal and the oblique line from the observer's eye to some object above eye level

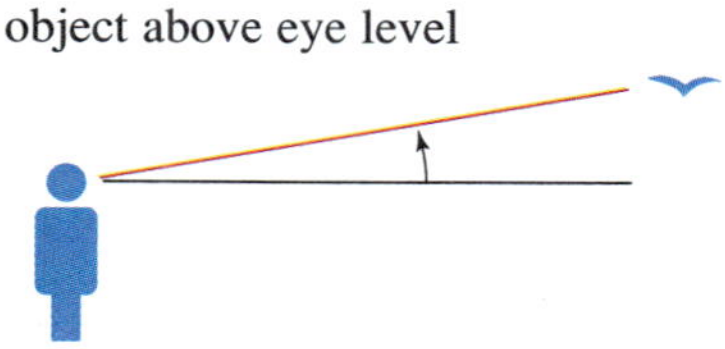

angle of inclination of a line segment: the acute angle, measured from the horizontal to the line segment

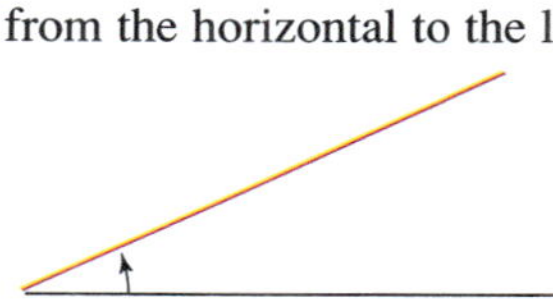

annuity: a series of regular, equal payments paid into, or out of, an account

approximation: a number close to the exact value of a quantity or an expression; the symbols $\doteq$ and $\cong$ mean "is approximately equal to"

3.14 is an approximation for π.

arc: part of the circumference of a circle or other curve.

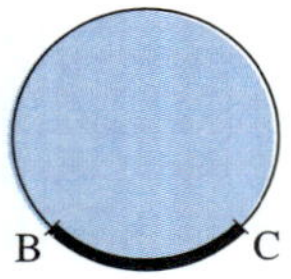

BC is an arc of this circle.

area: the number of square units needed to cover a surface; common units used to express area include cm^2, m^2, and hectares; see *circle*, *rectangle*, *square*, and *triangle*

arithmetic sequence: a sequence of numbers in which each term after the first is formed by adding a constant to the preceding term

The numbers 1, 4, 7, 10, ... form an arithmetic sequence, since each term after the first is formed by adding 3 to the preceding number.

average: a single number that is used to represent a set of numbers; to find the average, all the numbers in the data set are added together and the sum is divided by the number of entries in the data set; see *mean*

The data set 1, 3, 4, 7, 7, 8 has 6 entries.

$$\text{Average} = \frac{1+3+4+7+7+8}{6} = 5$$

average speed: the speed that, if the object travelled at that speed constantly, would result in the same total distance being travelled in the same total time. To calculate average speed, the total distance travelled during the given time period is divided by the total time.

In one hour, a car travels 100 km. The car stops for $\frac{1}{2}$ h and then travels 80 km in the next hour. Find the average speed of the car.

The total distance travelled is 180 km and the time required to travel this distance includes the stopping time of $\frac{1}{2}$ h.

$$\text{Average speed} = \frac{\text{Distance}}{\text{Time}}$$

$$v = \frac{d}{t} = \frac{180\text{ km}}{2.5\text{ h}} = 72\text{ km/h}$$

bar graph: a graph that displays data by using horizontal or vertical bars whose lengths are proportional to the number they represent

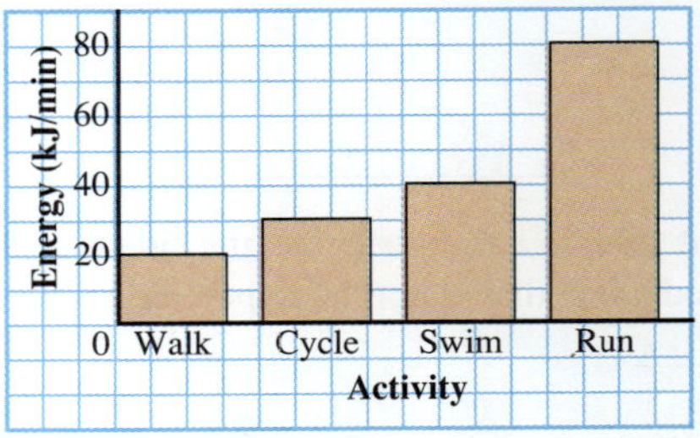

base: the side of a polygon, or the face of a solid, from which the height is measured; also, the factor repeated in a power

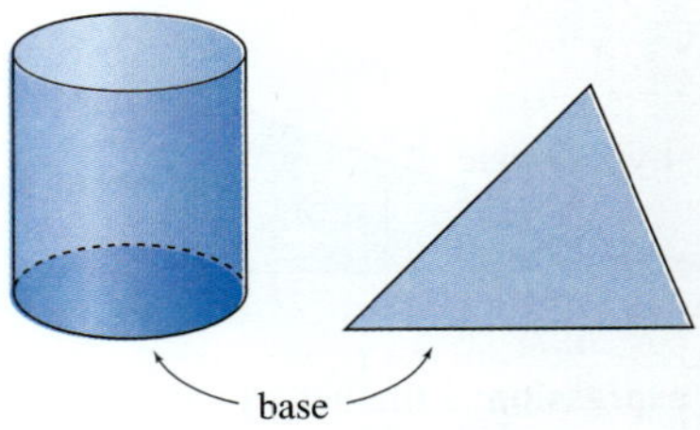

In the expression 5^3, 5 is the base.

bear market: a term used to describe the stock market when stock prices are falling

bearing: the 3-digit angle, measured in a clockwise direction, between the north line and a given direction

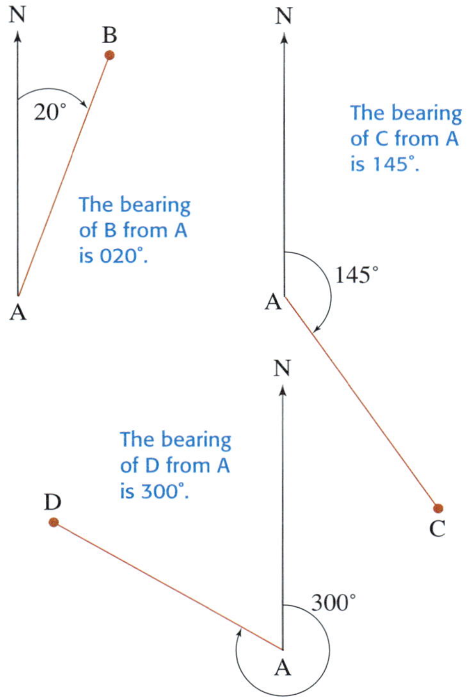

bias: an emphasis on characteristics that are not typical of the entire population

biased sample: a sample containing members of the population that are not representative of the whole population

binomial distribution: the probability distribution for a binomial experiment; see *Tutorial 3.1*

binomial experiment: an experiment with a fixed number of independent trials in which the outcomes can be classified as success or failure and the probabilities remain constant for each trial; see *Tutorials 1.1* and *3.1*

blueprints: the initial drawings used in a construction project

board lot: a unit of trading in the stock market

broken-line graph: a graph that displays data by using points that are joined by line segments

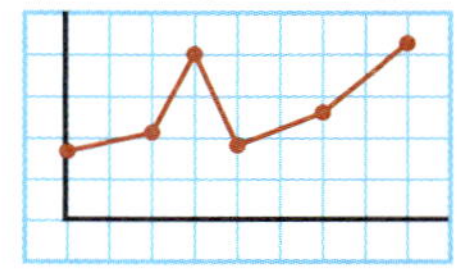

budget: a written plan to outline how money will be spent

bull market: a term used to describe the stock market when stock prices are rising

CBL: the Calculator-Based Laboratory; a data collection device used with the TI-83 calculator

CBR: the Calculator-Based Ranger; a sonic motion detector used with the TI-83 calculator

cash investments: short-term investments that are easily accessible; include bank accounts, term deposits, and money market funds; see *Tutorial 4.3*

cell: a rectangle in a spreadsheet into which data may be entered

cell reference: the name of a cell in a spreadsheet, given by indicating the column and row to which it belongs

Cell B3 is the cell in column B and row 3 of the spreadsheet document.

central angle: the angle whose vertex is at the centre of a circle between two radii

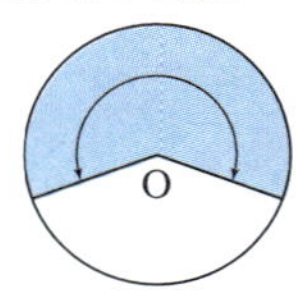

central angle subtended by an arc: the angle at the centre of a circle between the radii from the ends of an arc of the circle

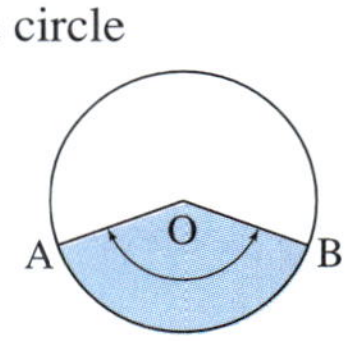

∠AOB is subtended by the arc AB.

circle: the set of points in a plane that are a given distance from a fixed point (the centre)

The area of a circle with radius *r* and diameter *d* is:

$$A = \pi r^2 = \frac{\pi d^2}{4}$$

The circumference *C* of a circle is:

$C = 2\pi r$, where r is the radius
$= \pi d$, where d is the diameter

The area of a circle with radius 5 cm is:

$$A = \pi(5\text{ cm})^2 = \pi(25\text{ cm}^2) \doteq 78.5\text{ cm}^2$$

The circumference of a circle whose radius is 5 cm is:

$$C = 2\pi(5\text{ cm}) \doteq 31.4\text{ cm}$$

circle graph: a diagram that uses parts of a circle to display data

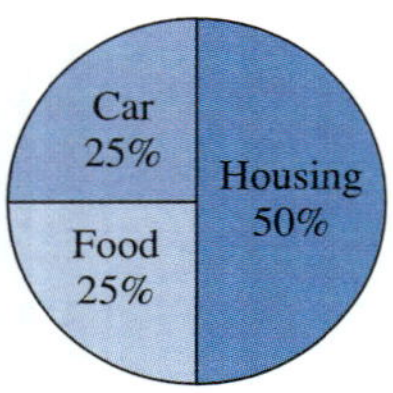

circumference: the distance around a circle; the boundary of any region whose boundary is a simple closed curve; see *circle*

cluster sample: a sample in which every member of a randomly chosen section of the population is selected

collecting like terms: putting together terms with exactly the same variable expressions, then simplifying by addition or subtraction

Collect like terms: $4(3x - 1) + 5x - 2$
$= 12x - 4 + 5x - 2$
$= 12x + 5x - 4 - 2$
$= 17x - 6$

column matrix: a matrix with only one column; see *Tutorial 2.1*

common difference: the number obtained by subtracting any term from the next term in an arithmetic sequence

For the arithmetic sequence 1, −5, −11, −17, . . . , the common difference $= (-5) - 1 = -6$

common ratio: the ratio of one term in a geometric sequence to the preceding term; see *Tutorial 6.1*

complement of event A: the entire set of outcomes that are not favourable to A; see *Tutorial 1.2*

complementary angles: two angles whose measures add up to 90°

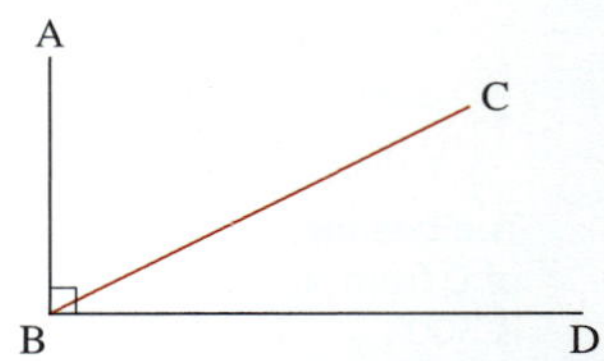

∠ABC and ∠CBD are complementary angles.

compound interest: when the interest due is added to the original amount invested and thereafter earns interest; the interest earned is compound interest. The formula for calculating the future amount is $A = P(1 + i)^n$.

Calculate the amount of interest on an investment of $100 after 4 years at 6.25%, compounded annually.

$P = 100 \quad i = 0.0625 \quad n = 4 \quad A = 100(1 + 0.0625)^4$
$= 100(1.274\ 429\ 321)$
$\doteq 127.44$

The $100 is worth $127.44 after 4 years so the amount of interest earned is $27.44.

compounding: the process of converting interest into principal. After a certain length of time, the interest becomes part of the money that earns interest.

conditional probability: the probability that an event will occur given that another event has occurred; see *Tutorial 1.4*

cone: a solid that is formed by a region (the base of the cone) and all the line segments joining points on the boundary of the region to a point not on the region

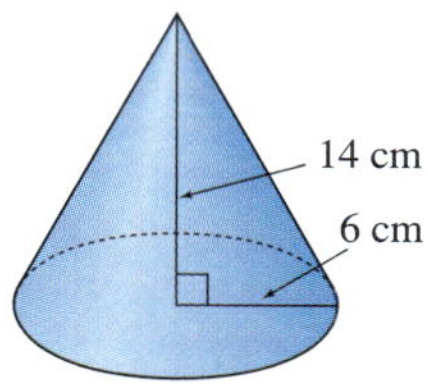

The volume of a cone can be found using the formula:

Volume = $\frac{1}{3}$(Base area)(Height)

= $\frac{1}{3}\pi r^2 h$, where the circular base has radius r, and the height of the cone is h.

The volume of a cone whose circular base has a radius of 6 cm and a height of 14 cm is:

$V = \frac{1}{3}\pi(6 \text{ cm})^2(14 \text{ cm})$

$\doteq 527.8 \text{ cm}^3$

congruent: figures that have the same size and shape, but not necessarily the same orientation

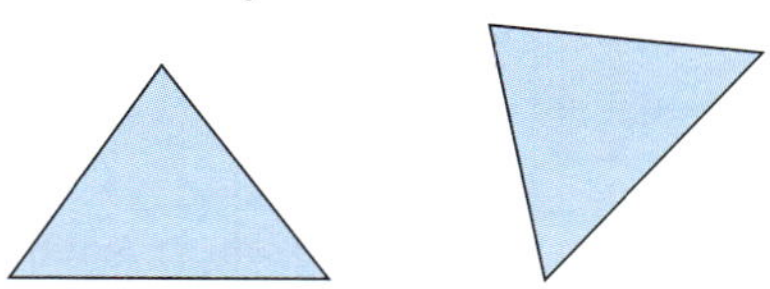

conjecture: a conclusion based on examples

consecutive integers: integers that come one after the other without any integers missing

34, 35, 36 are consecutive integers; so are −2, −1, 0, 1.

consistent system of equations: a system of equations with at least one solution

$x + y = 3$ and $x - y = 1$ is a consistent system since (2, 1) solves both equations.

constant: a particular number

7 is a constant.

constant term: a number

In the equation $y = 4x + 2$, the constant term is 2.

continuity correction: an adjustment of 0.5 added to or subtracted from the endpoints when the normal distribution is used to approximate the binomial distribution in probability calculations; see *Tutorial 3.5*

continuous data: data that can assume any value without a break; data about the time it takes for a plant to grow is continuous because in-between values have meaning

control chart: used in statistical process control to track the value of some aspect of a product over time

convenience sample: a sample whose members are selected based on convenience

coordinate axes: the horizontal and vertical number lines on a grid that represents a plane

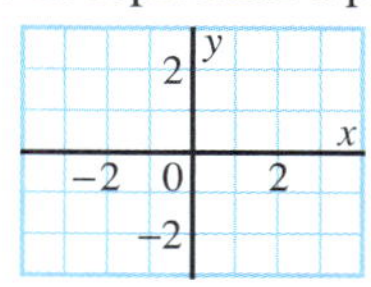

coordinate plane: a two-dimensional surface on which a coordinate system has been set up

coordinates: also called Cartesian coordinates; the numbers in an ordered pair that locate a point in the coordinate plane

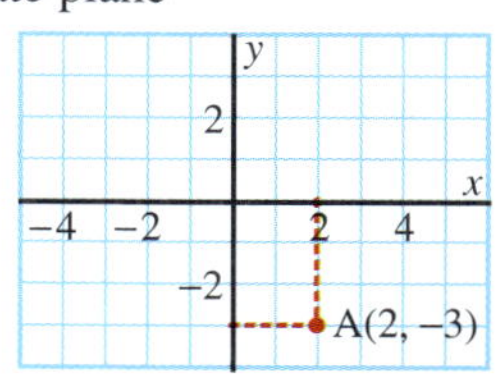

The coordinates of point A are (2, −3).

correlation coefficient: a measure of how closely data can be described by a certain type of function; the closer the value of the correlation coefficient to 1 or −1, the closer the data fit the function

corresponding angles: see *parallel lines*

corresponding angles in similar triangles: two angles, one in each triangle, that are equal

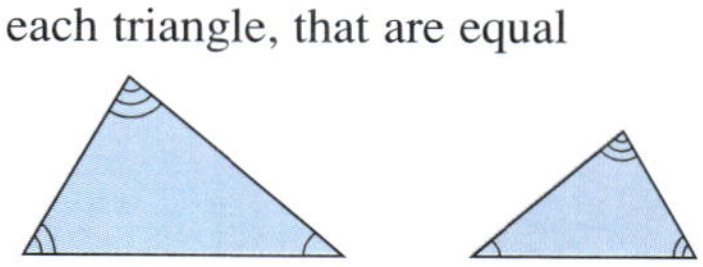

cosine: for an acute angle ∠A in a right triangle, the ratio of the length of the side adjacent to the angle to the length of the hypotenuse

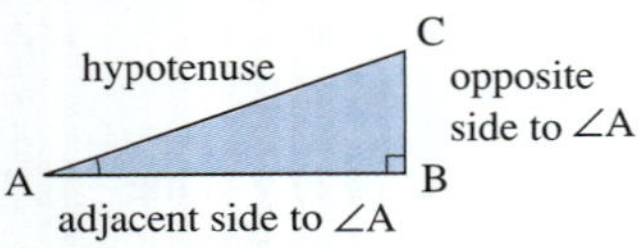

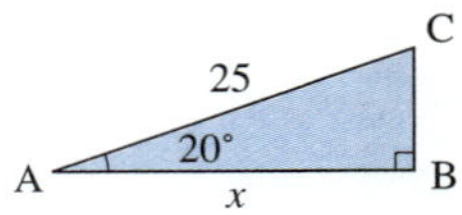

To determine AB in the diagram above, right:

$\cos\angle A = \frac{\text{Length of adjacent side}}{\text{Length of the hypotenuse}}$	**Substitute the known quantities.**
$\cos 20^\circ = \frac{x}{25}$	**Use a calculator to determine cos 20°.**
$0.9397 \doteq \frac{x}{25}$	**Multiply both sides by 25.**
$23.5 \doteq x$	

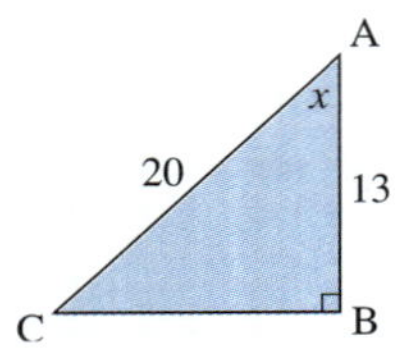

To determine ∠A in the diagram above:

$\cos\angle A = \frac{\text{Length of adjacent side}}{\text{Length of the hypotenuse}}$	**Substitute the known quantities.**
$\cos x^\circ = \frac{13}{20}$	**Convert the fraction into a decimal.**
$\cos x^\circ = 0.65$	**Use the $\cos^{-1}$ ("the angle whose cosine is") function on the calculator.**
$x \doteq 49.5^\circ$	

Cosine Law: a trigonometric law used to solve triangles that are not necessarily right triangles

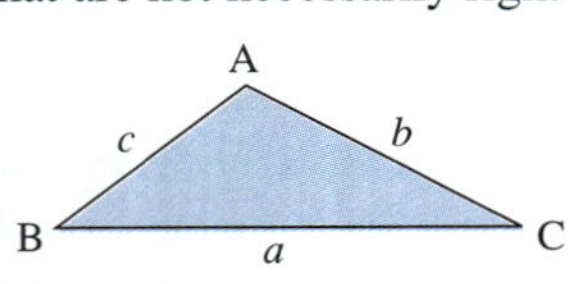

In any △ABC,

$a^2 = b^2 + c^2 - 2bc\cos A$, so $\cos A = \frac{b^2 + c^2 - a^2}{2bc}$

$b^2 = a^2 + c^2 - 2ac\cos B$, so $\cos B = \frac{a^2 + c^2 - b^2}{2ac}$

$c^2 = a^2 + b^2 - 2ab\cos C$, so $\cos C = \frac{a^2 + b^2 - c^2}{2ab}$

Use the Cosine Law to solve problems when you know all 3 sides of a triangle and wish to find an angle, or 2 sides and the included angle and wish to find the other side.

1.

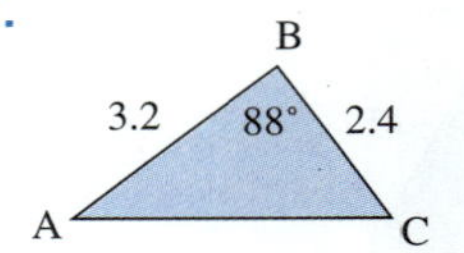

2.

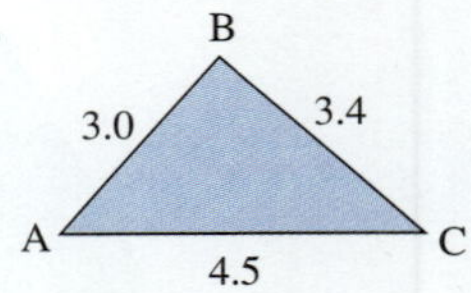

1. In △ABC, above left, find AC.

$b^2 = a^2 + c^2 - 2ac\cos B$

$b^2 = 2.4^2 + 3.2^2 - 2(2.4)(3.2)\cos 88^\circ$

$b^2 = 15.463\,943\,73$

$b \doteq 3.9$

2. In △ABC, above right, find ∠A.

$\cos A = \frac{b^2 + c^2 - a^2}{2bc}$

$\cos A = \frac{4.5^2 + 3.0^2 - 3.4^2}{2 \times 4.5 \times 3.0}$

$\cos A = 0.655\,185\,185\,2$ **Use $\cos^{-1}$ to find A.**

$A \doteq 49.1^\circ$

cost of financing: the difference between the cash price and the sum of the payments made for an item; see *Tutorial 4.4*

counterexample: an example that shows a conjecture is false

cube: a rectangular solid whose length, width, and height are all equal

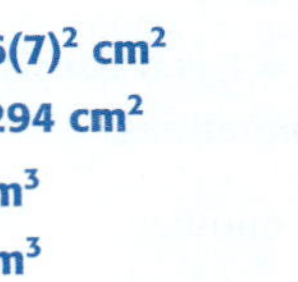

For a cube with sides of length *s*, the surface area is $S = 6s^2$; the volume of the cube is $V = s^3$.

If $s = 7$ cm

Surface area $= 6(7)^2\ \text{cm}^2$

$= 294\ \text{cm}^2$

Volume $= (7)^3\ \text{cm}^3$

$= 343\ \text{cm}^3$

cube root: a number that, when raised to the power 3, results in the given number

$\sqrt[3]{8} = 2$, since $2^3 = 8$

cubic units: units that measure volume; common cubic units include cm^3 and m^3

cycle: a cycle of a periodic function is a part of its graph from any point to the first point where the graph starts repeating; see *Tutorial 5.1*

cylinder: a solid with two parallel, congruent, circular bases

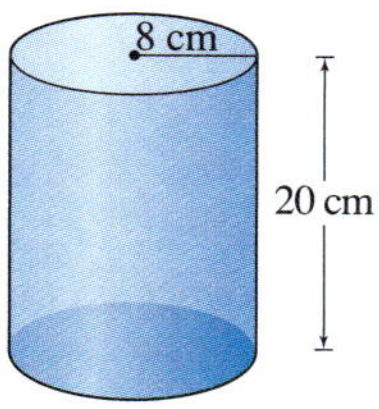

Surface area $= 2\pi rh + 2\pi r^2$
$= 2\pi(8)(20) + 2\pi(8)^2$
$\doteq 1005.3 + 402.1$
$\doteq 1407.4\ cm^2$

Volume $= \pi r^2 h$
$= \pi(8)^2(20)$
$\doteq 4021.2\ cm^3$

data: numeric or non-numeric facts or information

debt: money owing

declination of the sun: see *angle of declination of the sun*

dependent events: the occurrence of one event is affected by the occurrence of another event; see *Tutorial 1.4*

dependent variable: the output of a relation, often denoted as *y*; also called the responding variable

depreciation: the decrease in value of an asset; see *Tutorial 4.4*

diagonal: a line that joins two vertices of a figure, but is not a side

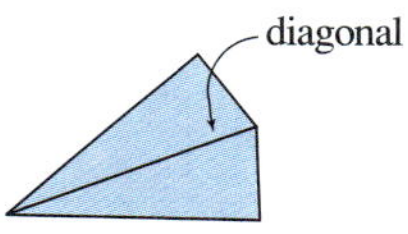

diameter: a line segment that joins two points on a circle (or sphere) and passes through its centre; the diameter of a circle is twice the length of the radius; see *circle*

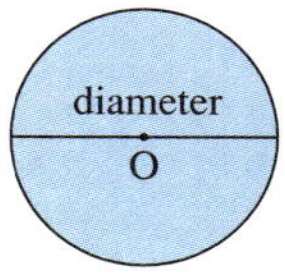

dimensions of a matrix: the number of rows and columns of a matrix; an $m \times n$ matrix has *m* rows and *n* columns; see *Tutorial 2.1*

direct variation: when the ratio of two variable quantities remains constant

If *y* varies directly as *x*, the equation that relates *y* to *x* is $y = mx$, where *m* is a constant. The graph of a direct variation is a straight line that passes through the origin.

discrete data: distinct data; data about the number of oranges in a crate are discrete because intermediate values have no meaning

displacement: a vector that describes the distance and direction an object moves; see *Tutorial 7.1*

distance: the space between two points; also, the distance travelled by an object that is moving at a constant speed for a time is determined from the relation Distance = Speed × Time, where a consistent set of units must be used

If an object travels at a constant speed of 20 m/s for 2 min, the distance travelled is
$d = (20\ m/s)(120\ s)$
$= 2400\ m$

distance formula: a formula used to determine the distance between two points whose coordinates are known

If $A(x_A, y_A)$ and $B(x_B, y_B)$, then
$AB = \sqrt{(x_B - x_A)^2 + (y_B - y_A)^2}$

Find the distance between the points A(7, 2) and B(−4, 4) $AB = \sqrt{(x_B - x_A)^2 + (y_B - y_A)^2}$
$= \sqrt{(-4 - 7)^2 + (4 - 2)^2}$
$= \sqrt{125}$
$\doteq 11.2$

dividends: a portion of a company's earnings paid to shareholders

domain of a relation or function: the set of all possible x-values (or valid input values) represented by the graph or equation

dominant trait: in genetics, only one dominant allele needs to be present for the trait to be expressed

effective annual interest rate: the rate that with annual compounding has the same effect as the stated rate; see *Tutorial 4.5*

elements: the entries in a matrix; see *Tutorial 2.1*

equal vectors: have the same magnitude and direction; see *Tutorial 7.1*

equation: a mathematical statement indicating that two expressions are equal

$2x + 5y = -4$ is an equation.

equation of a line: an equation that gives the relationship between the coordinates of every point on the line; see *linear equation and its graph*

equidistant: the same distance apart

Points A and B are equidistant from the y-axis since they are both 3 units from the y-axis.

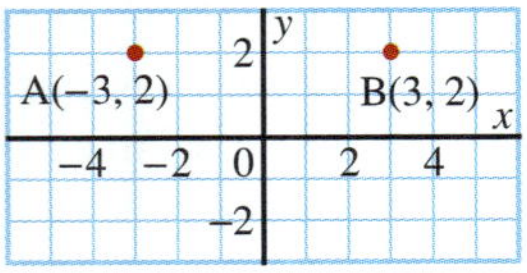

equilibrant: a force, equal and opposite to the resultant of forces acting on an object so that the object does not move; see *Tutorial 7.2*

equity: the difference between the market value of real estate and the amount still owing; see *Tutorial 4.6*

equity investments: stocks and mutual funds; see *Tutorial 4.3*

evaluate an expression: substitute a value for each variable in the expression, then calculate the resulting arithmetic expression applying the order of operations rules

Evaluate $2x^2 + 3y - 4$, if $x = -3$ and $y = 5$.

Replace each letter with its value, placing each number in parentheses to prevent errors with signs.

$$\begin{aligned} 2x^2 + 3y - 4 &= 2(-3)^2 + 3(5) - 4 \\ &= 2(9) + 3(5) - 4 \\ &= 18 + 15 - 4 \\ &= 29 \end{aligned}$$

event: any outcome, or set of outcomes, of an experiment; see *Tutorial 1.1*

expected value: the number that would be expected to be the average when an experiment is repeated many times; the mean of a probability distribution; see *Tutorial 3.1*

expenses: items that must be paid from income; for example, food, shelter, transportation

experiment: a procedure, carried out under controlled conditions, that is used to test a hypothesis

experimental probability: probability determined using sampling or a simulation; see *Tutorial 1.1*

exponent: a number, shown in a smaller size and raised, that tells us how many times the number before it is used as a factor

2 is the exponent in 6^2.

The laws of exponents are given below left, with examples to the right.

$a^m \cdot a^n = a^{m+n}$	$x^3 \cdot x^5 = x^{3+5} = x^8$
$(a^m)^n = a^{mn}$	$(x^3)^4 = x^{3 \cdot 4} = x^{12}$
$(ab)^m = a^m b^m$	$(4x)^3 = 4^3 \cdot x^3 = 64x^3$
$\left(\frac{a}{b}\right)^m = \frac{a^m}{b^m}$	$\left(\frac{x}{2}\right)^4 = \frac{x^4}{2^4} = \frac{x^4}{16}$
$\frac{a^m}{a^n} = a^{m-n}$	$\frac{x^8}{x^2} = x^{8-2} = x^6$
$a^{-m} = \frac{1}{a^m}$	$2^{-4} = \frac{1}{2^4} = \frac{1}{16}$
$a^0 = 1$	$7x^0 = 7 \cdot 1 = 7$

expression: a meaningful combination of mathematical symbols, such as a polynomial.

$3x - 2$ is an expression, as is $-13x^2 - 5x + 6$.

extrapolate: estimate a value beyond the known values

To use the graph below to estimate the distance travelled after 10 h requires extrapolation; the last known value occurs when the time is 8 h.

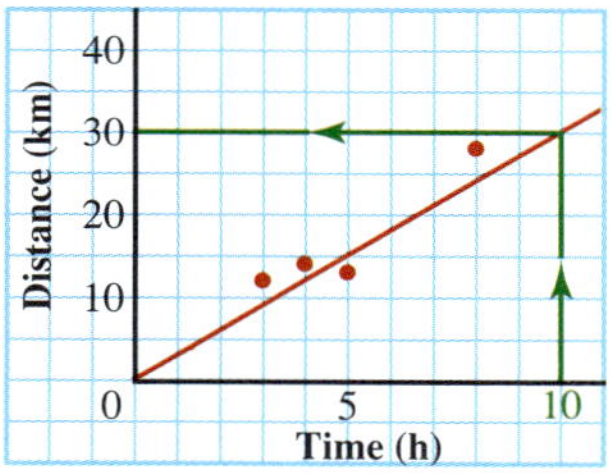

extremes: the highest and lowest values in a set of numbers

fair market value: an expert's determination of the value of a saleable item

Fibonacci sequence: a sequence of numbers in which each term after the second is formed by adding the two preceding terms; see *Tutorial 6.1*

1, 1, 2, 3, 5, 8, 13, 21, ...

fixed income investments: provide a source of regular income with limited risk; includes bonds, debentures, and guaranteed investment certificates; see *Tutorial 4.3*

force: a push or a pull on an object in a certain direction

A force of 1 Newton (N) will cause a 1-kg object to accelerate at a rate of 1 m/s^2.

formula: an equation that is used to describe the relationship between two or more quantities

The formula that describes how the volume, V, of a sphere is related to its radius, r, is $V = \frac{4}{3}\pi r^3$.

formula rearrangement: changing a formula to an equivalent form using the rules of equality

1. Rearrange the equation $y = mx + b$ to isolate x.

$$y = mx + b$$

Subtract b from both sides of the equation.

$$y - b = mx$$

Divide by m to isolate x.

$$\frac{y-b}{m} = \frac{mx}{m}$$

$$\frac{y-b}{m} = x$$

2. Make d the subject of the formula $F = \frac{12g}{d^2}$.

$$F = \frac{12g}{d^2}$$

Clear the fraction by multiplying both sides of the equation by d^2.

$$Fd^2 = 12g$$

Divide both sides of the equation by F to isolate d^2.

$$\frac{Fd^2}{F} = \frac{12g}{F}$$

$$d^2 = \frac{12g}{F}$$

Take the square root of both sides of the equation to isolate d.

$$d = \pm\sqrt{\frac{12g}{F}}$$

fractals: geometric figures that can be generated by repeating the same process many times; see *Tutorial 6.3*

frequency: the number of times an event occurs in an experiment; the number of times that something occurs in a given time; a common unit is the Hertz (Hz), which is the number of cycles that occur in 1 s

If you eat breakfast, lunch, and dinner every day, your meal frequency is 3 times a day.

If a pendulum swings back and forth 10 times in 5 s, its frequency would be

10 cycles/5 s = 2 cycles/s

= 2 Hz

function: a rule that gives a single output number for each input number

$y = 3x + 2$ and $y = 5x^2$ are functions.

function notation: the use of the function name, such as *f*, to indicate the output value for a particular input

$f(x) = 3x + 2$

Fundamental Counting Principle: if one item can be selected in *m* different ways and a second item can be selected in *n* different ways, then the two items can be selected in $m \times n$ different ways; see *Tutorial 1.3*

genotype: the pair of alleles that determine a genetic characteristic

geometric sequence: a sequence of numbers in which each term after the first is formed by multiplying the preceding term by a constant; see *Tutorial 6.1*

geometric vectors: a directed line segment used to represent a vector quantity; the length represents the magnitude and the arrowhead indicates the direction; see *Tutorial 7.1*

gradien: a unit of angle measure; 400 grad = 360°; see *Tutorial 5.2*

gross income: the amount of money earned before deductions

gross profit: the difference between the cost price and selling price of an item; also called the *markup*

Lori's Fashions buys a coat for $29.00 and sells it for $49.99.

The gross profit on the coat is $(49.99 – 29.00), or $20.99.

Guaranteed Investment Certificates: a type of investment offered at banks and trust companies which usually offers a higher rate of return than a bank account, but is for a fixed period of time and is not cashable until the expiration of the fixed period

half-life: the time taken for an item to reduce its quantity by half

head-to-tail: two vectors are drawn head-to-tail if the second vector begins where the first vector ends; see *Tutorial 7.2*

heading: the direction toward which a vehicle is steered; see *Tutorial 7.4*

hectare: a metric unit used for land area; the area of a square of side 100 m, so 1 hectare = 10 000 m^2

histogram: a graph that uses bars, where each bar represents a range of values

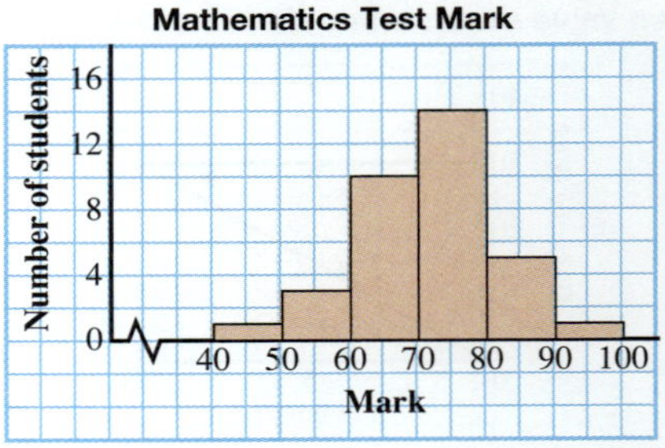

horizontal intercept: the horizontal coordinate of the point(s) where the graph of the line or function intersects the horizontal axis; see *intercepts*

hypotenuse: the side opposite the right angle in a right triangle

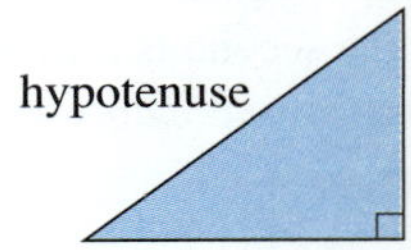

imperial system: a system of measures that was used in Canada prior to 1976; a variant is still used in the U.S.A.; measuring devices using this system often have each unit subdivided by halving, then halving the subdivisions

Relationship between common imperial units	Relationship between common imperial units and SI units
Length	
1 mile = 1760 yards	1 mile = 1.609 km
1 yard = 3 feet	1 yard = 0.9144 m
1 foot = 12 inches	1 inch = 2.54 cm
Capacity (Volume)	
1 gallon = 4 quarts	1 gallon = 4.546 L
1 quart = 2 pints	
Mass (Weight)	
1 ton = 2000 pounds	1 pound = 0.454 kg
1 pound = 16 ounces	1 ounce = 28.35 g

income tax: money paid as tax to the federal and provincial governments based on the amount of income earned

inconsistent system of equations: a system of linear equations with no solutions

$x + y = 7$ and $x + y = 14$ is an inconsistent system. The solution represents distinct parallel lines.

independent events: two or more events for which the occurrence or nonoccurrence of one does not affect the occurrence of the others; see *Tutorial 1.4*

independent variable: the input variable in a relation, often called x; also called the manipulated variable

index: used to measure the performance of the stock market; common indices are the TSE 300 and the Dow Jones

inferential statistics: statistics calculated for a set of data collected from a sampling of the population; used to draw conclusions about the population; see *Tutorial 1.1*

initial probability matrix: a row matrix containing the probabilities at the beginning of an experiment; see *Tutorial 2.4*

integers: see *number systems*

intercepts: the horizontal and vertical coordinates of the points at which a graph crosses the horizontal and vertical axes

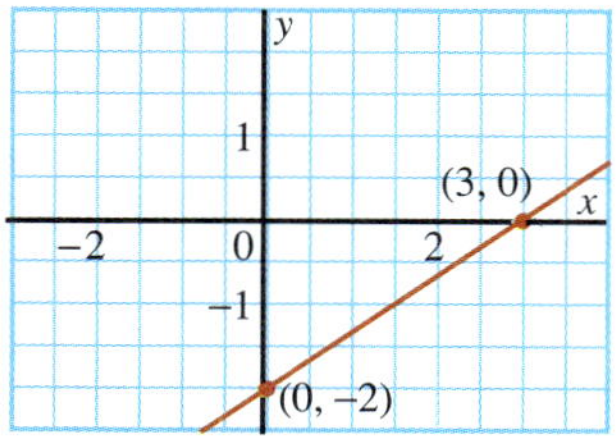

The x-intercept is 3 and the y-intercept is −2.

interest (simple): money paid for the use of money, usually at a predetermined percent. If P is the amount invested or borrowed, r, the rate of interest per annum, and t, the time in years, then I, the interest, is given by the formula $I = Prt$.

Juanita purchased a $500 bond at an interest rate of 6.5% per annum. After 6 months, she receives the following interest:

$$\begin{aligned} I &= \$500 \times 6.5\% \times \tfrac{6}{12} \\ &= \$500 \times 0.065 \times 0.5 \\ &= \$16.25 \end{aligned}$$

interpolate: estimate a value that lies between known values

To use the graph below to estimate the distance travelled after 6 h requires interpolation.

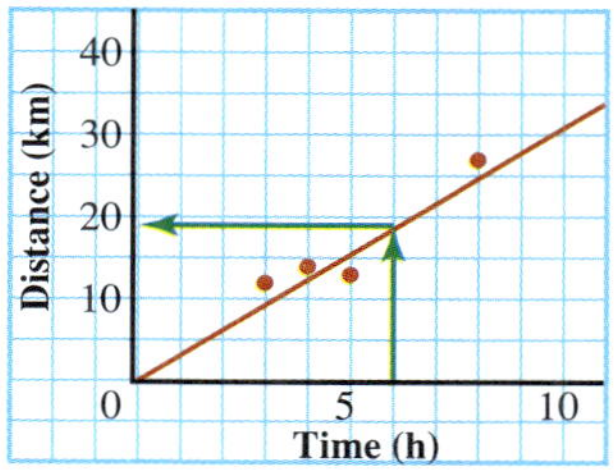

irrational numbers: see *number systems*

iterative procedure: a procedure in which a sequence of steps is repeated many times; see *Tutorial 6.3*

lease: to rent an item from the owner; the lease payments cover the depreciation of the item over the course of the lease plus interest on the outstanding balance of the full purchase price; see *Tutorial 4.4*

least squares method: for a set of data, a method that is used by many calculators and computers to determine a line or curve of best fit

limiting factor: a factor that restricts the number of possible outcomes of an experiment

line of best fit: a line that passes as close as possible to a set of plotted points

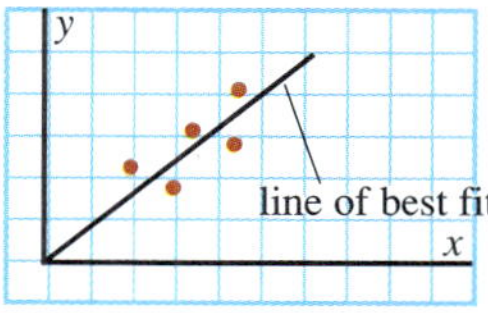

line segment: the part of a line between two points on the line, including the two points

linear equation and its graph: an equation that represents a straight line; can be written in the form $y = mx + b$, where m is the slope and b is the y-intercept, or $Ax + By + C = 0$, where A, B, and C are numbers

Slope–Intercept method of graphing: Graph the line $y = 3x + 2$.

This line has a slope of 3 and a y-intercept of 2. Mark the point (0, 2) on the y-axis. From that point, move up 3 for every 1 to the right. Mark this point, and then draw the line joining these points.

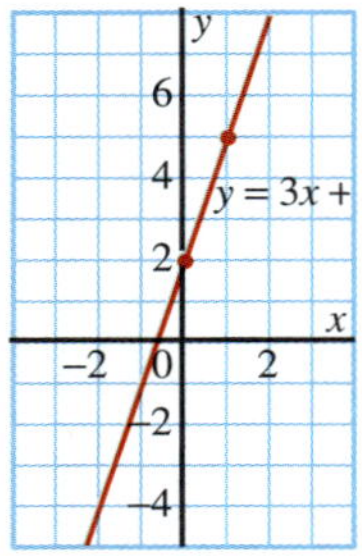

Intercept method of graphing: Calculate the points at which the line intersects the axes.

y-intercept: when $x = 0$, $y = 3(0) + 2 = 2$

x-intercept: when $y = 0$, $0 = 3x + 2$

$$-2 = 3x$$

$$-\frac{2}{3} = x$$

Plot, then join the intercepts $(-\frac{2}{3}, 0)$ and $(0, 2)$.

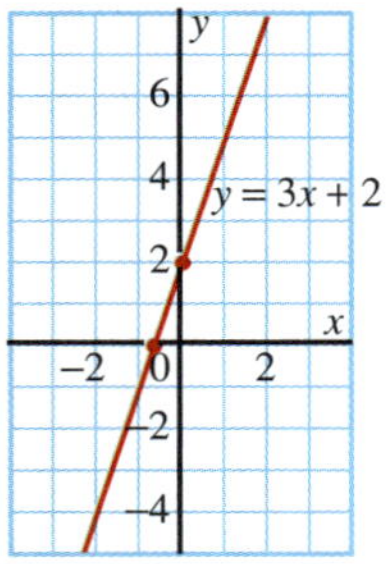

linear function: a function whose equation can be written in the form $y = mx + b$, and whose graph is a non-vertical line

linear system: two or more linear equations in the same variables

$5x - 2y = 7$ and $y = \frac{2}{3}x + 1$ form a linear system.

maintenance level: the maximum level of an ingested substance that will be absorbed into the bloodstream; see *Tutorial 6.2*

margin of error: the proportion that we add to and subtract from the result to construct the confidence interval; see *Tutorial 3.6*

marginal tax rate: the rate of income tax charged on the last dollar earned; as your income increases, the rate of tax you pay increases; see *Tutorial 4.2*

Mary paid a total of \$26 000 tax on her \$81 000 salary. She received a final bonus of \$1000, and paid \$450 tax on these last thousand dollars. This means that her marginal tax rate is \$450 per \$1000, or 45%.

markup: see *gross profit*

mass: a measure of the amount of material in an object; common units are grams or kilograms

matrix: a rectangular array of numbers; see *Tutorial 2.1*

mean: the average of a set of numbers; see *average*

measurement error: the difference between a measure and its true value

measurement standards: the standard set by the General Conference on Weights and Measures in 1889 so that all persons would be dealing in the same measurements

median: the middle number of a set of numbers arranged in numerical order; if there are two middle numbers, their average is the median of the data set

For the data 2, 4, 8, 9, and 11 the median is 8.

For the data 2, 4, 6, 8, 9, and 11 the median is 7, since $\frac{6+8}{2} = 7$.

meiosis: the process of cell division in which sperm and egg cells are produced

metric system: also called the *SI system*; based on a decimal system, with each unit subdivided into tenths and prefixes showing the relation of a unit to the base unit; commonly used base units are:

Metre (m) for length	Gram (g) for mass
Litre (L) for capacity	Second (s) for time

Prefix	Multiplying Factor	Symbol
mega	1 000 000	M
kilo	1000	k
hecto	100	h
deca	10	da
	1	
deci	0.1	d
centi	0.01	c
milli	0.001	m
micro	0.000 001	μ
nano	0.000 000 001	n

Note: 1 L = 1000 cm^3, so 1 mL = 1 cm^3

1 tonne (sometimes called a metric ton) is used for mass: 1 t = 1000 kg

Convert 34.6 cm to m:

from the table above, 1 cm = 0.01 m

$$34.6 \times 1 \text{ cm} = 34.6 \times 0.01 \text{ m}$$
$$34.6 \text{ cm} = 0.346 \text{ m}$$

Convert 246 cm^2 to mm^2:

from the table above, 1 cm = 10 mm

$$(1 \text{ cm})^2 = (10 \text{ mm})^2$$
$$1 \text{ cm}^2 = 100 \text{ mm}^2$$
$$246 \times 1 \text{ cm}^2 = 246 \times 100 \text{ mm}^2$$
$$246 \text{ cm}^2 = 24\,600 \text{ mm}^2$$

Convert 35 mL to L:

from the table above, 1 mL = 0.001 L

$$35 \times 1 \text{ mL} = 35 \times 0.001 \text{ L}$$
$$35 \text{ mL} = 0.035 \text{ L}$$

midpoint: the point that divides a line segment into two equal parts

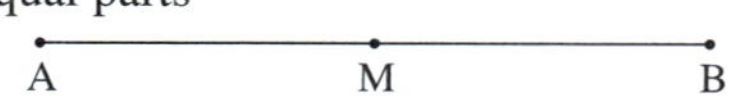

On a coordinate grid, if the endpoints are A(x_A, y_A) and B(x_B, y_B), the coordinates of M are:

$$x_M = \frac{x_A + x_B}{2} \qquad y_M = \frac{y_A + y_B}{2}$$

Find the midpoint of the line segment with endpoints P(−2, 8) and Q(3, 0).

$$\text{Midpoint} = \left(\frac{-2+3}{2}, \frac{8+0}{2}\right)$$
$$= \left(\frac{1}{2}, 4\right)$$

mill rate: the rate (in thousandths of a dollar) at which property tax is to be paid on the assessed value of the property

mode: the most frequently occurring value in a set of data

In the data set {5, 12, 8, 7, 3, 5, 3, 10, 5}, the mode is 5.

momentum: a vector quantity found by multiplying an object's mass by its velocity; see *Chapter 7 Project*

mortgage: a long-term loan on real estate that gives the person or firm providing the money a claim on the property if the loan is not repaid

mutual fund: an investment in which your money is pooled with the money of many other investors; a professional fund manager invests the monies in a variety of securities; see *Tutorial 4.3*

mutually exclusive events: two or more events that do not share any common outcomes; see *Tutorial 1.5*

natural numbers: see *number systems*

negative number: a number less than 0

95% confidence interval: the range of values that lie within 1.96 standard deviations of the mean; the probability a particular data value lies in that range is 0.95; see *Tutorial 3.6*

nominal rate of interest: the stated rate of interest for the given period; see *Tutorial 4.5*

non-linear systems: two or more equations in which at least one is not a linear equation

The system $y = 2x^2 + 3$ and $4x + 3y = 12$ is a non-linear system.

normal distribution: a probability distribution with mean μ and standard deviation σ; the bell-shaped graph is symmetrical about the mean; obeys the 68-95-99 rule; see *Tutorial 3.3*

number systems:

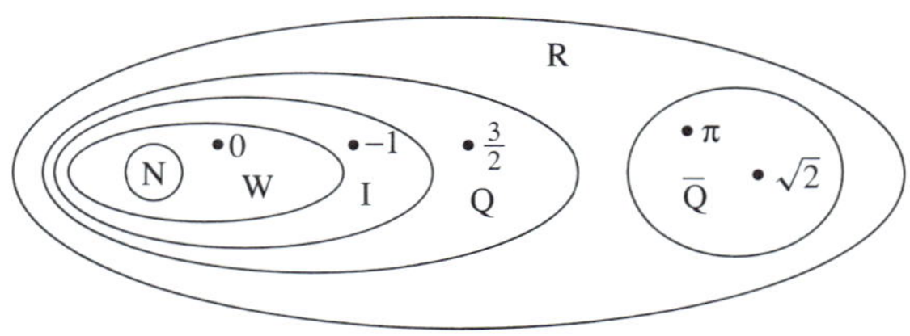

The *Natural numbers*, N, also called the counting numbers, are 1, 2, 3, 4, 5, ...

If the number 0 is included, we get the *Whole numbers*, W: 0, 1, 2, 3, 4, 5, ...

The *Integers*, I, contain all of the whole numbers and all of their negatives: ..., −3, −2, −1, 0, 1, 2, 3, ...

The set of numbers that can be written in the form $\frac{m}{n}$, where m and n are integers and $n \neq 0$ are the *Rational numbers, Q*. Integers are rational numbers, since they can be expressed as fractions with denominator 1. All terminating or repeating decimals are rational numbers.

Irrational numbers, $\overline{Q}$, cannot be expressed as fractions involving integers. They are non-terminating, non-repeating decimals. Numbers such as $\sqrt{2}$ and π are irrational.

The *Real number* system, R, consists of all rational numbers together with all irrational numbers. It can be represented by all positions on a number line.

oblique triangle: a triangle that does not contain a 90° angle

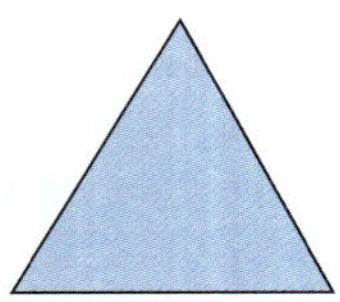

obtuse angle: an angle greater than 90° but less than 180°

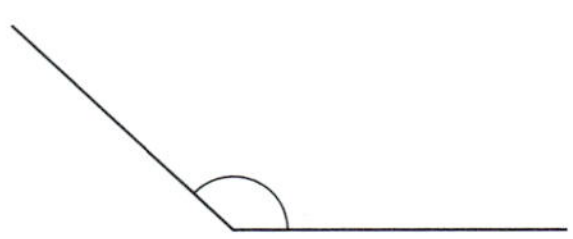

obtuse triangle: a triangle with one obtuse angle

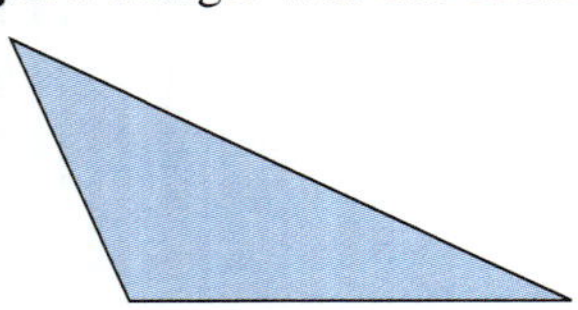

opposite side: the side opposite a given angle in a right triangle

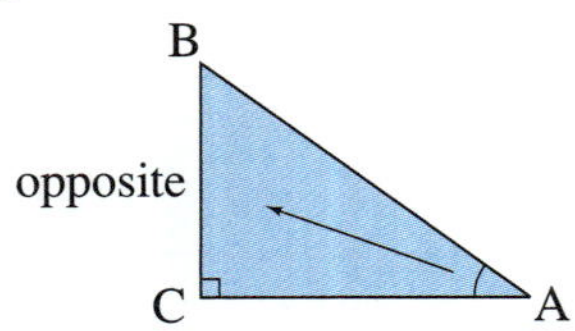

opposite vectors: have the same magnitude but act in opposite directions; see *Tutorial 7.1*

order of operations: the rules that are followed when simplifying or evaluating an expression:

Complete all operations within brackets following the order of operations.

Evaluate all exponents.

Complete all multiplication and division in the order they appear from left to right.

Complete all addition and subtraction in the order they appear from left to right.

Evaluate the following:

$5 - 2(4 + 2^3 \div 4)$	**Begin with the exponent within the brackets.**
$= 5 - 2(4 + 8 \div 4)$	**Complete the division within the brackets.**
$= 5 - 2(4 + 2)$	**Complete the addition within the brackets.**
$= 5 - 2(6)$	**Complete the multiplication.**
$= 5 - 12$	**Finish by completing the subtraction.**
$= -7$	

ordered pair: a pair of numbers, written as (x, y) that represent a point on the coordinate plane; see *coordinates*

orthographic diagram: a diagram commonly used in industry that shows at least three points of view: the top, front and side; also shows the internal features

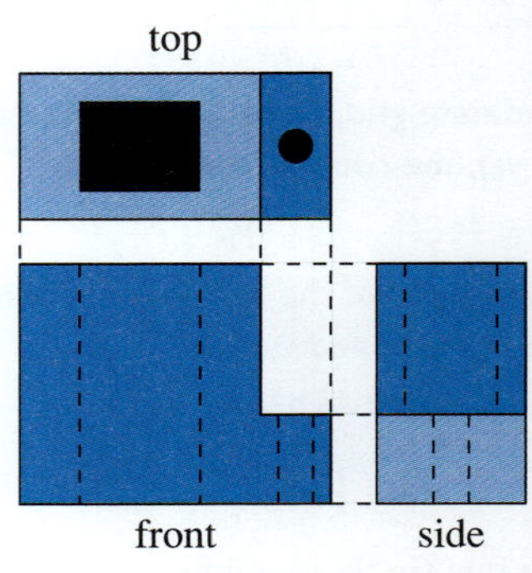

outcome: a possible result of an experiment; a possible answer to a survey question

For the experiment of tossing a six-sided die, the possible outcomes are rolling a 1, 2, 3, 4, 5, or 6.

outlier: an observed value that differs markedly from the pattern established by most of the data

parallel lines: lines in the same plane that do not intersect

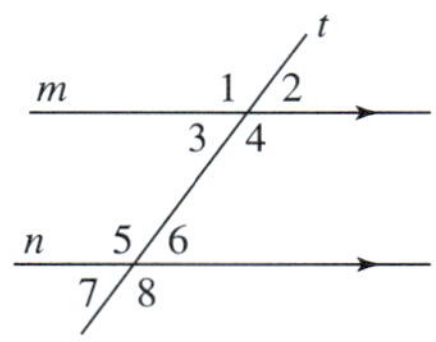

Lines *m* and *n* are parallel with a transversal, *t*. If two lines are parallel and cut by a transversal, then the following will be true:

The alternate interior angles will have equal measures (congruent). $\angle 3 = \angle 6$ and $\angle 4 = \angle 5$

The corresponding angles will have equal measures (congruent).
$\angle 1 = \angle 5$ and $\angle 2 = \angle 6$ $\quad \angle 3 = \angle 7$ and $\angle 4 = \angle 8$

The same-side interior angles add up to 180° (supplementary).
$\angle 3 + \angle 5 = 180°$ and $\angle 4 + \angle 6 = 180°$

If any one of these conditions is true then the lines must be parallel. Lines that are parallel will have the same slope.

parallelogram: a quadrilateral with opposite sides parallel

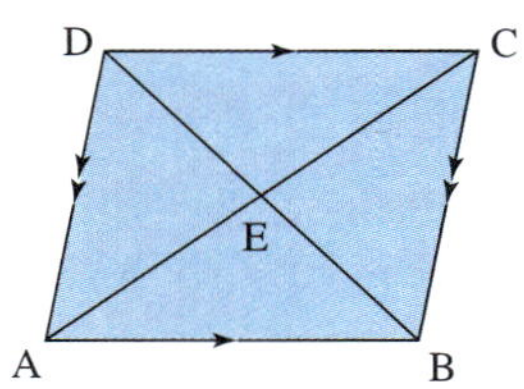

A parallelogram has these properties:

The opposite sides have equal lengths.
AB = CD and AD = BC

The opposite angles have equal measures (congruent). $\angle A = \angle C$ and $\angle B = \angle D$

The diagonals bisect each other (cut each other into equal lengths).
AE = EC and DE = EB

parallelogram method of vector addition: a method for finding the sum of two vectors arranged tail-to-tail; see *Tutorial 7.2*

parameters: the constants in the equation of a particular function; see *Tutorial 5.4*

percent: means "out of 100"; symbolized by %

A percent can be written as a fraction with denominator 100, or as a decimal.

$$45\% = \frac{45}{100} = 0.45 \text{ and } 150\% = \frac{150}{100} = 1.50$$

1. What percent is 33 out of 48?

We must determine *x*, where $\frac{33}{48} = \frac{x}{100}$.

To isolate *x*, multiply both sides by 100.

$$100 \times \frac{33}{48} = 100 \times \frac{x}{100}$$
$$68.75 = x$$

Thus, 33 out of 48 is 68.75%.

2. Find 7% of 45.95.

7% of 45.95 = 0.07 × 45.95, changing the percent to its decimal form
= 3.2165

Thus, 7% of 45.95 is 3.2165.

3. 7.5 is 20% of what number?

Let the unknown number be *x* and write the question as:

$$7.5 = 20\% \text{ of } x$$

Use the decimal form of the percent.

$$7.5 = 0.20 \times x$$

Solve for *x* by dividing both sides of the equation by 0.20.

$$\frac{7.5}{0.20} = \frac{0.20x}{0.20}$$
$$37.5 = x$$

Thus, 7.5 is 20% of 37.5.

perimeter: the distance around a closed figure; see *square*, *rectangle*, *triangle*, and *circle*

period: the length of the shortest part of the graph that repeats, measured along the horizontal axis; see *Tutorials 5.1*, *5.2*, *5,3*, and *5.4*

periodic data: data that contain cycles that repeat at regular intervals; see *Tutorial 5.1*

perpendicular: at right angles

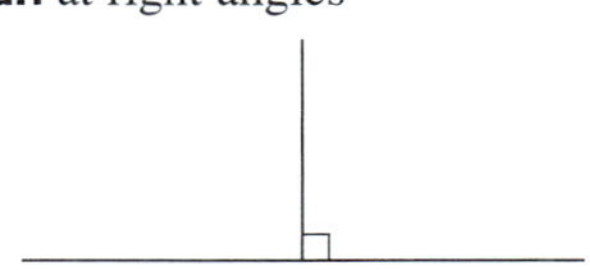

Two lines are perpendicular if their slopes are negative reciprocals of one another.

phenotype: the appearance of a genetic trait

pi (π): the ratio of the circumference of a circle and its diameter; $\pi \doteq 3.1416$

pictorial diagram: a two dimensional representation of a three dimensional figure drawn from one perspective

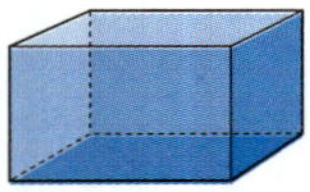

polygon: a closed figure that consists of line segments that only intersect at their endpoints

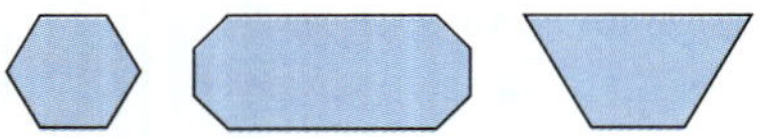

The above figures are polygons.

The table below gives the names of some common polygons.

Number of Sides	Polygon
3	Triangle
4	Quadrilateral
5	Pentagon
6	Hexagon
8	Octagon
10	Decagon
n	n-gon

population: the entire set of objects, people, or processess being studied

positive number: a number greater than 0

power: see *exponent*

pressure: a measure of the amount of force acting on a unit area of a surface; calculated by dividing the force acting on a surface by the area of the surface

If a force of 20 N is acting on a surface whose area is 4 m², what is the pressure on the surface?

$$\text{Pressure} = \frac{\text{Force}}{\text{Area}}$$

$$P = \frac{20}{4}$$

$$= 5 \text{ N/m}^2$$

prime numbers: a whole number with exactly two factors, itself and 1; for example, 3, 5, 7, 11, 29, 31, and, 43

principal: the original amount of a loan or investment

Suppose you invest \$5000 at 7%, compounded annually, for 5 years. The principal is \$5000.

prism: a solid with two congruent and parallel faces (bases), all other faces are parallelograms

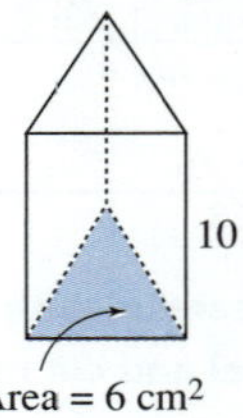

The volume of a prism is the area of the base multiplied by the height.

$$\text{Volume} = (\text{Area of base})(\text{Height})$$

$$V = A \times h$$

For the prism shown:

$$V = (6 \text{ cm}^2)(10 \text{ cm})$$

$$= 60 \text{ cm}^3$$

probability: an indication of the likelihood of an event occurring; if the outcomes of an experiment are equally likely, it is the ratio of the number of favourable outcomes to the total number of outcomes

Find the probability of getting a 5 in a single roll of one die.

There are 6 outcomes, only one of which is favourable (getting a 5), so the probability is $\frac{1}{6}$.

probability distribution: a probability distribution gives the theoretical probability for each outcome of an experiment; see *Tutorial 3.1*

property taxes: property owners pay tax to their municipal government; the amount of property tax is calculated using the formula $\frac{\text{Assessed value} \times \text{Mill rate}}{1000}$ where the assessed value is a percent of the fair market value and the mill rate is determined locally

Determine the property taxes for a property that has a fair market value of \$265 000 if the assessed value is 75% of the fair market value and the mill rate is 18 mills.

$$\text{Assessed value} = 0.75 \times 265\,000$$

$$= \$198\,750$$

Property tax = $\frac{198\,750 \times 18}{1000}$

= \$3577.50

The property tax is \$3577.50

proportion: a statement that two ratios are equal

To solve for an unknown quantity within a proportion, first clear the fraction by multiplication.

$$\frac{x}{12} = \frac{2}{5}$$
$$12 \times \frac{x}{12} = 12 \times \frac{2}{5}$$
$$x = 4.8$$

$$\frac{3}{x} = \frac{4}{5}$$
$$5x \times \frac{3}{x} = 5x \times \frac{4}{5}$$
$$15 = 4x$$
$$\frac{15}{4} = \frac{4x}{4}$$
$$3.75 = x$$

Punnett Square: a table showing the sample space for the possible genotypes of the offspring of parents with given genotypes

pyramid: a solid with one face that is a polygon (base) and other faces that are triangles with a common vertex

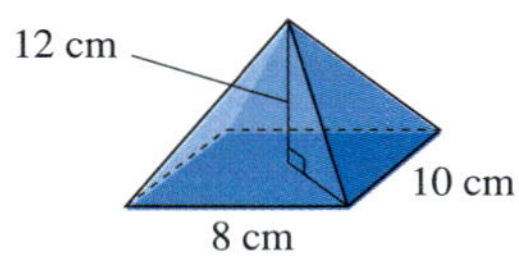

The formula to calculate the volume, V, of a pyramid is:

Volume = $\frac{1}{3}$(Base area)(Height)

For the pyramid shown:

$$V = \frac{1}{3}(8\text{ cm} \times 10\text{ cm})(12\text{ cm})$$
$$= 320\text{ cm}^3$$

Pythagorean Theorem: for any right triangle, the area of the square on the hypotenuse is equal to the sum of the areas of the squares on the other two sides

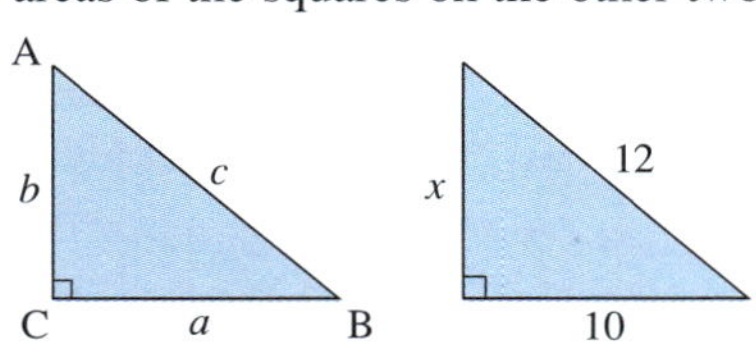

In right triangle ABC: $a^2 + b^2 = c^2$.

Using the values for the right triangle shown, substitute into the above formula.

$$10^2 + x^2 = 12^2$$
$$100 + x^2 = 144 \quad \text{Simplify each side.}$$
$$100 + x^2 - 100 = 144 - 100 \quad \text{Isolate the variable.}$$
$$x^2 = 44 \quad \text{To solve, take the}$$
$$x \doteq 6.6 \quad \text{square root of both sides.}$$

quadrant: one of the four regions into which the coordinate axes divide the plane, usually numbered as shown in the diagram

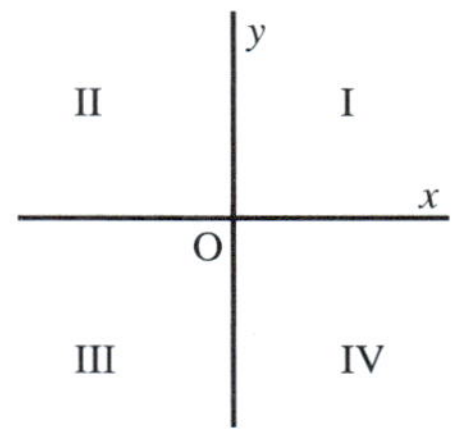

quadrilateral: a four-sided polygon; see *polygon*

radian: a unit for angle measure; the measure of the central angle of a sector with arc length of the sector equal to the radius; see *Tutorial 5.2*

π **radians = 180°**

radical: the root of a number

$\sqrt{5}$, $\sqrt[3]{5}$, $\sqrt{2.6}$

radius: the distance from the centre of a circle to any point on the circumference; also, a line segment joining the centre of a circle to any point on the circumference

The radius of a circle is half the length of its diameter.

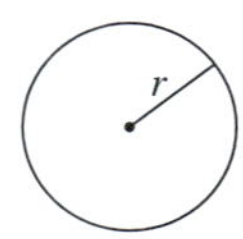

random sample: a sampling in which all members of the population have an equal chance of being selected

range: the difference between the highest and the lowest data values

For the data set {5, 12, 8, 7, 3, 5, 3, 10, 5}, the range is 12 – 3 = 9.

range of a function or a relation: the set of output numbers of a function or a relation

rate: a certain quantity of one thing considered in relation to one unit of another

Speed is the rate at which distance travelled changes in relation to one unit of time.

The slope of a line is the rate at which the line rises for one unit of horizontal run.

ratio: a comparison of two or more quantities with the same unit

A rectangular box has length 18 cm, width 5 cm, and height 8 mm. To give the ratio of the lengths of its sides, first change all lengths to the same units, mm, then reduce by dividing by common factor:

$$180 : 50 : 8 = 90 : 25 : 4$$

rational numbers: see *number systems*

real numbers: see *number systems*

recessive trait: in genetics, both alleles must be recessive for the trait to be expressed

rectangle: a quadrilateral that has four right angles

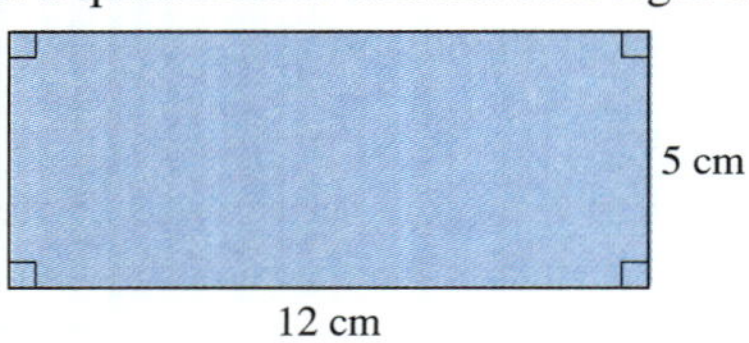

Perimeter = 2 × Length + 2 × Width

$$P = 2l + 2w$$

Area = Length × Width

$$A = lw$$

For the rectangle shown:

$$P = 2(12\text{ cm}) + 2(5\text{ cm}) = 24\text{ cm} + 10\text{ cm} = 34\text{ cm}$$

$$A = (12\text{ cm})(5\text{ cm}) = 60\text{ cm}^2$$

rectangular prism: a prism that has rectangular faces; see *prism*

recursively generated sequence: a sequence in which each term, after the first, is generated using the terms that come before it

The sequence 3, 6, 12, 24 is recursive, since each successive term after the first is twice the previous one.

Registered Retirement Savings Plan (RRSP): a savings plan for individuals who earn income, where funds contributed and interest earned are not taxed until the funds are withdrawn

regression: a process by which a curve or line of best fit is fitted to a set of data using a least squares process

regular polygon: a polygon with all sides and all angles equal

The polygons below are regular polygons.

relation: a rule that produces one or more output numbers for every valid input number

Here are two relations: $25 = x^2 + y^2$ and $f(x) = 3x - 7$.

reliability: for a sample statistic, if nearly the same result is obtained in subsequent studies; for a test, if, when the test is repeated, the results are nearly the same

remove fractions from an equation: to make an equation easier to simplify by getting rid of the fractions

Solve:

$$\frac{5}{2}x - 3x + \frac{2}{7}x = 9$$

$$2\left(\frac{5}{2}x\right) - (2)3x + 2\left(\frac{2}{7}x\right) = 2(9)$$ **Multiply each term by one denominator.**

$$5x - 6x + \frac{4}{7}x = 18$$ **Simplify.**

$$-x + \frac{4}{7}x = 18$$

$$7(-x) + 7\left(\frac{4}{7}x\right) = 7(18)$$ **Multiply each term by another value on the denominator.**

$$-7x + 4x = 126$$ **Simplify.**

$$-3x = 126$$

$$x = -42$$

residual: the vertical distance between a plotted point and a line of best fit

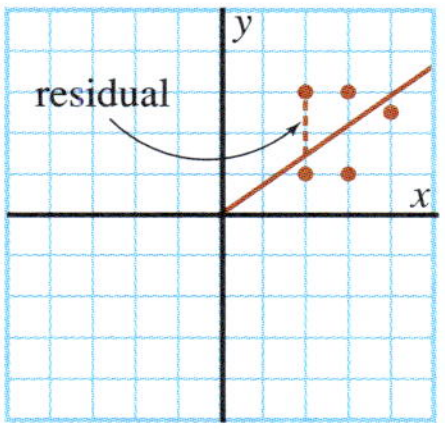

residual value: the value of a leased item at the end of the lease period; see *Tutorial 4.4*

resultant: the vector that results from adding two or more vectors; see *Tutorial 7.2*

revenue: money brought in from the sale of goods or services

The sale of 158 tickets for a show costing $20 each produces revenue of $3160.

rhombus: a parallelogram with four equal sides

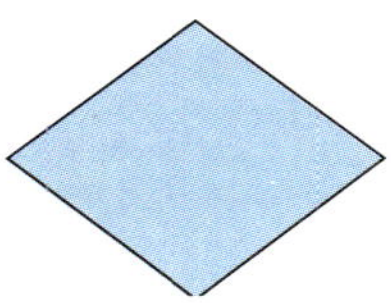

right angle: a 90° angle

right triangle: a triangle with one right angle

rise: the vertical distance between two points; see *slope*

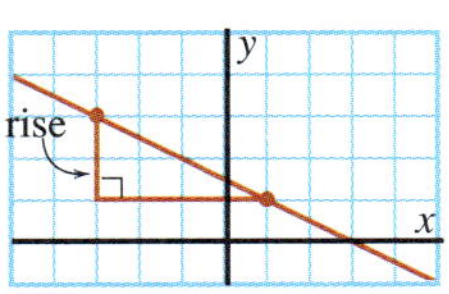

rough sketch: a drawing that is not accurate, but is representative of the data being considered

row matrix: a matrix with only one row; see *Tutorial 2.1*

Rule of 72: the number of years it takes money to double in value is approximately $\frac{72}{\text{Annual interest rate}}$

Susan invests a sum of money at 10%/a. She will double her money in about $\frac{72}{10}$ or 7.2 years.

run: the horizontal distance between two points; see *slope*

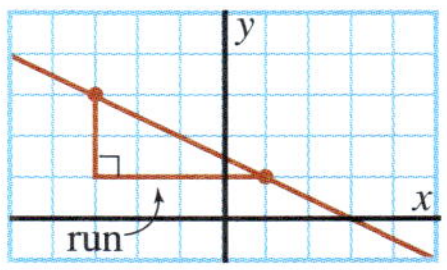

SI system: see *metric system*; SI stands for Système Internationale des unités

salary: a fixed amount of money earned by an individual over a specific amount of time

sample: part of a population chosen to represent the total population in a study

sample space: the list of all possible outcomes in an experiment; see *Tutorial 1.2*

sample statistic: a numerical value that is used to describe a sample, such as the *mean*

sampling: the process used to choose part of a population to represent the total population in a study

scalar: a constant; see *Tutorials 2.1* and *7.3*

scalar quantities: quantities that can be described by specifying their magnitude only; see *Tutorial 7.1*

scale: for a map, model, or diagram, the ratio of the distance between two points to the distance between the actual locations; also the numbers on the coordinate axes

scale break: a small zig-zag mark on a graph's scale to indicate a break in the scale

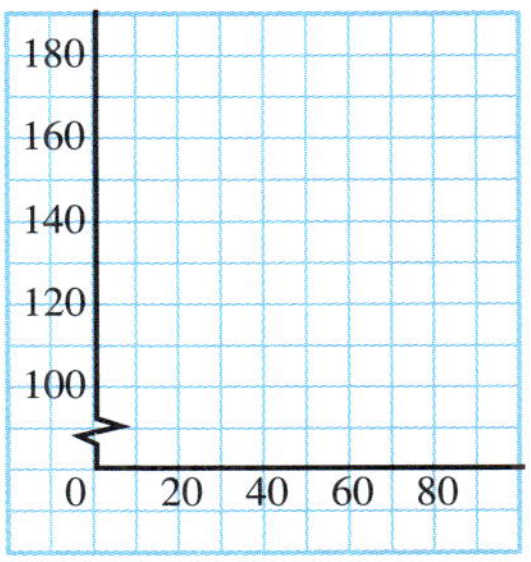

scale drawing: a drawing which either enlarges or reduces the original by a particular factor; see *scale factor*

scale factor: the ratio of corresponding sides on similar figures

The triangles in the diagram are similar. To find the scale factor, find the ratio of the corresponding sides.

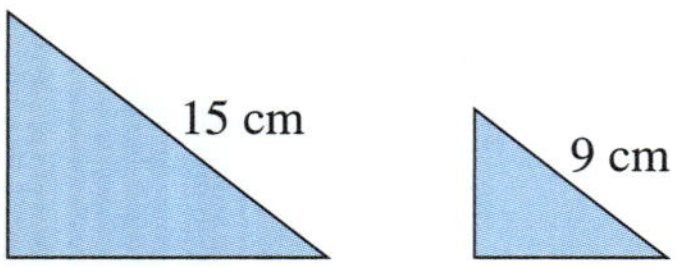

Scale factor $= \frac{9 \text{ cm}}{15 \text{ cm}}$

$= \frac{3}{5}$ **Divide by the common factor 3.**

$= 0.6$

scatterplot: a graph of data that are a series of points

Height (cm)	154	162	172	178
Mass (kg)	56.3	60.1	72.2	64.3

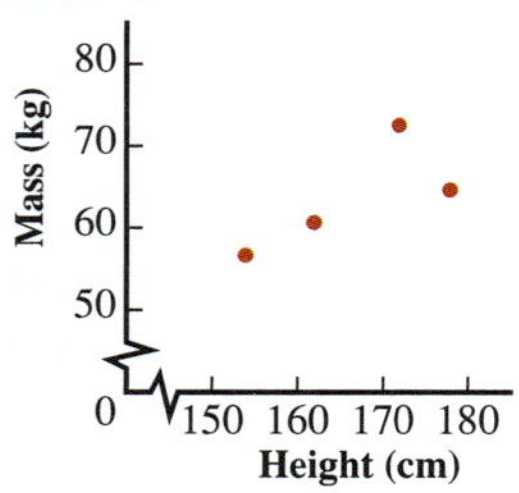

scientific notation: a number expressed as the product of a number greater than –10 and less than –1, or greater than 1 and less than 10, and a power of 10

$47\,000 = 4.7 \times 10^4$, **and** $-26 = -2.6 \times 10^1$

sector: the figure formed by an arc of a circle, the radii at the ends of the arc, and all the enclosed points; see *Tutorial 5.2*

sector angle: the angle at the centre of the circle between two radii; see *central angle*; see *Tutorial 5.2*

self-selected sample: a sample in which only interested members of the population will participate

self-similarity: a part of an object is geometrically similar to the whole object; see *Tutorial 6.3*

semicircle: half a circle

significant digits: the meaningful (non-zero) digits of a number

The measurement 120.7 cm has 4 significant digits. The measurement 0.004 mm has 1 significant digit; the zeros are acting as placeholders.

similar figures: figures that have the same shape but are not necessarily the same size

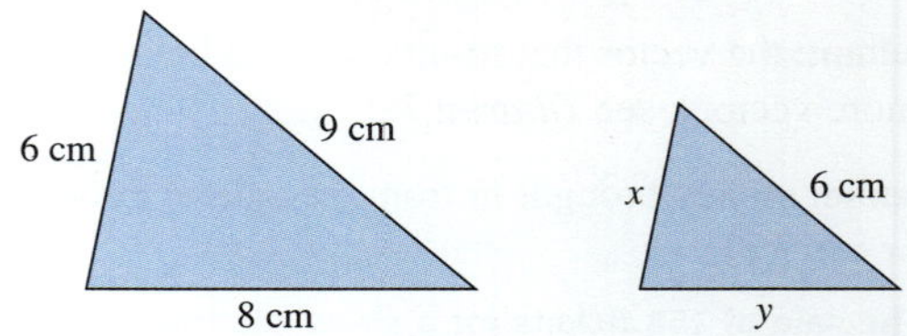

When two figures are similar, their corresponding angles will have equal measures, and their corresponding sides are in proportion (all have the same scale factor). The symbol ~ is used to indicate two figures are similar.

To find an unknown side of one similar figure, use a proportion.

$$\frac{9}{6} = \frac{8}{y} = \frac{6}{x}$$

$$\text{Scale factor} = \frac{9}{6} = \frac{3}{2} = 3:2$$

To find x, solve the following proportion.

$$\frac{3}{2} = \frac{6}{x}$$
$$2x \times \frac{3}{2} = 2x \times \frac{6}{x}$$
$$3x = 12$$
$$x = 4$$

To find y, solve the following proportion.

$$\frac{3}{2} = \frac{8}{y}$$
$$2y \times \frac{3}{2} = 2y \times \frac{8}{y}$$
$$3y = 16$$
$$y \doteq 5.33$$

similar objects: objects that have the same shape but are not necessarily the same size

All the corresponding angles will have equal measures and all dimensions will be proportional.

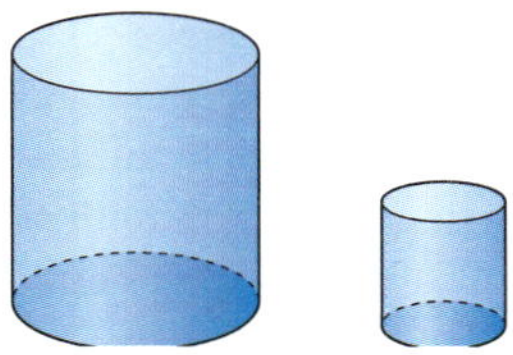

simple interest: see *interest* (simple)

sine: for an acute angle ∠A in a right triangle, the ratio of the length of the opposite side to the length of the hypotenuse

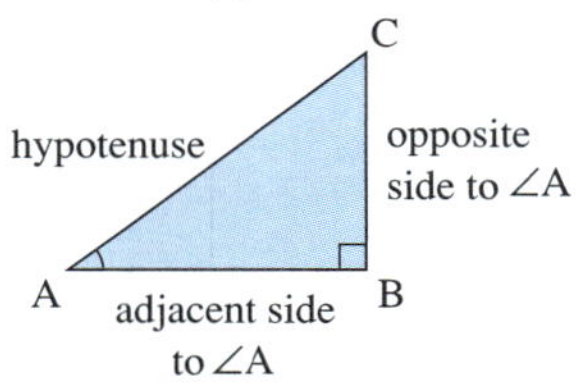

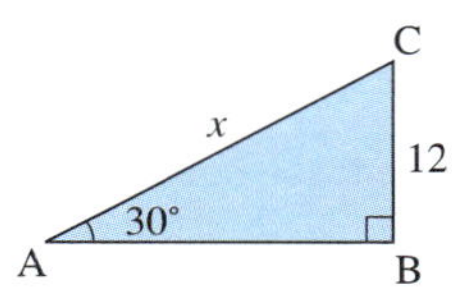

$$\sin A = \frac{\text{Length of opposite side}}{\text{Length of hypotenuse}}$$

$\sin 30^\circ = \frac{12}{x}$ — **Substitute in the known quantities.**

$0.5 = \frac{12}{x}$ — **Use a calculator to determine the sine ratio.**

$x \times 0.5 = x \times \frac{12}{x}$ — **Multiply to clear the fraction.**

$0.5x = 12$

$\frac{0.5x}{0.5} = \frac{12}{0.5}$ — **Divide to isolate *x*.**

$x = 24$

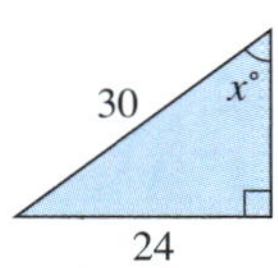

$\sin A = \frac{\text{Length of opposite side}}{\text{Length of hypotenuse}}$ — **Substitute in the known quantities.**

$\sin x^\circ = \frac{24}{30}$ — **Convert the fraction into an equivalent decimal.**

$\sin x^\circ = 0.8$

$x \doteq 53.1^\circ$ — **Use $\sin^{-1}$ ("the angle whose sine is") function.**

Sine Law: a trigonometric law used to solve triangles

Use the Sine Law in an oblique triangle where 2 angles and one side are known.

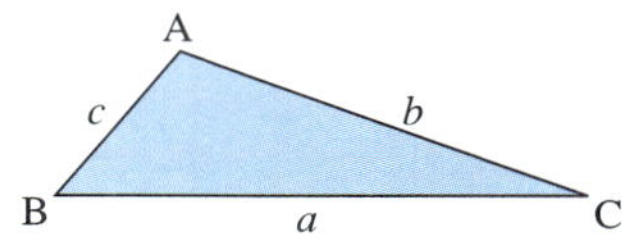

In any triangle ABC: $\frac{\sin A}{a} = \frac{\sin B}{b} = \frac{\sin C}{c}$ **and** $\frac{a}{\sin A} = \frac{b}{\sin B} = \frac{c}{\sin C}$

Use the Sine Law to find the length of AB.

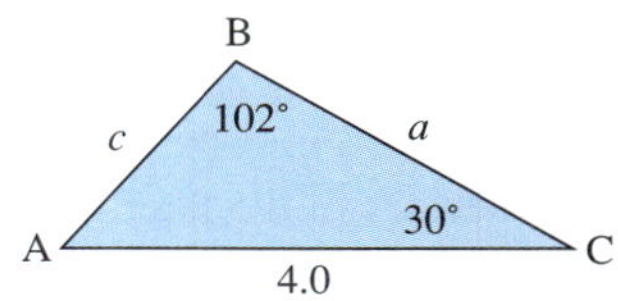

$\frac{b}{\sin B} = \frac{c}{\sin C}$

$\frac{4.0}{\sin 102^\circ} = \frac{c}{\sin 30^\circ}$

$\frac{4.0 \times \sin 30^\circ}{\sin 102^\circ} = c$

$2.0 \doteq c$

sinusoid: see *sinusoidal function*

sinusoidal data: periodic data that, when graphed, looks like a sine curve; see *Tutorial 5.1*

sinusoidal function: functions of curves that can be described by an equation of the form $y = a\sin(bx + c) + d$; see *Tutorial 5.3*

sinusoidal regression: determining the equation of the sinusoid of best fit for given data; see *Tutorial 5.5*

68-95-99 rule: about 68% of the population are within 1 standard deviation of the mean; about 95% of the population are within 2 standard deviations of the mean; about 99.7% of the population are within 3 standard deviations of the mean; see *Tutorials 3.3* and *3.4*

slope: a measure of the steepness of a line; the tangent of the angle made by the line with the *x*-axis

The slope of a line segment joining A(x_A, y_A) and B(x_B, y_B) is:

$$\text{Slope} = \frac{\text{Rise}}{\text{Run}} = \frac{y_A - y_B}{x_A - x_B}$$

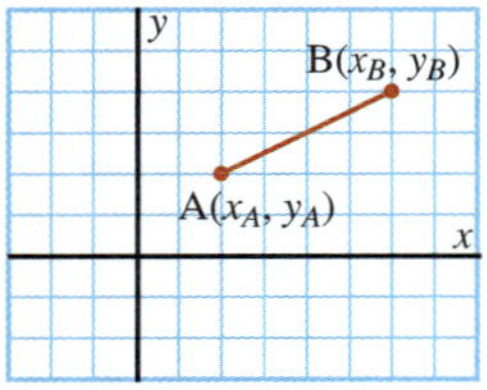

Find the slope of the line that passes through the points M(5, –1) and N(1, 4)

$$\text{Slope} = \frac{-1-4}{5-1}$$
$$= \frac{-5}{4}$$

slope, *y*-intercept form: the equation of a line in the form $y = mx + b$, where m is the slope of the line and b is the y-intercept of the line; see *linear equation and its graph*

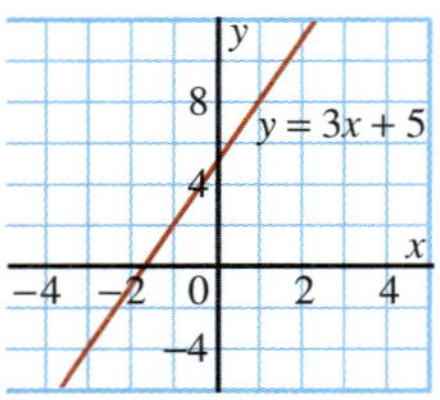

The equation $y = 3x + 5$ is that of a line with a slope of 3 and a y-intercept of 5.

solve a linear system: to find all the ordered pairs that satisfy both equations in the linear system

Solve the system: $2x + y = 5$ ①
$4x - y = 7$ ②

$2x + y = 5$ **Add the two equations.**
$+ \quad 4x - y = 7$
$2x + 4x = 5 + 7$
$6x = 12$
$x = 2$

Substitute $x = 2$ in either of the original equations.
$4(2) - y = 7$
$8 - y = 7$
$8 - 7 = y$
$1 = y$

Therefore, (2, 1) solves the system.

solve an equation: to determine the value(s) of the unknown that, when substituted for the unknown in the equation, result(s) in a true statement

Solve the equation $3x + 5 = 5x - 4$.

Use inverse operations to move all the terms in the variable to one side of the equation.

$3x + 5 - 3x = 5x - 4 - 3x$ **Subtract $3x$ from both sides.**
$3x - 3x + 5 = 5x - 3x - 4$ **Rearrange to collect like terms.**
$5 = 2x - 4$ **Simplify.**
$5 + 4 = 2x - 4 + 4$ **Isolate the term in x; add 4 to both sides.**
$9 = 2x$ **Simplify.**
$\frac{9}{2} = \frac{2x}{2}$ **Isolate x; divide both sides by 2.**
$4.5 = x$

speed: see *average speed*

sphere: the set of points in space that are a given distance (radius) from a fixed point (centre)

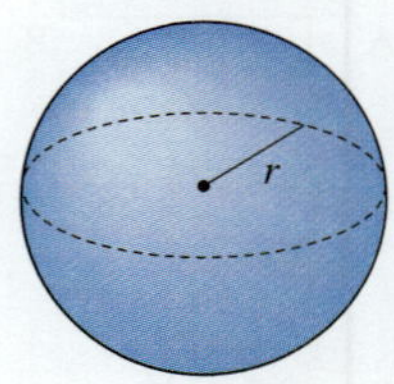

Surface Area, $S = 4\pi r^2$
$= \pi d^2$

Volume, $V = \frac{3}{4}\pi r^3$
$= \frac{\pi d^3}{6}$

spreadsheet: a computer-generated arrangement of data in rows and columns, where a change in one value can result in appropriate calculated changes in the other values

square: a rectangle with four equal sides

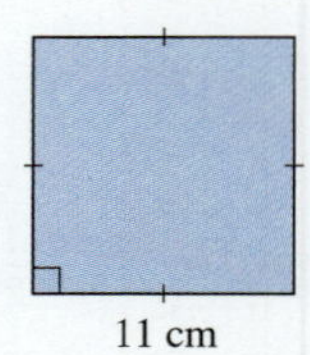

Perimeter = 4 × Length of side
$P = 4s$

Area = (Length of side)^2
$A = s^2$

For the square shown:

$P = 4(11 \text{ cm})$ $A = (11 \text{ cm})^2$
$= 44 \text{ cm}$ $= 121 \text{ cm}^2$

square matrix: a matrix with the same number of rows as columns; see *Tutorial 2.1*

square root: a number which, when multiplied by itself, results in the given number

5 and –5 are the square roots of 25, since $5^2 = 25$ and $(-5)^2 = 25$. The notation $\sqrt{25}$ is reserved for the positive square root only.

standard deviation: a measure of the extent to which data cluster around the mean; see *Tutorial 3.2*

standard form: for the equation of a line, the standard form is $Ax + By + C = 0$, where A, B, and C are integers; see *equation of a line*

standard normal distribution: a normal distribution with mean 0 and a standard deviation of 1; the bell-shaped graph is symmetrical about the mean; obeys the 68-95-99 rule; see *Tutorial 3.4*

start point of a sine curve: the middle point on the increasing portion of the sine curve closest to the y-axis; see *Tutorial 5.4*

statistic: a quantity used to describe a set of data; see *Tutorial 3.2*

statistics: the branch of mathematics that deals with the collection, organization, and interpretation of data

steady state: the probability matrix does not change when multiplied by the transition matrix; in other words, $P_{n+1} = P_n$; see *Tutorial 2.4*

strata: the non-overlapping subgroup obtained when a population is separated into subpopulations

stratified random sample: a sample in which all members of different segments of the population have an equal chance of being selected

substituting into an expression: in an algebraic expression, replacing the letters with the indicated numbers or expressions each time they occur and simplifying according to the order of operations

Substitute $r = 4.5$ and $t = 3v$ into the expression $8r^2 - 2t$.

$$\begin{aligned} 8r^2 - 2t &= 8(4.5)^2 - 2(3v) \\ &= 8(20.25) - 6v \\ &= 162 - 6v \end{aligned}$$

supplementary angles: two angles whose sum is 180°

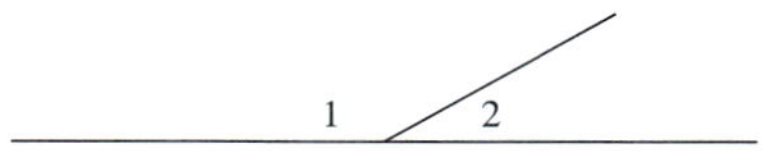

∠1 and ∠2 are supplementary.

surface area: a measure of the area on the surface of a three-dimensional object; see *cube*, *cylinder*, and *sphere*

survey: an investigation of a topic to find out people's views

systematic sample: a sample in which every nth member of a population is selected

tail-to-tail: two vectors are drawn tail-to-tail if they have the same start point; see *Tutorial 7.2*

tangent: for an acute ∠A in a right triangle, the ratio of the length of the opposite side to the length of the adjacent side

tangent of ∠A = $\dfrac{\text{Length of side opposite to } \angle A}{\text{Length of side adjacent to } \angle A}$

1.

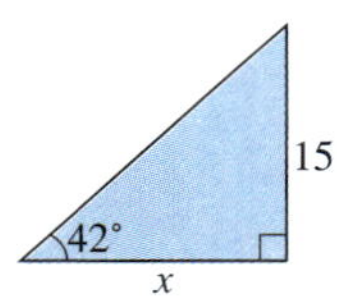

2.

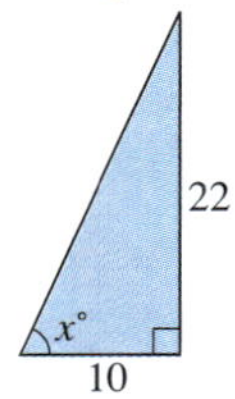

1. In the triangle shown above left,

$$\tan A = \frac{\text{Length of opposite side}}{\text{Length of adjacent side}}$$

$\tan 42^\circ = \frac{x}{15}$	**Substitute the known values.**
$0.9004 \doteq \frac{x}{15}$	**Use a calculator to obtain the value of tan 42°.**
$0.9004 \times 15 = 15 \times \frac{x}{15}$	**Clear the fraction.**
$13.506 = x$	

2. In the triangle shown above right,

$$\tan A = \frac{\text{Length of opposite side}}{\text{Length of adjacent side}}$$

$\tan x^\circ = \frac{22}{10}$	**Substitute in the known values.**
$\tan x^\circ = 2.2$	**Convert the fraction into an equivalent decimal.**
$x \doteq 65.6^\circ$	**Use $\tan^{-1}$ ("the angle whose tangent is") function.**

template: a pattern used to create spreadsheets or documents; see *Tutorial 4.1*

term deposit: an account whose rate of interest is guaranteed for a specified term where withdrawal before the end of the term may result in the loss of interest

theoretical probability: probability determined using the sample space; see *Tutorial 1.1*

three-dimensional: having length, width, and depth or height

Toronto Stock Exchange 300 Composite Index (TSE 300): tracks the composite price of 300 of the largest companies traded on the Toronto Stock Exchange; see *Tutorial 4.3*

transition matrix: a square matrix that shows how the probability of one event is dependent on the probability of another event occurring; see *Tutorial 2.4*

translation: a transformation that moves a point or a figure in a straight line to another position in the same plane; see *Tutorial 5.3*

tree diagram: a branching diagram used to show all possible outcomes of an experiment

This tree diagram shows all the outcomes of an experiment in which a die is rolled and a coin is tossed.

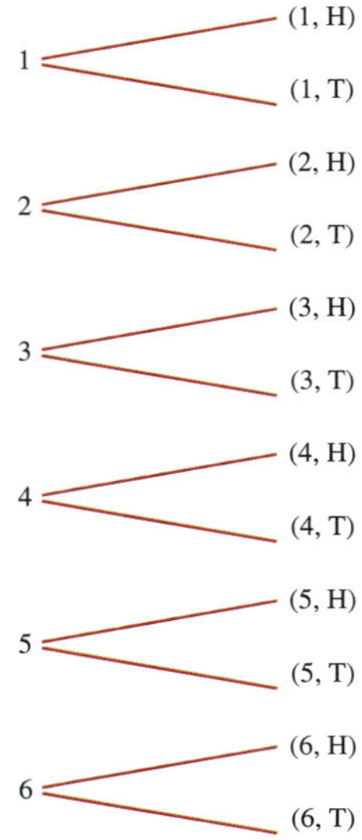

triangle: a three-sided polygon

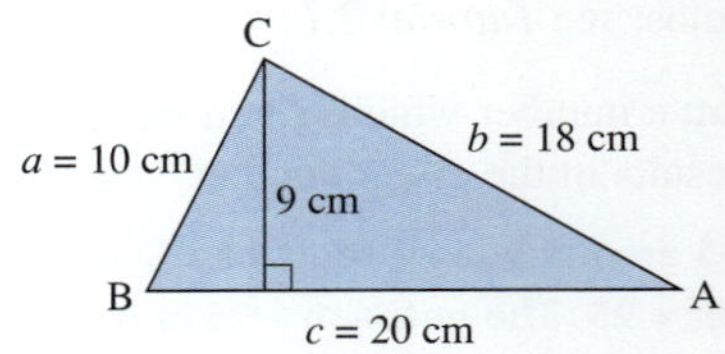

Perimeter = Sum of the three sides

$$= a + b + c$$
$$= 10 \text{ cm} + 18 \text{ cm} + 20 \text{ cm}$$
$$= 48 \text{ cm}$$

$$\text{Area} = \frac{1}{2}(\text{Base})(\text{Height})$$
$$= \frac{1}{2}bh$$
$$= \frac{1}{2}(20 \text{ cm})(9 \text{ cm})$$
$$= 90 \text{ cm}^2$$

The interior angles in any triangle add up to 180°. If the measures of two angles within a triangle are known, the third can be calculated by adding the measures of the two known angles and subtracting the result from 180°.

If $\angle A = 35°$ and $\angle B = 48°$,
then $\angle C = 180° - (35° + 48°)$
$= 97°$

triangle method of vector addition: a method for finding the sum of two vectors arranged head-to-tail; see *Tutorial 7.2*

trigonometric ratios: see *cosine*, *sine*, and *tangent*

two-dimensional: having length and width, but no thickness, height, or depth

uniform distribution: a probability distribution in which all the probabilities are equal; see *Tutorial 3.1*

unit price: the price of one item, or the price for a particular mass or volume of an item

validity: an indicator of how well a test really measures what it is suppose to measure; a sample statistic is valid if it closely approximates the quantity for the population

variable: a letter or symbol used to represent a quantity that can vary

vector quantities: quantities that are described by specifying both magnitude and direction; see *Tutorial 7.1*

velocity: a vector quantity describing the speed and direction of a moving object; see *Tutorial 7.1*

Venn diagram: a diagram where the sample space is represented by a rectangle, and event sets are represented by closed curves within the rectangle; see *Tutorials 1.2*, *1.4*, and *1.5*

vertex: the corner of a figure or solid

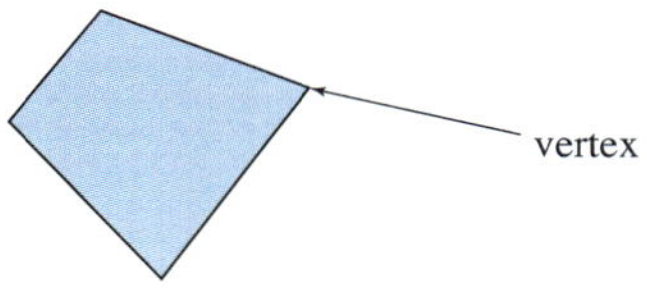

vertical intercept: the vertical coordinate of the point at which the graph of the line or function intersects the vertical axis; see *intercepts*

volume: the amount of space occupied by an object; see *cone*, *cube*, *cylinder*, *prism*, *pyramid*, *sphere*

wage: monies that are earned hourly, daily, or by piecework; this may include gratuities or overtime pay

whole numbers: see *number systems*

wholesale price: the price paid by a retailer to purchase an item for resale

***x*-axis:** the horizontal number line on a coordinate grid

***x*-intercept:** the x-coordinate where the graph of a line or function intersects the x-axis; see *intercepts* and *horizontal intercept*

***y*-axis:** the vertical number line on a coordinate grid

***y*-intercept:** the y-coordinate where the graph of a line or function intersects the y-axis; see *intercepts* and *vertical intercept*

zero vector: a vector with zero magnitude; parallel to any vector; the result of adding a vector and its opposite; see *Tutorial 7.2*

zeros of a function: the values of x for which a function $y = f(x)$ has the value 0

***z*-score:** the number of standard deviations a datum is from the mean in a normal distribution; $z = \frac{x - \mu}{\sigma}$; see *Tutorial 3.4*

Answers

CHAPTER 1 PROBABILITY

TUTORIAL 1.1

Investigation, page 4

1. Answers may vary.

2. Answers may vary.

3.

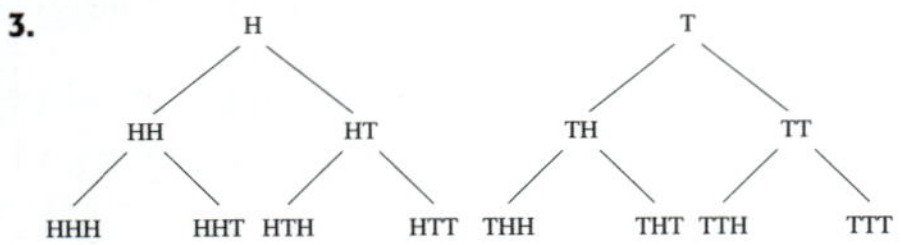

EXERCISES, page 10

Checking Your Skills

1. a) $\frac{40}{60}$ or 0.67 **b)** $\frac{1}{2}$ or 0.5

2. a) $\frac{1}{250}$ or 0.004 **b)** $\frac{5}{250}$ or 0.02 **c)** $\frac{10}{250}$ or 0.04

3. a) $\frac{4}{52}$ or 0.077 **b)** $\frac{13}{52}$ or 0.25 **c)** $\frac{26}{52}$ or 0.5
d) $\frac{2}{52}$ or 0.038

4. Answers may vary.

5. $\frac{30}{80}$ or 0.375

6. $\frac{18}{38}$ or 0.474

7. a) There is 1 pathway to exit A and G. There are 6 pathways to exit B and F. There are 15 pathways to exit C and E. There are 20 pathways to exit D.
b) $\frac{20}{64}$ or 0.3125

8. a) There are 28 pathways to A. There are 34 pathways to B. There are 21 pathways to C. There are 6 pathways to D.
b) i) $\frac{28}{89}$ or 0.315 **ii)** $\frac{21}{89}$ or 0.236

9. a) Answers may vary. **b)** $\frac{1}{4}$ or 0.25
c) Answers may vary.

10. a) i) $\frac{12}{50}$ or 0.24 **ii)** $\frac{24}{50}$ or 0.48
iii) 0 **iv)** $\frac{4}{50}$ or 0.08
b) Answers may vary. There are only 50 cards remaining in the deck when the third card is dealt.

11. a) The lighter side is more likely to appear because the heavier side is more likely to face down.
b) Answers may vary. The assumption is reasonable although it is theoretically possible for the coin to land on its edge.

TUTORIAL 1.2

Investigation, page 13

1.

Black Die

Green Die	1	2	3	4	5	6
1	(1, 1)	(2, 1)	(3, 1)	(4, 1)	(5, 1)	(6, 1)
2	(1, 2)	(2, 2)	(3, 2)	(4, 2)	(5, 2)	(6, 2)
3	(1, 3)	(2, 3)	(3, 3)	(4, 3)	(5, 3)	(6, 3)
4	(1, 4)	(2, 4)	(3, 4)	(4, 4)	(5, 4)	(6, 4)
5	(1, 5)	(2, 5)	(3, 5)	(4, 5)	(5, 5)	(6, 5)
6	(1, 6)	(2, 6)	(3, 6)	(4, 6)	(5, 6)	(6, 6)

2. 36

3.

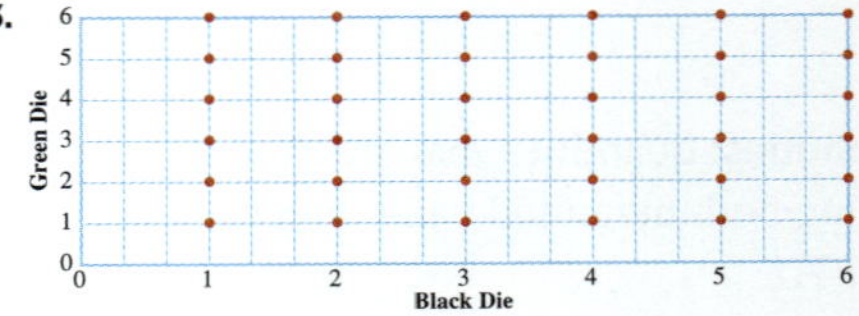

4. a)

Black Die

Green Die	1	2	3	4	5	6
1	(1, 1)	(2, 1)	(3, 1)	(4, 1)	(5, 1)	(6, 1)
2	(1, 2)	(2, 2)	(3, 2)	(4, 2)	(5, 2)	(6, 2)
3	(1, 3)	(2, 3)	(3, 3)	(4, 3)	(5, 3)	(6, 3)
4	(1, 4)	(2, 4)	(3, 4)	(4, 4)	(5, 4)	(6, 4)
5	(1, 5)	(2, 5)	(3, 5)	(4, 5)	(5, 5)	(6, 5)
6	(1, 6)	(2, 6)	(3, 6)	(4, 6)	(5, 6)	(6, 6)

b) $\frac{6}{36}$ or 0.167 **c)** $\frac{30}{36}$ or 0.833

5. a)

Black Die

Green Die	1	2	3	4	5	6
1	(1, 1)	(2, 1)	(3, 1)	(4, 1)	(5, 1)	(6, 1)
2	(1, 2)	(2, 2)	(3, 2)	(4, 2)	(5, 2)	(6, 2)
3	(1, 3)	(2, 3)	(3, 3)	(4, 3)	(5, 3)	(6, 3)
4	(1, 4)	(2, 4)	(3, 4)	(4, 4)	(5, 4)	(6, 4)
5	(1, 5)	(2, 5)	(3, 5)	(4, 5)	(5, 5)	(6, 5)
6	(1, 6)	(2, 6)	(3, 6)	(4, 6)	(5, 6)	(6, 6)

b) $\frac{4}{36}$ or 0.11 **c)** $\frac{32}{36}$ or 0.89

6. a)

Black Die

Green Die	1	2	3	4	5	6
1	(1, 1)	(2, 1)	(3, 1)	(4, 1)	(5, 1)	(6, 1)
2	(1, 2)	(2, 2)	(3, 2)	(4, 2)	(5, 2)	(6, 2)
3	(1, 3)	(2, 3)	(3, 3)	(4, 3)	(5, 3)	(6, 3)
4	(1, 4)	(2, 4)	(3, 4)	(4, 4)	(5, 4)	(6, 4)
5	(1, 5)	(2, 5)	(3, 5)	(4, 5)	(5, 5)	(6, 5)
6	(1, 6)	(2, 6)	(3, 6)	(4, 6)	(5, 6)	(6, 6)

b) $\frac{6}{36}$ or 0.167 **c)** $\frac{30}{36}$ or 0.833

7. Yes

8. The sum of the two probabilities is 1.

EXERCISES, page 17

Checking Your Skills

1. $\frac{9}{10}$

2. a) Either 3, 4, 5, 6, 7, 8, 9, 10, 11, or 12 are broken.
b) Either 0, 1, or 2 are broken.
c) At least 2 eggs are broken, that is, either 2, 3, 4, 5, 6, 7, 8, 9, 10, 11, or 12 are broken.

3. a) $\frac{4}{36}$ or 0.11 **b)** $\frac{6}{36}$ or 0.167 **c)** $\frac{3}{36}$ or 0.083

d) $\frac{6}{36}$ or 0.167

4. a)

		Die					
		1	2	3	4	5	6
Coin	H	1, H	2, H	3, H	4, H	5, H	6, H
	T	1, T	2, T	3, T	4, T	5, T	6, T

b) $\frac{1}{4}$ or 0.25 **c)** $\frac{1}{3}$ or 0.33

5. a)

		6-Sided Die					
		1	2	3	4	5	6
4-Sided Die	1	(1, 1)	(2, 1)	(3, 1)	(4, 1)	(5, 1)	(6, 1)
	2	(1, 2)	(2, 2)	(3, 2)	(4, 2)	(5, 2)	(6, 2)
	3	(1, 3)	(2, 3)	(3, 3)	(4, 3)	(5, 3)	(6, 3)
	4	(1, 4)	(2, 4)	(3, 4)	(4, 4)	(5, 4)	(6, 4)

b) $\frac{4}{24}$ or 0.167 **c)** $\frac{20}{24}$ or 0.833 **d)** $\frac{6}{24}$ or 0.25

e) The complement of event A, $P(\overline{A})$ is the event of the sum of the two dice being 7 or less.

f) $\frac{18}{24}$ or 0.75

6. a)

		Bus					
		8	9	10	11	12	13
Train	11	(8, 11)	(9, 11)	(10, 11)	(11, 11)	(12, 11)	(13, 11)
	12	(8, 12)	(9, 12)	(10, 12)	(11, 12)	(12, 12)	(13, 12)
	13	(8, 13)	(9, 13)	(10, 13)	(11, 13)	(12, 13)	(13, 13)
	14	(8, 14)	(9, 14)	(10, 14)	(11, 14)	(12, 14)	(13, 14)
	15	(8, 15)	(9, 15)	(10, 15)	(11, 15)	(12, 15)	(13, 15)

b) 30 **c)** 3 **d)** 3 **e)** $\frac{3}{30}$ or 0.1

7. a) (F, F, F) (Sa, Sa, Sa) (Su, Su, Su)
(F, F, Sa) (Sa, Sa, Su) (Su, Su, F)
(F, F, Su) (Sa, Sa, F) (Su, Su, Sa)
(F, Sa, F) (Sa, Su, Sa) (Su, F, Su)
(F, Sa, Sa) (Sa, Su, Su) (Su, F, F)
(F, Sa, Su) (Sa, Su, F) (Su, F, Sa)
(F, Su, F) (Sa, F, Sa) (Su, Sa, Su)
(F, Su, Sa) (Sa, F, Su) (Su, Sa, F)
(F, Su, Su) (Sa, F, F) (Su, Sa, Sa)

b) 27 **c)** $\frac{3}{27}$ or 0.11 **d)** $\frac{19}{27}$ or 0.704

8. a) (B, B, B, B) (B, B, G, G) (G, B, B, G) (G, G, B, G)
(B, B, B, G) (B, G, G, G) (G, B, G, B) (G, G, G, B)
(B, B, G, B) (B, G, G, B) (B, G, G, G) (G, G, G, G)
(B, G, B, B) (G, G, B, B) (G, B, G, G) (G, B, B, B)

b) $\frac{1}{16}$ or 0.0625 **c)** $\frac{15}{16}$ or 0.9375 **d)** $\frac{6}{16}$ or 0.375

9. A+, A–, B+, B–, AB+, AB–, O+, O–

Extending Your Thinking

10. a) 32. There are 5 questions, each with 2 possibilities. Thus, $2 \times 2 \times 2 \times 2 \times 2 = 32$.

b) $\frac{1}{32}$ or 0.3125.

P (getting all 5 questions correct)

$= \frac{1}{2} \times \frac{1}{2} \times \frac{1}{2} \times \frac{1}{2} \times \frac{1}{2} = \frac{1}{32}$

c) $\frac{31}{32}$ or 0.96875.

P (getting at least 1 correct) = 1 – P (getting all 5 wrong).

P (getting all 5 wrong) $= \frac{1}{2} \times \frac{1}{2} \times \frac{1}{2} \times \frac{1}{2} \times \frac{1}{2} = \frac{1}{32}$

P (getting at least 1 correct) $= 1 - \frac{1}{32} = \frac{31}{32}$

TUTORIAL 1.3

Investigation, page 20

1. a) (L1, S1) (L1, S2) (L1, S3) (L1, S4) (L1, S5)
(L2, S1) (L2, S2) (L2, S3) (L2, S4) (L2, S5)
(L3, S1) (L3, S2) (L3, S3) (L3, S4) (L3, S5)

b) 15 **c)** 3 **d)** 5

e) Multiply the number of literature class choices by the number of science class choices; $3 \times 5 = 15$.

2. a) (S1, P1, H1) (S1, P2, H1) (S1, P1, H2)
(S1, P2, H2) (S1, P1, H3) (S1, P2, H3)
(S2, P1, H1) (S2, P2, H1) (S2, P1, H2)
(S2, P2, H2) (S2, P1, H3) (S2, P2, H3)
(S3, P1, H1) (S3, P2, H1) (S3, P1, H2)
(S3, P2, H2) (S3, P1, H3) (S3, P2, H3)

b) 18

c) Multiply the number of shirt choices by the number of pants choices by the number of shoes choices; $3 \times 2 \times 3 = 18$.

3. a) 15

b) (P1, D1) (P1, D2) (P1, D3)
(P2, D1) (P2, D2) (P2, D3)
(P3, D1) (P3, D2) (P3, D3)
(P4, D1) (P4, D2) (P4, D3)
(P5, D1) (P5, D2) (P5, D3)

4. $m \times n$

EXERCISES, page 23

Checking Your Skills

1. a) (P1, T1) (P1, T2) (P1, T3) (P1, T4) (P1, T5)
(P2, T1) (P2, T2) (P2, T3) (P2, T4) (P2, T5)
(P3, T1) (P3, T2) (P3, T3) (P3, T4) (P3, T5)
There are 15 outcomes.

b) $3 \times 5 = 15$

2. 624

3. a) 80 **b)** 100

4. a) 25 **b)** 20 **c)** 125 **d)** 60

5. a) 6 **b)** 8 **c)** 30

6. a) 17 576 000. Assume all letters and digits can be repeated.

b) 156 000 000

c) 12 654 720

d) $\frac{1\,976\,000}{17\,576\,000}$ or 0.1124

e) $\frac{4\,921\,280}{17\,576\,000}$ or 0.28

7. a) 18 **b)** 10 **c)** 12

8. a) 216 000 **b)** 205 320 **c)** $\frac{1}{8}$ or 0.125

d) $\frac{5\,832}{216\,000}$ or 0.27

9. a) 4 **b)** 1024

c) $\frac{1}{1024}$ or 0.000 977

10. a) 2^{50} or about 1.126×10^{15} **b)** about 187 650

Extending Your Thinking

11. Answers may vary.

12. **a)** **i)** (A, B), (A, C), (B, C)
   ii) (A, B), (A, C), (A, D), (B, C), (B, D), (C, D)
   iii) (A, B), (A, C) (A, D), (A, E), (B, C), (B, D), (B, E), (C, D), (C, E), (D, E)
   b) **i)** 6 **ii)** 12
   iii) 20; Answers in part a are one-half answers in part b. The same CD is not chosen twice.
   c) 45

TUTORIAL 1.4

Investigation 1, page 28

1.

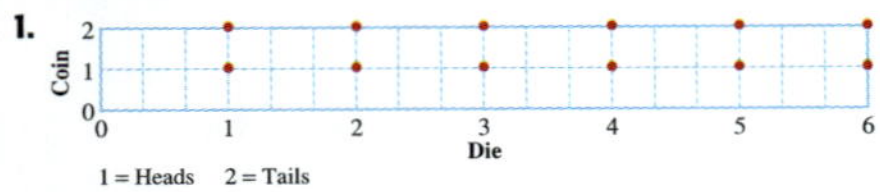

2. **a)** $\frac{6}{12}$ or 0.5 **b)** $\frac{2}{12}$ or 0.167 **c)** $\frac{1}{12}$ or 0.083
   d) P(A and B) = P(A) × P(B)

3. **a)** $\frac{6}{12}$ or 0.5 **b)** $\frac{6}{12}$ or 0.5 **c)** $\frac{3}{12}$ or 0.25
   d) P(M and N) = P(M) × P(N)

4. **a)** $\frac{6}{12}$ or 0.5 **b)** $\frac{8}{12}$ or 0.67 **c)** $\frac{4}{12}$ or 0.33
   d) P(S and T) = P(S) × P(T)

Investigation 2, page 29

1. No. The probability of choosing the second marble is affected by the first marble chosen.

2. 6

3.

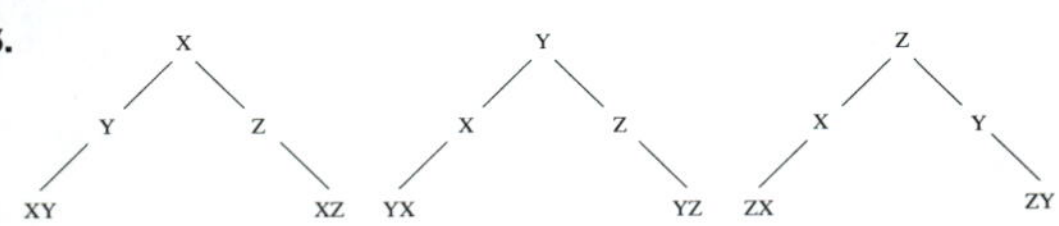

4. **a)** $\frac{5}{6}$ or 0.83 **b)** $\frac{5}{6}$ or 0.83 **c)** $\frac{2}{6}$ or 0.33
   d) No. The events are not independent.

EXERCISES, page 31

Checking Your Skills

1. **a)** Independent **b)** Dependent **c)** Independent
   d) Independent **e)** Dependent **f)** Independent

2. No, dependent. P(T) = 0.6, P(A) = 0.25, P(T and A) = 0.15; P(T) × P(A) ≠ P(T and A)

3. Yes, independent. P(T) = 0.4, P(D) = 0.5 P(T and D) = 0.2; P(T) × P(D) = P(T and D)

4. **a)** $\frac{1}{16}$ or 0.0625 **b)** $\frac{1}{4}$ or 0.25 **c)** $\frac{9}{169}$ or 0.053
   d) $\frac{1}{169}$ or 0.006 **e)** $\frac{9}{16}$ or 0.5625

5. **a)** $\frac{1}{16}$ or 0.0625 **b)** $\frac{1}{16}$ or 0.0625 **c)** $\frac{15}{16}$ or 0.9375

6. $\frac{1}{2}$ or 0.5

7. **a)** 0.24 **b)** 0.24

8. **a)** $\frac{1}{32}$ or 0.031 25 **b)** $\frac{31}{32}$ or 0.968 75 **c)** 7

9. **a)** $\frac{1}{10\,000}$ or 0.0001 **b)** 0.0488
   c) $\frac{1}{10\,000}$ or 0.0001 **d)** 0.3936

Extending Your Thinking

10. **a)** $\frac{2}{3}$ **b)** $\frac{1}{2}$ **c)** $\frac{1}{3}$
   d) P(A and B) = P(A) × P(B|A)

11. **a)** $\frac{156}{2652}$ or 0.0588 **b)** $\frac{650}{2652}$ or 0.2451 **c)** $\frac{132}{2652}$ or 0.498
   d) $\frac{12}{2652}$ or 0.0045 **e)** $\frac{1482}{2652}$ or 0.5588

TUTORIAL 1.5

Investigation, page 34

1.

Die 2 \ Die 1	1	2	3	4	5	6
1	(1, 1)	(2, 1)	(3, 1)	(4, 1)	(5, 1)	(6, 1)
2	(1, 2)	(2, 2)	(3, 2)	(4, 2)	(5, 2)	(6, 2)
3	(1, 3)	(2, 3)	(3, 3)	(4, 3)	(5, 3)	(6, 3)
4	(1, 4)	(2, 4)	(3, 4)	(4, 4)	(5, 4)	(6, 4)
5	(1, 5)	(2, 5)	(3, 5)	(4, 5)	(5, 5)	(6, 5)
6	(1, 6)	(2, 6)	(3, 6)	(4, 6)	(5, 6)	(6, 6)

2. **a)** $\frac{4}{36}$ **b)** $\frac{5}{36}$ **c)** 0 **d)** $\frac{9}{36}$
   e) P(X or Y) = P(X) + P(Y) − P(X and Y)

3. **a)** $\frac{4}{36}$ **b)** $\frac{6}{36}$ **c)** 0 **d)** $\frac{10}{36}$
   e) P(X or Z) = P(X) + P(Z) − P(X and Z)

4. **a)** $\frac{5}{36}$ **b)** $\frac{6}{36}$ **c)** $\frac{1}{36}$

Die 2 \ Die 1	1	2	3	4	5	6
1	(1, 1)	(2, 1)	(3, 1)	(4, 1)	(5, 1)	(6, 1)
2	(1, 2)	(2, 2)	(3, 2)	(4, 2)	(5, 2)	(6, 2)
3	(1, 3)	(2, 3)	(3, 3)	(4, 3)	(5, 3)	(6, 3)
4	(1, 4)	(2, 4)	(3, 4)	(4, 4)	(5, 4)	(6, 4)
5	(1, 5)	(2, 5)	(3, 5)	(4, 5)	(5, 5)	(6, 5)
6	(1, 6)	(2, 6)	(3, 6)	(4, 6)	(5, 6)	(6, 6)

   d) $\frac{10}{36}$
   e) P(Y or Z) = P(Y) + P(Z) − P(Y and Z)

5. Answers may vary.

EXERCISES, page 37

Checking Your Skills

1. a) Yes **b)** No; 7 of clubs
c) No; 91, 93, 95, 97, or 99 **d)** Yes
e) Yes

2. a) 1; The probability of an event occurring or not occurring
b) 0; The probability of an event occurring and not occurring

3. a) $\frac{3}{5}$ or 0.6 **b)** $\frac{3}{5}$ or 0.6 **c)** $\frac{3}{5}$ or 0.6
d) $\frac{7}{10}$ or 0.7 **e)** $\frac{4}{5}$ or 0.8

4. a) $\frac{26}{52}$ or 0.5 **b)** $\frac{16}{52}$ or 0.308 **c)** $\frac{22}{52}$ or 0.423
d) $\frac{16}{52}$ or 0.308

5. a) $\frac{21}{36}$ or 0.583 **b)** $\frac{12}{36}$ or 0.333 **c)** 1
d) $\frac{8}{36}$ or 0.222 **e)** $\frac{12}{36}$ or 0.333

6. a) $\frac{30}{200}$ or 0.15 **b)** $\frac{80}{200}$ or 0.4
c) $\frac{20}{200}$ or 0.1

7. 0.133

8. $\frac{34}{64}$ or 0.531 25

9. a) $\frac{1}{8}$ or 0.125 **b)** $\frac{688}{2704}$ or 0.254
c) $\frac{160}{2704}$ or 0.0592 **d)** $\frac{304}{2704}$ or 0.1124

10. a) $\frac{72}{1296}$ or 0.0556 **b)** $\frac{648}{1296}$ or 0.5
c) $\frac{44}{1296}$ or 0.034 **d)** $\frac{344}{1296}$ or 0.2654

Extending Your Thinking

11. a) 3 and 5 **b)** 5 and 6
c) 5 and 7 **d)** 2, 3, 4, 5, 6, and 7
e) 3, 4, 5, 6, 7, and 8 **f)** 2, 3, 5, 6, 7, and 8
g) 5 **h)** 2, 3, 4, 5, 6, 7, and 8

CONSOLIDATING YOUR SKILLS

What Should I Be Able To Do?, page 43

1. a)

		First Spin		
		Win	Win	Lose
Second Spin	Win	WW	WW	LW
	Win	WW	WW	LW
	Lose	WL	WL	LL

b) i) $\frac{4}{9}$ or 0.44 **ii)** $\frac{8}{9}$ or 0.89
c) i) $\frac{2}{3}$ **ii)** $\frac{2}{3}$ **iii)** $\frac{4}{9}$ **iv)** $\frac{8}{9}$
d) Yes
e) i) Answers may vary. **ii)** Answers may vary.
f) Answers may vary.
g) Answers may vary. As the number of trials increases, the experimental probability gets closer to the theoretical probability.

2. a)

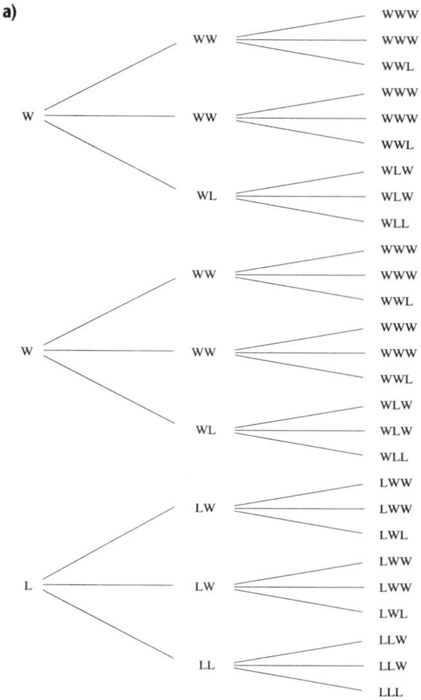

b) $\frac{8}{27}$ or 0.296
c) Answers may vary. The spinner lands on Lose at least once.
d) $\frac{19}{27}$ or 0.704

3. a) 17 576 000 **b)** 676 000 **c)** 67 600

4. a) 6561 **b)** 3024 **c)** $\frac{3537}{6561}$ or 0.5391

5. a) $\frac{1}{64}$ or 0.016 **b)** $\frac{1}{8}$ or 0.125 **c)** $\frac{27}{2197}$ or 0.012
d) $\frac{37}{64}$ or 0.578

6. No. Answers may vary. Since $P(A \text{ and } B) \neq P(A) \times P(B)$, taking Vitamin C and having a cold are not independent events.

7. a) $\frac{30}{40}$ or 0.75 **b)** $\frac{13}{40}$ or 0.325 **c)** $\frac{11}{40}$ or 0.275

8. 0.375

9. Answers may vary.

10. Answers may vary.

CHAPTER 2 MATRICES

TUTORIAL 2.1

Exercises, page 55

Checking Your Skills

1. a) $\begin{bmatrix} 4 & 8 & 3 \\ 12 & 7 & -2 \end{bmatrix}$ **b)** not possible

c) $\begin{bmatrix} 9 \\ 6 \\ -15 \end{bmatrix}$ **d)** $\begin{bmatrix} 5 & -14 & 15 \\ 12 & -4 & 2 \end{bmatrix}$

2. a) $\begin{bmatrix} 5 & 10 & 3 \\ 4 & 6 & 8 \\ 9 & 2 & 7 \end{bmatrix}$; Yes, each row, column, and diagonal sum to 18.

b) $\begin{bmatrix} 6 & 16 & 2 \\ 4 & 8 & 12 \\ 14 & 0 & 10 \end{bmatrix}$; Yes, each row, column, and diagonal sum to 24.

c) $\begin{bmatrix} 8 & 18 & 4 \\ 6 & 10 & 14 \\ 16 & 2 & 12 \end{bmatrix}$; Yes, each row, column, and diagonal sum to 30.

3. $\begin{bmatrix} 5 & 0 & 2 & 10 \\ 5 & 0 & 2 & 10 \\ 2 & 0 & 5 & 4 \\ 2 & 0 & 5 & 4 \end{bmatrix}$

4. a) $\begin{bmatrix} -1 \\ -1 \\ -1 \\ -7 \\ -18 \end{bmatrix}, \begin{bmatrix} -3 \\ -2 \\ 1 \\ -2 \\ -5 \end{bmatrix}, \begin{bmatrix} 1 \\ 0 \\ 2 \\ -1 \\ -2 \end{bmatrix}, \begin{bmatrix} -2 \\ -1 \\ 0 \\ 0 \\ 1 \end{bmatrix}, \begin{bmatrix} 1 \\ 2 \\ 1 \\ 1 \\ -1 \end{bmatrix}, \begin{bmatrix} -5 \\ -3 \\ -2 \\ -1 \\ -2 \end{bmatrix}$

b) $\begin{bmatrix} -9 \\ -5 \\ 1 \\ -10 \\ -27 \end{bmatrix}$

5. a) Answers will vary. **b)** Answers will vary.

6. a) $A = \begin{bmatrix} 59 & 37 & 14 \\ 101 & 88 & 31 \\ 29 & 18 & 56 \\ 83 & 121 & 6 \end{bmatrix}, B = \begin{bmatrix} 62 & 45 & 18 \\ 127 & 75 & 42 \\ 18 & 7 & 68 \\ 79 & 103 & 21 \end{bmatrix},$

$C = \begin{bmatrix} 75 & 59 & 28 \\ 122 & 77 & 39 \\ 6 & 9 & 49 \\ 62 & 118 & 15 \end{bmatrix}$

b) $\begin{bmatrix} 196 & 141 & 60 \\ 360 & 240 & 112 \\ 53 & 34 & 173 \\ 224 & 342 & 42 \end{bmatrix}$

c) $\begin{bmatrix} 546.84 & 393.39 & 167.40 \\ 976.50 & 669.60 & 312.48 \\ 147.87 & 94.86 & 482.67 \\ 624.96 & 954.18 & 117.18 \end{bmatrix}$

7. a) $A = \begin{bmatrix} 3.75 & 8 & 8 & 8 & 8 & 8 & 8 \\ 0 & 8 & 8 & 8 & 8 & 8 & 6.5 \\ 3 & 8 & 8 & 8 & 8 & 8 & 8 \end{bmatrix},$

$B = \begin{bmatrix} 0 & 1.5 & 0 & 0 & 1 & 1.75 & 2 \\ 0 & 0 & 1 & 1 & 0.75 & 1 & 0 \\ 0 & 0 & 0 & 0 & 0.5 & 0.5 & 0.5 \end{bmatrix}$

b) $\begin{bmatrix} 65.63 & 179.37 & 140 & 140 & 166.25 & 185.94 & 192.5 \\ 0 & 140 & 166.25 & 166.25 & 159.69 & 166.25 & 113.75 \\ 52.5 & 140 & 140 & 140 & 153.13 & 153.13 & 153.13 \end{bmatrix}$

8. a) $\begin{bmatrix} 12 & 4 & 4 & 10 \\ 18 & 6 & 8 & 16 \\ 16 & 8 & 9 & 15 \end{bmatrix}$ **b)** $\begin{bmatrix} 4 & 2 & 1 & 0 \\ 6 & 1 & 4 & 3 \\ 5 & 3 & 3 & 6 \end{bmatrix}$

c) $\begin{bmatrix} 8 & 2 & 3 & 10 \\ 12 & 5 & 4 & 13 \\ 11 & 5 & 6 & 9 \end{bmatrix}$ **d)** $\begin{bmatrix} 16 & 6 & 5 & 10 \\ 24 & 7 & 12 & 19 \\ 21 & 11 & 12 & 21 \end{bmatrix}$

9. a) GST = \$223.88, PST = \$107.09

b) Answers may vary.

$\begin{bmatrix} 501.34 \\ 781.03 \end{bmatrix} + 1.07 \begin{bmatrix} 1250.56 \\ 417.79 \end{bmatrix} + 1.14 \begin{bmatrix} 822.60 \\ 707.28 \end{bmatrix} = \begin{bmatrix} \text{Sat.} \\ \text{Sun.} \end{bmatrix}$

c) $\begin{bmatrix} 501.34 \\ 781.03 \end{bmatrix} + 1.05 \begin{bmatrix} 1250.56 \\ 417.79 \end{bmatrix} + 1.17 \begin{bmatrix} 822.60 \\ 707.28 \end{bmatrix} = \begin{bmatrix} \text{Sat.} \\ \text{Sun.} \end{bmatrix}$

Extending Your Thinking

10. a) Answers may vary. Yes.

b) Yes. Since every number is added to or multiplied by the same factor, the rows, columns, and diagonals sum to the same result.

c) Yes. Taking the square root or squaring all the numbers would result in a matrix that is not a magic square.

11. \$770.92, \$942.82, \$1169.17

TUTORIAL 2.2

Investigation 1, page 60

1.

Team	Wins	Ties	Losses	Points
Blazers	2 × 7 = 14	1 × 2 = 2	0 × 4 = 0	14 + 2 + 0 = 16
Hawks	2 × 5 = 10	1 × 1 = 1	0 × 7 = 0	10 + 1 + 0 = 11
Tigers	2 × 8 = 16	1 × 0 = 0	0 × 5 = 0	16 + 0 + 0 = 16
Eagles	2 × 4 = 8	1 × 1 = 1	0 × 8 = 0	8 + 1 + 0 = 9

2. Total points = (Wins × 2 points) + (Ties × 1 point) + (Losses × 0 points)
The process is the same for all three teams.

3. $S = \begin{bmatrix} 7 & 2 & 4 \\ 5 & 1 & 7 \\ 8 & 0 & 5 \\ 4 & 1 & 8 \end{bmatrix}$ **4.** $P = \begin{bmatrix} 2 \\ 1 \\ 0 \end{bmatrix}$

5. $T = \begin{bmatrix} 16 \\ 11 \\ 16 \\ 9 \end{bmatrix}$ **6.** $S \times P = \begin{bmatrix} 16 \\ 11 \\ 16 \\ 9 \end{bmatrix}$

7. $S \times P = T$. Answers may vary.

8. $S \times P$ is a 4×1 matrix. The number of rows in S and the number of columns in P are the same.

Investigation 2, page 61

1. $S = \begin{bmatrix} 0 & 14 & 103 & 32 \\ 67 & 29 & 45 & 18 \\ 0 & 0 & 86 & 54 \end{bmatrix}$

2. $A = \begin{bmatrix} 5 \\ 6 \\ 8 \\ 6 \end{bmatrix}$

3. $\begin{bmatrix} 1100 \\ 977 \\ 1012 \end{bmatrix}$; 3×1; the same number of rows as S and the same number of columns as A.

4. Each element shows the total sales for each movie.

5. Cannot be done.

6. The number of columns in the first matrix must equal the number of rows in the second to be able to multiply the two matrices.

7. Answers will vary.

Exercises, page 64

Checking Your Skills

1. a) $\begin{bmatrix} -175 \\ 180 \end{bmatrix}$ **b)** $\begin{bmatrix} 11 & 6 \\ 3 & 1 \end{bmatrix}$

2. a) $\begin{bmatrix} 16.75 & 3.225 \\ 21.28 & -5.31 \\ 21.21 & -7.535 \end{bmatrix}$

b) $\begin{bmatrix} 14.675 & 14.525 & 6.5125 \\ 17.53 & 22.19 & 11.645 \\ 14.285 & 22.295 & 13.5375 \end{bmatrix}$

c) Not possible

3. a) $\begin{bmatrix} 43 & 14 & 32 \\ 76 & 26 & 59 \\ 52 & 19 & 51 \end{bmatrix}$ **b)** $\begin{bmatrix} 78 \\ 47 \\ 65 \end{bmatrix}$

c) \$6092, \$10 985, \$8264

4. a) $P = [2.65 \quad 3.73 \quad 3.15 \quad 2.99]$, $S = \begin{bmatrix} 5 \\ 8 \\ 7 \\ 10 \end{bmatrix}$

b) \$95.04

5. a)

Team	Points
Tigers	70
Irish	59
Colts	57
Jets	63

b) $\begin{bmatrix} 30 & 2 & 8 & 2 \\ 24 & 9 & 2 & 7 \\ 25 & 7 & 0 & 10 \\ 26 & 1 & 10 & 5 \end{bmatrix} \begin{bmatrix} 2 \\ 1 \\ 1 \\ 0 \end{bmatrix} = \begin{bmatrix} 70 \\ 59 \\ 57 \\ 63 \end{bmatrix}$

c)

Team	Points		
	i	ii	iii
Tigers	100	92	156
Irish	83	81	147
Colts	82	82	146
Jets	89	79	133

d) iii **e)** ii

f) i **g)** Answers may vary.

6. Supershop

7. a) $A = \begin{bmatrix} 0.5 & 0.8 & 0.4 \\ 0.8 & 1.0 & 0.5 \\ 0.6 & 0.4 & 0.5 \end{bmatrix}$ **b)** $B = \begin{bmatrix} 1000 & 1100 \\ 600 & 850 \\ 800 & 725 \end{bmatrix}$

c) 9315 h

d)

	Panda	Kangaroo	Rabbit
Alberta	\$16.49	\$19.16	\$12.10
Saskatchewan	\$15.72	\$18.18	\$11.55
Manitoba	\$18.94	\$21.68	\$13.89

8. a) $A = \begin{bmatrix} 20 & 3 & 15 & 21 & 3 \\ 19 & 1 & 13 & 19 & 4 \\ 18 & 2 & 10 & 18 & 5 \\ 19 & 2 & 11 & 19 & 4 \\ 18 & 2 & 10 & 21 & 5 \\ 16 & 1 & 11 & 20 & 4 \\ 17 & 0 & 12 & 18 & 3 \end{bmatrix}$

b) $B = \begin{bmatrix} 1.15 & 0 & 0 & 0 & 0 \\ 0 & 1 & 0 & 0 & 0 \\ 0 & 0 & 1.05 & 0 & 0 \\ 0 & 0 & 0 & 0.90 & 0 \\ 0 & 0 & 0 & 0 & 0.90 \end{bmatrix}$

c)

	Abbotsford	Cranbrook	Hope	Surrey	Trail
Monday	23	3	15.75	18.9	2.7
Tuesday	21.85	1	13.65	17.1	3.6
Wednesday	20.7	2	10.5	16.2	4.5
Thursday	21.85	2	11.55	17.1	3.6
Friday	20.7	2	10.5	18.9	4.5
Saturday	18.4	1	11.55	18	3.3
Sunday	19.55	0	12.6	16.2	2.7

9. a) $[200 \quad 300 \quad 40]$ **b)** $\begin{bmatrix} 0.96 & 0 & 0 \\ 0 & 1.1 & 0 \\ 0 & 0 & 1.3 \end{bmatrix}$

c) 574

Extending Your Thinking

10. a) 40 business calculators, 60 scientific calculators

b) \$920

TUTORIAL 2.3

Investigation, page 73

1.

Number of Routes Between Cities with Exactly One Stopover					
FROM City \ TO City	Yellowknife	Inuvik	Norman Wells	Rankin Inlet	Whitehorse
Yellowknife	4	1	1	0	0
Inuvik	1	2	1	1	1
Norman Wells	1	1	2	1	1
Rankin Inlet	0	1	1	1	1
Whitehorse	0	1	1	1	1

$B = \begin{bmatrix} 4 & 1 & 1 & 0 & 0 \\ 1 & 2 & 1 & 1 & 1 \\ 1 & 1 & 2 & 1 & 1 \\ 0 & 1 & 1 & 1 & 1 \\ 0 & 1 & 1 & 1 & 1 \end{bmatrix}$

2. $A^2 = \begin{bmatrix} 4 & 1 & 1 & 0 & 0 \\ 1 & 2 & 1 & 1 & 1 \\ 1 & 1 & 2 & 1 & 1 \\ 0 & 1 & 1 & 1 & 1 \\ 0 & 1 & 1 & 1 & 1 \end{bmatrix} = B$

3. $C = \begin{bmatrix} 2 & 5 & 5 & 4 & 4 \\ 5 & 2 & 3 & 1 & 1 \\ 5 & 3 & 2 & 1 & 1 \\ 4 & 1 & 1 & 0 & 0 \\ 4 & 1 & 1 & 0 & 0 \end{bmatrix}$

4. $A^3 = \begin{bmatrix} 2 & 5 & 5 & 4 & 4 \\ 5 & 2 & 3 & 1 & 1 \\ 5 & 3 & 2 & 1 & 1 \\ 4 & 1 & 1 & 0 & 0 \\ 4 & 1 & 1 & 0 & 0 \end{bmatrix} = C$

5. Calculate A^5.

6. $\begin{bmatrix} 4 & 2 & 2 & 1 & 1 \\ 2 & 2 & 2 & 1 & 1 \\ 2 & 2 & 2 & 1 & 1 \\ 1 & 1 & 1 & 1 & 1 \\ 1 & 1 & 1 & 1 & 1 \end{bmatrix}$; The entries represent the number of routes between cities with at most 1 stopover.

7. $A + A^2 + A^3$

Exercises, page 76

Checking Your Skills

1. a)

From \ To	A	B	C	D
A	0	1	0	1
B	1	0	0	1
C	1	0	0	0
D	0	1	1	1

b)

From \ To	W	X	Y
W	0	1	1
X	0	0	1
Y	1	0	0

2. a)

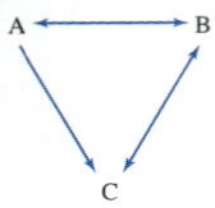

b)

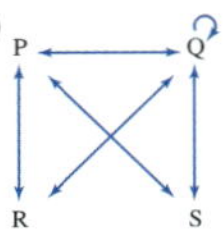

3. a)

From \ To	V	C	E	W
V	0	1	0	0
C	1	0	1	1
E	0	1	0	0
W	0	1	0	0

b)

From \ To	V	C	E	W
V	1	0	1	1
C	0	3	0	0
E	1	0	1	1
W	1	0	1	1

These elements represent the number of flights with exactly one stopover.

c) Explanations may vary. To fly from Calgary to Calgary with one stopover occurs 3 times (it passes through each city). Each of the other 3 cities must pass through Calgary to reach any other city, so there are 9 possibilities with 1 stopover. Six elements equal 0 because you cannot get to Calgary with exactly one stopover from the other cities.

4. a) $C =$

From	A	B	C	D	E
A	0	1	1	1	0
B	1	0	0	0	1
C	1	0	0	1	0
D	1	1	0	0	1
E	0	0	0	1	0

b) $C^2 = \begin{bmatrix} 3 & 1 & 0 & 1 & 2 \\ 0 & 1 & 1 & 2 & 0 \\ 1 & 2 & 1 & 1 & 1 \\ 1 & 1 & 1 & 2 & 1 \\ 1 & 1 & 0 & 0 & 1 \end{bmatrix}$

c) $C + C^2 = \begin{bmatrix} 3 & 2 & 1 & 2 & 2 \\ 1 & 1 & 1 & 2 & 1 \\ 2 & 2 & 1 & 2 & 1 \\ 2 & 2 & 1 & 2 & 2 \\ 1 & 1 & 0 & 1 & 1 \end{bmatrix}$

5. a)

From \ To	CEO	VPS	VPM	VPF	RSR	RMR	RA
CEO	0	1	1	1	0	0	0
VPS	0	0	1	0	1	1	0
VPM	0	1	0	1	1	1	0
VPF	0	0	1	0	0	0	1
RSR	0	0	0	0	0	0	0
RMR	0	0	0	0	0	0	0
RA	0	0	0	0	0	0	0

b) 2 **c)** 5

6. a)

From \ To	L	M	CiH	WP	G	CoH	TI
L	0	1	0	0	1	0	0
M	0	0	1	0	0	1	0
CiH	0	0	0	1	1	1	1
WP	0	0	0	0	0	0	1
G	1	1	0	0	0	0	0
CoH	0	1	1	0	1	0	0
TI	0	0	0	1	0	1	0

b) 3

7. a) 2 **b)** 4 **c)** 6 **d)** 1

Extending Your Thinking

8. a) Dangerous Goods

From \ To	A	B	C	D	E
A	0	1	0	1	0
B	1	0	1	0	0
C	0	1	0	0	1
D	1	0	0	0	1
E	0	0	1	1	0

Other Goods

From \ To	A	B	C	D	E
A	0	1	1	1	1
B	1	0	1	1	1
C	1	1	0	1	1
D	1	1	1	0	1
E	1	1	1	1	0

b) 4 **c)** 17

d) Dangerous Goods

From \ To	A	B	C	D	E
A	0	1	1	1	1
B	1	0	0	0	0
C	1	0	0	0	0
D	1	0	0	0	0
E	1	0	0	0	0

Other Goods

From \ To	A	B	C	D	E
A	0	1	1	1	1
B	1	0	1	1	1
C	1	1	0	1	1
D	1	1	1	0	1
E	1	1	1	1	0

TUTORIAL 2.4

Investigation, page 80

1. $P_1 = [0.475 \quad 0.525]$; the probability Dana will win the 2nd game and the probability Dana will lose the 2nd game

2. $P_2 = [0.466\,25 \quad 0.533\,75]$; the probability that Dana will win the 3rd game and the probability that Dana will lose the 3rd game

3. $P_3 = [0.463 \quad 0.537]$; the probability that Dana will win the 4th game and the probability that Dana will lose the 4th game

4. $P_4 = [0.462 \quad 0.538]$; the probability that Dana will win the 5th game and the probability that Dana will lose the 5th game

5. $[0.466\ 25 \quad 0.533\ 75]$; equal to P_2.

6. $[0.463 \quad 0.537]$, equal to P_3; $[0.462 \quad 0.538]$, equal to P_4

7. Use $P_{20} = P_{19} \times T$ or $P_{20} = P_0 \times T^{20}$. The second method is easier because P_0 and T are known.

Exercises, page 83

Checking Your Skills

1. a) $[0.548 \quad 0.175 \quad 0.277]$ b) $[0.474 \quad 0.221 \quad 0.306]$ c) $[0.476 \quad 0.219 \quad 0.305]$

2. a) $P_0 = [0.5 \quad 0.5]$, $T = \begin{bmatrix} 0.75 & 0.25 \\ 0.45 & 0.55 \end{bmatrix}$ b) 0.4

3. $\begin{bmatrix} 0.9 & 0.1 \\ 0.85 & 0.15 \end{bmatrix}$

4. a) $P_0 = [0.95 \quad 0.95]$, $T = \begin{bmatrix} 0.97 & 0.03 \\ 0.85 & 0.15 \end{bmatrix}$ b) 96.6%

5. a) $P_0 = [0.3 \quad 0.2 \quad 0.5]$
b) $T = \begin{bmatrix} 0.85 & 0.13 & 0.02 \\ 0.05 & 0.91 & 0.04 \\ 0.03 & 0.21 & 0.76 \end{bmatrix}$
c) 28% full-size, 32.6% midsize, 39.4% compact

6. a) $P_0 = [0.6 \quad 0.4]$, $T = \begin{bmatrix} 0.8 & 0.2 \\ 0.25 & 0.75 \end{bmatrix}$ b) 58% c) 56.9%

7. a) $P_0 = [0.22 \quad 0.78]$, $T = \begin{bmatrix} 0.25 & 0.75 \\ 0.10 & 0.90 \end{bmatrix}$ b) 42

8. a) 44% 6 L; 56% 10 L b) 37.6% 6 L; 62.4% 10 L

9. 61.6% Liberal; 38.4% Conservative

Extending Your Thinking

10. a) $P_0 = [0.57 \quad 0.43]$, $T = \begin{bmatrix} 0.8 & 0.2 \\ 0.4 & 0.6 \end{bmatrix}$ b) $[0.666 \quad 0.333]$

Consolidating Your Skills

What Should I Be Able To Do?, page 89

1. a) $[0.14 \quad 0.07 \quad 0.14 \quad 0.14 \quad 0.15 \quad 0.145 \quad 0.15 \quad 0.15 \quad 0.17 \quad 0.15]$
b) $\begin{bmatrix} 2082 \\ 2479 \\ 1873 \\ 1905 \\ 2439 \\ 1986 \\ 1873 \\ 1874 \\ 1870 \\ 2219 \end{bmatrix}$
c) \$291.48, \$173.53, \$262.22, \$266.70, \$365.85, \$287.97, \$280.95, \$281.10, \$317.90, \$332.85
d) No.

2. a) $\begin{bmatrix} -12.5 & 5 & -22.5 \\ 15 & -7.5 & 7.5 \\ 0 & 2.5 & 20 \end{bmatrix}$ b) $\begin{bmatrix} -35 & -116 \\ 15 & 69 \\ 30 & 72 \end{bmatrix}$
c) $\begin{bmatrix} 37 & -25 & -21 \\ -48 & 24 & -39 \\ 6 & 5 & 67 \end{bmatrix}$ d) Not possible

3. a) 77.5%, 65.3%, 64.6%, 89.1%, 52%

4. a) $\begin{bmatrix} 0.04 & 0.09 & 0.04 & 0.02 \\ 0.13 & 0.2 & 0.05 & 0.03 \\ 0.52 & 0.49 & 0.26 & 0.08 \\ 0.21 & 0.16 & 0.38 & 0.17 \\ 0.07 & 0.04 & 0.16 & 0.19 \\ 0.03 & 0.02 & 0.11 & 0.51 \end{bmatrix}$
b) $\begin{bmatrix} 130 \\ 110 \\ 310 \\ 450 \end{bmatrix}$
c) 36, 68, 238, 239, 149, 270

5. a) $\begin{bmatrix} 8.5 & 11.75 & 15.5 & 24.75 \\ 1.75 & 2 & 2.5 & 3.5 \end{bmatrix}$
b) $\begin{bmatrix} 350 \\ 300 \\ 150 \\ 75 \end{bmatrix}$ c) \$12 531.25

6. a) $[1\ 143\ 400 \quad 1\ 028\ 100 \quad 2\ 969\ 000 \quad 4\ 029\ 300]$
b) $\begin{bmatrix} 1.004\ 635 \\ 1.002\ 334 \\ 2.009\ 872 \\ 1.006\ 577 \end{bmatrix}$ c) 12 202 310
d) Answers may vary. Assume the population will increase at the same rate each year. This is not a valid process because it does not take into account the possibility of a decrease in population. Also, it is not possible to determine a pattern from one year.

7.

From \ To	A	B	C
A	0	1	0
B	1	0	1
C	1	1	0

8. a) A =

From \ To	AP	M	6th	7th	AS
AP	0	1	0	0	0
M	1	0	1	0	0
6th	0	1	0	1	1
7th	0	0	1	0	1
AS	0	0	1	1	0

b) $\begin{bmatrix} 1 & 0 & 1 & 0 & 0 \\ 0 & 2 & 0 & 1 & 1 \\ 1 & 0 & 3 & 1 & 1 \\ 0 & 1 & 1 & 2 & 1 \\ 0 & 1 & 1 & 1 & 2 \end{bmatrix}$ c) $\begin{bmatrix} 1 & 1 & 1 & 0 & 0 \\ 1 & 2 & 1 & 1 & 1 \\ 1 & 1 & 3 & 2 & 2 \\ 0 & 1 & 2 & 2 & 2 \\ 0 & 1 & 2 & 2 & 2 \end{bmatrix}$
d) The matrix represents the number of ways to travel from one station to another with at most 3 stops.

9. a)

$$\begin{array}{ccc} & & \text{To} \\ & & \begin{array}{ccccc} \text{C} & \text{CSR} & \text{M} & \text{CRO} & \text{PRVP} \end{array} \\ \text{From} & \begin{array}{c} \text{C} \\ \text{CSR} \\ \text{M} \\ \text{CRO} \\ \text{PRVP} \end{array} & \begin{bmatrix} 0 & 1 & 0 & 0 & 0 \\ 0 & 0 & 1 & 0 & 0 \\ 1 & 0 & 0 & 1 & 0 \\ 1 & 0 & 0 & 0 & 1 \\ 0 & 0 & 0 & 1 & 0 \end{bmatrix} \end{array}$$

b) $\begin{bmatrix} 1 & 0 & 0 & 1 & 0 \\ 1 & 1 & 0 & 0 & 1 \\ 0 & 1 & 1 & 1 & 0 \\ 1 & 0 & 1 & 0 & 1 \\ 0 & 1 & 0 & 1 & 0 \end{bmatrix}$

10. a) $\begin{bmatrix} 0.75 & 0.1 & 0.1 & 0 & 0.5 \\ 0.3 & 0.7 & 0 & 0 & 0 \\ 0.25 & 0 & 0.65 & 0.1 & 0 \\ 0.1 & 0 & 0.1 & 0.8 & 0 \\ 0.1 & 0 & 0 & 0 & 0.9 \end{bmatrix}$

b) 30% for A, 16% for B, 17% for C, 18% for D, and 28% for E

c) 36% for A, 14% for B, 16% for C, 16% for D, and 40% for E

Chapter 3 Statistics

Tutorial 3.1

Investigation, page 98

1.

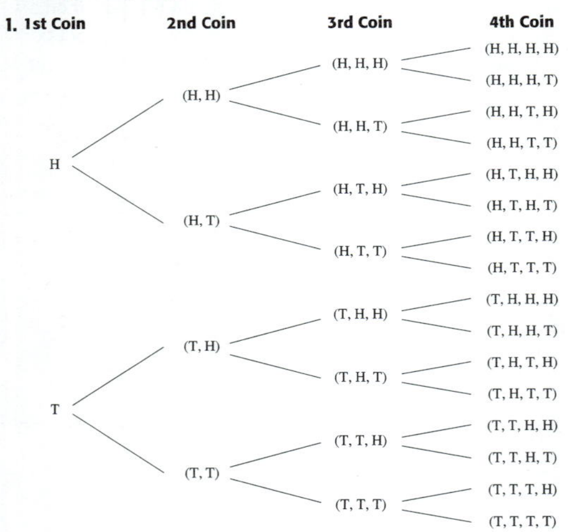

2.

Outcome	Probability
0 heads	$\frac{1}{16}$
1 head	$\frac{4}{16}$
2 heads	$\frac{6}{16}$
3 heads	$\frac{4}{16}$
4 heads	$\frac{1}{16}$

3. $\frac{1}{16}$; $\frac{6}{16}$; 0 and 1

4. 1

6. Answers will vary.

7. Answers may vary.

8. Answers may vary. 0 and 1; 1

9. Answers may vary.

10. Answers may vary. The probabilities are similar to those in exercise 2.

11. Answers may vary. 0 and 1; 1

12. As the number of repetitions increases, the probability distribution should approach that found in exercise 2.

Exercises, page 102

Checking Your Skills

1. Answers may vary.

2. a) Distribution 2; The probability has a constant value of $\frac{1}{7}$ for each day.

b) Distributions 1 and 4; The sum of the probabilities is equal to 1 for both graphs and the probabilities are all less than 1.

c) Distribution 3; The sum of the probabilities is less than 1.

3. a)

Outcome	Probability
Winning 1st Prize	$\frac{1}{5000}$
Not Winning 1st Prize	$\frac{4999}{5000}$

Outcome	Probability
Winning 2nd Prize	$\frac{1}{2500}$
Not Winning 2nd Prize	$\frac{2499}{2500}$

Outcome	Probability
Winning 3rd Prize	$\frac{1}{500}$
Not Winning 3rd Prize	$\frac{499}{500}$

b) Yes. There are two possible outcomes — win or lose.

4. a)

Die	1	2	3	4	5	6
1	2	3	4	5	6	7
2	3	4	5	6	7	8
3	4	5	6	7	8	9
4	5	6	7	8	9	10
5	6	7	8	9	10	11
6	7	8	9	10	11	12

b) No. Answers may vary. There are 6 outcomes. Binomial distribution must have exactly 2 outcomes.

5.

Outcome	Probability
50% discount	$\frac{1}{15}$
25% discount	$\frac{4}{15}$
10% discount	$\frac{10}{15}$

6. a)

Outcome	Probability
Green light	$\frac{75}{140}$
Amber light	$\frac{5}{140}$
Red light	$\frac{60}{140}$

b) No. There are 3 possible outcomes.

8. Answers may vary.

Extending Your Thinking

9. 2 heads

Tutorial 3.2

Investigation 1, page 105

2. Answers may vary. Both histograms have the same shape and the maximum occurs at 34. The first histogram is more spread out.
3. 33.65, 4.594; 33.65, 2.824
4. Data Set 1; Data Set 1; The higher the standard deviation, the more spread out the data is from the mean.
5. Yes.

Investigation 2, page 106

2. 7.5, 2.372
3. Same
4. Same
5. Answers may vary. To estimate the mean, multiply the number of trials by the probability of a success. To estimate the standard deviation, multiply the mean by the probability of a failure and then take the square root of the result.

Exercises, page 108

Checking Your Skills

1. Part b is more likely to have the greatest standard deviation since the data is spread out evenly and not clustered about the mean. Part c is likely to have the least standard deviation since the data is clustered more closely about the mean.
2. **a)** 15.92, 0.954 **b)** 16, 1.443 **c)** 15.92, 0.640
3. Graph a is more likely to have the greatest standard deviation since the data is spread evenly and not clustered about the mean. Graph c is more likely to have the least standard deviation since the data is closely clustered about the mean.
4. The group from the shopping mall is more likely to have the greatest standard deviation because they are not as skilled in this activity.
5. The group from the video arcade is more likely to have the lesser standard deviation because they are more skilled in this activity.
6. **a)** 354.23, 2.37
   b) Answers may vary. If the standard deviation is too high then either the cans are under-filled or overfilled. This could result in unsatisfied customers or damaged cans.
7. Answers may vary.
8. **a)** 5.6, 1.35; 5.6, 0.79
   b) Battery Y is better. Explanations may vary. Its lifetime is more reliable since the standard deviation is smaller.
9. **a)** 8.4, 2.86 **b)** 4.5, 2.11

Extending Your Thinking

10. Answers may vary. Since 6 is closer to 4.5 than 8.4 and within the standard deviation of the improved process, then the result supports the consultant's claim.

Tutorial 3.3

Exercises, page 116

Checking Your Skills

1. **a)** ii **b)** iii **c)** i
2. **b)** **i)** 99.85% **ii)** 0.15% **iii)** 2.5% **iv)** 2.5% **v)** 95% **vi)** 34%
3. **a)** 0.5 **b)** 0.84 **c)** 0.16 **d)** 0.68
4. **a)** 2σ **b)** 0.025 **c)** 12 cm
5. **a)** 2.5% **b)** 0.15% **c)** 75 **d)** 0.975
6. 0.815
7. **a)** 0.025 **b)** 11.4 cm
8. 2.3 kg
9. 5

Extending Your Thinking

10. **a)** 1σ or 0.833; 2σ or 1.25 **b)** **i)** Answers may vary.

Tutorial 3.4

Exercises, page 123

Checking Your Skills

1. **a)** 1 **b)** 0.583 **c)** −3 **d)** −8.5
2. **a)** 0.4192 **b)** 0.7699 **c)** 0.0047
3. 0.1587
4. The two heights are not equivalent. Explanations may vary.
5. **a)** 1.6 **b)** 0.4452 **c)** 0.9452 **d)** 0.0082
6. **a)** 35 **b)** 1
7. 0.0047
8. The student is in the top 45% in Physics, in the top 42% in Mathematics, and in the top 24% in English.
9. 1.96%

Extending Your Thinking

10. 1.6 kg

Tutorial 3.5

Investigation, page 126

1. **a)** 0.775 87
   b) **i)** 7.5, 1.3 **ii)** −0.365 **iii)** 0.642 44, 0.133 43
   c) 1.875
2. **a)** 0.803 42
   b) **i)** 22.5, 2.3717 **ii)** −0.6325 **iii)** 0.736 47, 0.066 95
3. **a)** 0.851 42
   b) **i)** 45, 3.35 **ii)** −0.8955 **iii)** 0.814 74, 0.036 68
   c) 11.25
4.

Exercise	Value of $n \times p \times (1 - p)$	Difference between Actual and Estimated Probability
1	1.875	0.133 43
2	5.625	0.066 95
3	11.25	0.036 68

5. a) 3; 1 b) 1; 3
c) The greater the difference between the actual and the estimated probability, the lower the value of $n \times p \times (1 - p)$.

Exercises, page 130

Checking Your Skills

1. a) normal b) binomial c) normal
2. 0.786 22
3. 0.000 007
4. 1
5. 0.9377
6. 0.036 33
7. 0.018 31
8. 0.002 05

Extending Your Thinking

9. 27

Tutorial 3.5

Investigation, page 134

1. 0.9544
2. 0.9534
3. a) −1.96, 1.96
b) The values are not exact because they are rounded to 2 decimal places. Answers may vary.
c) 1.96

Exercises, page 139

Checking Your Skills

1. a) 59.848, 76.312, ±8.948% b) 89.373, 115.519, ±7.223%
c) 55.331, 81.477, ±6.67% d) 85.983, 118.785, ±5.062%
2. a) 36%, 52% b) ±7.94%
3. a) 36%, 50% b) ±6.8%
4. 64% to 88% with a margin of error of ±6.04%
5. Between 12 and 28; ±4.1%
6. a) 32%, 60% b) ±13.8% c) ±9.8%
7. a) 91%, 93% b) Yes. Answers may vary.
8. a) 47%, 70% b) Yes. Answers may vary.
9. a) 65%, 84% b) Yes. Answers may vary.

Extending Your Thinking

10. a) ±3.3% b) 951
11. a) $-2.58 < z < 2.85$ b) $-1.64 < z < 1.64$
c) Answers may vary. The greater the level of confidence, the larger the margin of error.
d) For 95%: ±1.96σ, for 90%: ±1.64σ, for 99%: ±2.58σ

Consolidating Your Skills

What Should I Be Able To Do?, page 145

1. Answers will vary.
2. a) Uniform distribution b) Binomial distribution
3. a)

	Crispy Chips	Special Spuds
μ	80	80
σ	6.9	5.45

b) Special Spuds Company c) Answers may vary.
4. 1.86
5. 0.2781
6. 91.6%
7. a) 0.022 75 b) 0.0796
8. 0.9981
9. a) Answers will vary. b) Answers will vary.
c) Increase the sample size.

Chapter 4 Personal Finance

Tutorial 4.1

Practise Your Prior Skills, page 152

1. a) \$100, \$112.62, 0.02, 6 b) \$100, \$112.72, 0.005, 24
2. a) \$1790.85 b) \$2298.91 c) \$2165.19
3. a) i) 18 years ii) 9 years iii) 3 years
b) i) \$202.58 ii) \$199.90 iii) \$190.66
c) Answers may vary. The Rule of 72 is only an estimate.
4. \$6163.53

Investigation 1, page 153

2. Explanations may vary.
5. \$12 007.74
6. 3.25 years

Investigation 2, page 155

5. \$14 226.47
6. 12 years

Exercises, page 158

Checking Your Skills

1. a) An investment of \$275 is made at the end of each month for 2 years with an annual interest rate of 8%, compounded semi-annually. The future value of the investment will be S7122.55.
b) A deposit of \$5000 is made into an account with an annual interest rate of 8%, compounded monthly. A monthly withdrawal of \$226.14 will be made over 2 years resulting in a future value of \$0.
2. a) 7.2 or about 8 years. Yes, it is close to the Rule of 72.
b) No. c) The doubling period will be shorter.
3. 9% compounded semi-annually; \$22.86
5. b) \$28 302.68
6. \$30 088.68
7. a) \$496 737.85; \$486 275.71 b) \$43 592.22
c) \$30 044.34.

Extending Your Thinking

8. \$51 013.43

Tutorial 4.2

Investigation 1, page 160

2. D9: =B9+C9, E9: =D9*$D$4, F9: =D9+E9, G9: =$D$5*C9
5. A10: =A9+1, B10: =F9, C10: =G9, D10: =B10+C10, E10: =D10*$D$4, F10: =D10+E10, G10: =$D$10*C10
6. $22 455.90; $22 466.10
7. 14.96%; The rate of return on the RRSP is 9.46% better.
9. $10 794.35

Investigation 2, page 161

4. F9: =D9*B9, H9: =D9/12, I9: =D9/365
7. N = 148, I% = 8, PV = 0, FV = 1 000 000, P/Y = 4, C/Y = 4, PMT: END, PMT = −1127.2426...
8. Answers may vary.
9. Answers may vary.

Exercises, page 163

Checking Your Skills

3. **a)** $31 893.09 **b)** $352 642.60
4. **a)** $788.95 **b)** $1303.83
5. **b)** $2340.11 **c)** $6170.04 **d)** $25 504.90
6. **b)** $3222.58 **c)** $11 331.60 **d)** $59 698.32
7. **a)** The amount required to begin investing in later ages increases as the interest rate increases.
   b) Answers may vary.

Extending Your Thinking

8. **a)** Answers will vary.

Tutorial 4.3

Exercises, page 171

Checking Your Skills

1. 20.258%
2. **a)** 12.551% **b)** 5.746% **c)** 4.917%
3. **b)** Answers may vary.
4. **a)** 13.507%
   b) Each rate of return is weighted differently.
   c) Answers may vary.
   d)

Investment	Percentage of Total Investment
Savings account	2.09%
Money market funds	9.04%
Canada Savings Bonds	4.87%
Term deposits	6.49%
Guaranteed investment certificates	6.37%
Stock in a computer company	3.94%
Mutual funds	67.21%

Answers may vary. The electrician should consider reallocating some investments into stocks and GICs and moving some from the mutual funds, money market funds, and term deposits.

5. **b)** $17 900.38
6. **a)** $324 784.89 **b)** 7.39%
7. **a)** $6934.96 **b)** 3.344%

Extending Your Thinking

8. **a)** $4675.43 **b)** 4.188%

Tutorial 4.4

Investigation 1, page 174

1.

	Length of Lease (months)				
	12	24	36	48	60
Residual Values ($)	12 100	10 780	9240	7920	7040
Depreciation ($)	9900	11 220	12 760	14 080	14 960
Average Annual Cost (before interest and taxes)	9900	5610	4253.33	3520	2992

2. Explanations may vary. The average annual cost decreases when the length of the lease increases.
3. **a)** $15 325.21 **b)** $17 675.21
   c) $11 830 **d)** $20 330

Investigation 2, page 175

2. **a)** $11 155.46 **b)** $22 333.16 **c)** $21 713.46
3. **a)** $302.45 **b)** $311.56
4. **a)** $623.46 **b)** $1265.45

Exercises, page 178

Checking Your Skills

1. **a)** 1.54% **b)** $2005 **c)** $1948.95
2. **a)** It is $282.99 cheaper to lease and then purchase.
   b) The monthly lease payment is $146.69 more than the monthly loan payment. Answers may vary.
   c) $512.14, $518.46
   d) It is $17.27 cheaper to lease and then purchase. Answers may vary.
3. **a)** $30 150, $26 650
   b) New car: $791.22, $28 506; 1-year-old car: $437.08, $15 756.56; 2-year-old car: $370.21, $13 349.56; $354.14, $421.01; $12 749.04, $15 156.36
4. 4 one-year-old cars
5. **a)** Tyler should lease the sport utility vehicle because the lease costs are less.
   b) The $\frac{1}{2}$-ton truck is $457.23 less expensive.
   c) $795.85; $47 751.02
6. **a)** $267.55 **b)** $16 313.69
   c) The monthly lease payments increase to $342.06.
   d) $444.22 **e)** About 50%
7. **a)** $112.26; $5388.48; $137.63, $4954.69; $189.27, $4542.58
   b) $5003.60, $5430.41; $5850.66
8. **a)** $1594.86 **b)** $57.54
   c) 12.75%; $1907.28
   d) The cost to purchase is $312.42 less than the cost to lease. Answers may vary.

Extending Your Thinking

9. a) $2802.92 **b)** $2667.57 **c)** Answers may vary.

Tutorial 4.5

Practise Your Prior Skills, page 184

1. $3640 **2.** $37 700

Investigation, page 184

1. $5254.17

2. $1418.60 to $1681.33

3. $1944.04 to $2101.67

4. $1064.04

5. $151 717.04

6. $120 128.75

Exercises, page 187

Checking Your Skills

1. a) $88 851.84 **b)** $81 392.22 **c)** $70 162.75

2. $617.40

3. a) $1340 **b)** $191 314.94
c) $174 181.92; $150 014.09; $115 922.98
d) $211 314.94; $194 181.92; $170 014.09; $135 922.98

4. a) $1058 **b)** About 10 years
c) $782.44 **d)** Answers may vary.

5. a)

Amortization Period (years)	Monthly Payments	Total Payments
15	$866.37	$155 946.60
20	$740.50	$177 720.00
25	$669.82	$200 946.00

Comparing Amortization Periods (years)	Total Saving
20 vs. 25	$23 226.00
15 vs. 25	$44 999.40
15 vs. 20	$21 773.40

b)

Amortization Period (years)	Monthly Payments	Total Payments
15	$970.14	$174 625.20
20	$852.07	$204 496.80
25	$788.45	$236 535.00

Comparing Amortization Periods (years)	Total Saving
20 vs. 25	$32 038.20
15 vs. 25	$61 909.80
15 vs. 20	$29 871.60

c) Answers may vary.

6. a) $115 169.70; $47 130.90
b) $952.21; $87 014.09; $28 976.99

Extending Your Thinking

8. a) $124 581.60
b) 12 years, 1 month; $34 140.37; $25 441.23
c) $2900 **d)** $121 357
e) Answers may vary.

Tutorial 4.6

Practise Your Prior Skills, page 191

1. a) $3198, $266.50 **b)** $324

Investigation, page 192

2. $877.87

7. Answers may vary.

8. Answers may vary.

9. Answers may vary.

Exercises, page 194

Checking Your Skills

1. $1873.51

2. Answers may vary. It would be better for them to buy the condominium.

3. a) Answers may vary. It would be better for them to rent the house.
b) Answers may vary. It would be better for them to buy the house.

Extending Your Thinking

4. a) $1084.59; $325 377; $175 377
b) $85 000; 8 years 9 months **c)** $64 055.90
d) About 8 years 10 months **e)** $116 606.77
f) $221 746.20; about 11 years 11 months; $157 344
g) $289 344; yes
h) It is less expensive to buy a total of 3 homes; $21 033
i) Yes; Answers may vary; real estate, legal, moving, transfer taxes, etc.
j) Answers may vary.

Consolidating Your Skills

What Should I Be Able To Do?, page 201

1. a)

Age at Initial Investment	Number of Years to Invest	Investment per Year	Future Value, if Growth Rate is 6%	Future Value, if Growth Rate is 10%	Future Value, if Growth Rate is 14%
25	40	$1000.00	$154 761.97	$442 592.56	$1 342 025.10
30	35	$1142.86	$127 354.35	$309 742.91	$792 656.50
35	30	$1333.33	$105 410.65	$219 324.82	$475 714.61
40	25	$1600.00	$87 783.22	$157 355.30	$290 993.32
45	20	$2000.00	$73 571.18	$114 550.00	$182 049.86
50	15	$2666.67	$62 069.33	$84 762.72	$116 913.25
55	10	$4000.00	$52 723.18	$63 749.70	$77 349.18

b) $27 407.62, $132 849.65, $549 368.60
c) $39 571.13, $152 387.61, $501 663.18
d) Answers may vary. The larger the growth rate, the larger the difference in amounts.

2. a)

Year	Opening Balance	Rate of Return	Closing Balance
1	$1000.00	12.1%	$1121.00
2	$1121.00	8.3%	$1214.04
3	$1214.04	(6.8%)	$1131.49
4	$1131.49	19.8%	$1355.52
5	$1131.49	55.1%	$2102.42

b) 36.3%, 19.2%, 16.0%

4. $114 046.00; 0.110 516 809

5. a) $13 027

c)

Investment Amount	Interest Rate	Marginal Tax Rate	Net Interest Rate	Number of Years of Investment	Total Value of Investment
$1000	12%	25%	9%	10	$1411.71
$2500	8.5%	39%	5.185%	15	$2887.76
$4000	14.5%	44%	8.12%	25	$25 258.43

6. $14 249.50, 20.444%

7. a) Answers may vary. b) $11 382.49, $9852.53

8. a) $28 800, $27 000, $23 400 b) $852.85, $651.59, $761.43
c) Answers may vary. d) Answers may vary.

9. a) $612 b) About 19 years 5 months

10. Answers may vary. They should buy the condominium.

Chapter 5 Sinusoidal Data

Tutorial 5.1

Investigation 1, page 210

2. 30°

3. a) 30° b) 60° c) 120° d) 165° e) 306°

4.

Time	Position-angle	Time	Position-angle
12:00 noon	0°	12:00 midnight	360°, 0°
1:00 pm	30°	1:00 am	30°
2:00 pm	60°	2:00 am	60°
3:00 pm	90°	3:00 am	90°
4:00 pm	120°	4:00 am	120°
5:00 pm	150°	5:00 am	150°
6:00 pm	180°	6:00 am	180°
7:00 pm	210°	7:00 am	210°
8:00 pm	240°	8:00 am	240°
9:00 pm	270°	9:00 am	270°
10:00 pm	300°	10:00 am	300°
11:00 pm	330°	11:00 am	330°
		12:00 noon	0°

5. Yes. The position-angle of the hour hand can be determined for the times in between the hours.

6. Starting at noon, the graph increases in a linear manner, 30 degrees for every hour, until midnight. The graph then repeats for the next 12 h.

Investigation 2, page 211

2. No. The times of the sunrise are discrete. Sunrise only occurs once a day.

3. The points of the graph form a repeating pattern that is wave-like.

4. About 6:55 am

5. About 6:53 am. The estimates are very close.

6. 6:44 am

Investigation 3, page 212

4. The graph is a straight line.

7. The graph fluctuates up and down. The graph appears to have a repeating pattern.

8. Answers may vary. The light sources are different with different light intensities.

9. No. Only the fluorescent light is periodic. The incandescent light is linear.

10. No. The light of the candle fluctuates randomly.

Exercises, page 216

Checking Your Skills

1. a) Periodic; 0.0025 b) Not periodic
c) Periodic; 35 years

2. Answers may vary. The graph is periodic and shows the repeated beating of the heart. Because the heart is a muscle, the voltage increases when the heart contracts and decreases when it expands.

3. a) Answers may vary. The time between heartbeats would be shorter and the peaks of the graph would be higher.
b) Answers may vary. The graph would remain periodic.

4. a) 5 s
b) Answers may vary. The person is involved in a low level activity or is resting.

6. a) 125 m, 125 m
b) 8 h
c) Answers may vary. The vertex of the graph would touch the horizontal axis and the graph would be narrower.
d) Answers may vary. The curves would open up more and the gap between the curves would be smaller.

7. b) About 4:18 am c) About 8:25 am

Extending Your Thinking

8. a) Periodic
b) Answers may vary. One cycle of the graph would correspond to the time it takes the pendulum to reach its maximum height after being released. One full swing forward and back (the period).
c) Answers may vary. The maximum horizontal displacement would correspond to the pendulum's swing back to its start position.

Tutorial 5.2

Investigation 1, page 222

5. The results are about the same with $\angle BCA \doteq 57.3°$.

6. No.

7. There are approximately 57.3° in 1 rad.

8. $C = 2\pi r$ or $C = \pi d$

9. About 6.28 or 2π arc lengths will fit into the circumference.

10. About 6.28 or 2π radians will fit into one complete revolution.

Investigation 2, page 223

5. $-2\pi, -\pi, 0, \pi, 2\pi$

6. Maximum: $\left(-\frac{3\pi}{2}, 1\right), \left(\frac{\pi}{2}, 1\right)$; Minimum: $\left(-\frac{\pi}{2}, -1\right), \left(\frac{3\pi}{2}, -1\right)$

7. 1 8. 2π rad

Tutorial 5.3

Investigation 1, page 225

2. $y \doteq 1.94 \sin(0.017x + 1.7) + 6.53$

3. Maximum = 8.47, minimum = 4.59, amplitude = 1.94, period = 369

4. 1.94; It is equal to the amplitude.

5. 0.017; About 369; It is equal to the period.

6. 6.53

7. 6.53; It is equal to the median time of sunrise.

8. 8:27 am

Investigation 2, page 226

3. Answers may vary. The period is about 0.004 s.

4. Answers may vary. $y \doteq 1.3 \sin(1068.4x - 0.7) + 2.54$

5. 3.57, 1.5, 2.54, 1.3, 0.004; The value of a is 1.3 which is the amplitude, the value of $\frac{2\pi}{b}$ is 0.004 which is the period, and the value of d is 2.54 which is the median value.

6. 250 Hz

7. Answers may vary. The frequency stamped on the tuning fork because it has gone through many tests to ensure its frequency and there could be errors in obtaining the values with the CBL.

Exercises, page 229

Checking Your Skills

1. a) $y \doteq 2.3 \sin(0.0172x - 1.328) + 19.193$
b) About 18:32 or 6:32 pm **c)** About 19:12 or 7:12 pm
d) About 2 h 18 min

2. a) $y \doteq 1.072 \sin(0.222x - 1.27) + 2$
b) 2 m **c)** About 2.4 m
d) Between 11:10 am and 2:22 pm

3. a) $y \doteq 1.56 \sin(0.46x + 0.703) + 2.52$
b) About 4 m **c)** About 2.52 m **d)** About 1.56 m

4. a) $y \doteq 11.45 \sin(0.152x - 1.63) + 13.36$
b) About 41.3 s **c)** About 24.8 m
d) About 22.9 m **e)** About 24.5 m

5. a) About 640 d
b) $y \doteq 39.35 \sin(0.0098x - 0.29) - 17.24$
c) The highest maximum temperature is about 22.11°C on about July 8, 1999. The lowest maximum temperature is about –56.69°C on about May 23, 2000.
d) About 17.24°C **e)** About 39.35°C
f) Answers may vary. The equation fits the data very well. However, more data are needed to obtain more accurate values.

Extending Your Thinking

6. a) i) $y \doteq 2.03 \sin(0.0165x + 1.8) + 6.335$
ii) $y \doteq 2.1 \sin(0.0164x - 1.2) + 18.41$
iii) $y \doteq 4.08 \sin(0.0167x - 1.32) + 12.14$
b) About 81 d. Answers may vary.

Tutorial 5.4

Investigation, page 233

1.

Revolution	0	$\frac{1}{4}$	$\frac{1}{2}$	$\frac{3}{4}$	1	$1\frac{1}{4}$	$1\frac{1}{2}$	$1\frac{3}{4}$	2
Time (s)	0	1.57	3.14	4.71	6.28	7.85	9.42	10.99	12.56
Height (m)	0	1	0	–1	0	1	0	–1	0

2. $y \doteq 1.00 \sin[1.00x + (-2.55 \times 10^{-14})] + (-2.73 \times 10^{-14})$
$\doteq 1.00 \sin(1.00x)$

4.

Revolution	0	$\frac{1}{4}$	$\frac{1}{2}$	$\frac{3}{4}$	1	$1\frac{1}{4}$	$1\frac{1}{2}$	$1\frac{3}{4}$	2
Time (s)	0	1.57	3.14	4.71	6.28	7.85	9.42	10.99	12.56
Height (m)	0	–1	0	1	0	–1	0	1	0

5. $y \doteq 1.00 \sin(1.00x - 3.14) + (2.73 \times 10^{-14})$
$\doteq 1.00 \sin(1.00x - 3.14)$

7. a) 1 **b)** 6.28 **c)** 0 **d)** 1, –1 **e)** 3.14

8. a) 1 **b)** 6.28 **c)** 0 **d)** 1, –1
e) 3.14; The values are the same as the result in exercise 7.

9. Answers may vary. The c values are different. In the equation from exercise 2, the c value is (2.55×10^{-14}) which is very close to 0. In the equation from exercise 5, the c value is –3.14. The d values differ in their sign. The graph of the equation from exercise 5 is shifted to the right by 3.14.

10.

Revolution	0	$\frac{1}{4}$	$\frac{1}{2}$	$\frac{3}{4}$	1	$1\frac{1}{4}$	$1\frac{1}{2}$	$1\frac{3}{4}$	2
Time (s)	0	1.57	3.14	4.71	6.28	7.85	9.42	10.99	12.56
Height (m)	0	2	0	–2	0	2	0	–2	0

11. $y \doteq 2.00 \sin[1.00x + (-2.55 \times 10^{-14})] + (-5.45 \times 10^{-14})$
$\doteq 2.00 \sin(1.00x)$;
amplitude = 2, period = 6.28, median value = 0,
maximum = 2, minimum = –2, x-coordinate = 0;
$a = 2, \frac{2\pi}{b} = 6.28, d = 0, d + a = 2, d - a = -2, -\frac{c}{b} = 0,$
The values are the same.; The a values are different. In the equation from exercise 2, the a value is 1. In the equation from exercise 11, the a value is 2. The graph of the equation from exercise 11 is stretched vertically by a factor of 2.

12.

Revolution	0	$\frac{1}{4}$	$\frac{1}{2}$	$\frac{3}{4}$	1	$1\frac{1}{4}$	$1\frac{1}{2}$	$1\frac{3}{4}$	2
Time (s)	0	3.14	6.28	9.42	12.56	15.70	18.84	21.98	25.12
Height (m)	0	1	0	–1	0	1	0	–1	0

13. $y \doteq 1.00 \sin[0.50x + (-4.53 \times 10^{-14})] + (2.64 \times 10^{-14})$
$\doteq 1.00 \sin(0.50x)$;
amplitude = 1, period = 12.56, median value = 0,
maximum = 1, minimum = –1, x-coordinate = 0;
$a = 1, \frac{2\pi}{b} = 12.56, d = 0, d + a = 1, d - a = -1, -\frac{c}{b} = 0,$
The values are the same.; The b values are different. In the equation from exercise 2, the b value is 1. In the equation from exercise 13, the b value is 0.50. The graph of the equation from exercise 13 is stretched horizontally by a factor of 2.

14.

Revolution	0	$\frac{1}{4}$	$\frac{1}{2}$	$\frac{3}{4}$	1	$1\frac{1}{4}$	$1\frac{1}{2}$	$1\frac{3}{4}$	2
Time (s)	0	1.57	3.14	4.71	6.28	7.85	9.42	10.99	12.56
Height (m)	1	2	1	0	1	2	1	0	1

15. $y \doteq 1.00 \sin[1.00x + (-3.5 \times 10^{-13})] + 1$
$\doteq 1.00 \sin(1.00x) + 1$;
amplitude = 1, period = 6.28, median value = 1,
maximum = 2, minimum = 0, x-coordinate = 0;

$a = 1, \frac{2\pi}{b} = 6.28, d = 1, d + a = 2, d - a = 0, -\frac{c}{b} = 0,$
The values are the same.; The d values are different. In the equation from exercise 2, the d value is 0. In the equation from exercise 15, the d value is 1. The graph of the equation from exercise 15 is shifted up by 1.

16.

Revolution	0	$\frac{1}{4}$	$\frac{1}{2}$	$\frac{3}{4}$	1	$1\frac{1}{4}$	$1\frac{1}{2}$	$1\frac{3}{4}$	2
Time (s)	0	1.57	3.14	4.71	6.28	7.85	9.42	10.99	12.56
Height (m)	0	1	2	1	0	1	2	1	0

17. $y \doteq 1.00 \sin(1.00x - 1.57) + 1$
amplitude = 1, period = 6.28, median value = 1, maximum = 2, minimum = 0, x-coordinate = 0;
$a = 1, \frac{2\pi}{b} = 6.28, d = 1, d + a = 2, d - a = 0, -\frac{c}{b} = 1.57,$
The values are the same.; The c and d values are different. In the equation from exercise 2, the c value is 0 and the d value is 0. In the equation from exercise 17, the c value is –1.57 and the d value is 1. The graph of the equation from exercise 17 is shifted up by 1 and to the right by 1.57.

Exercises, page 238

Checking Your Skills

1. a) **i)** 1 **ii)** –3 **iii)** 2
iv) 2π **v)** –1 **vi)** (0, –1)
b) **i)** 1 **ii)** –1 **iii)** 1
iv) π **v)** 0 **vi)** $\left(-\frac{\pi}{6}, 0\right)$
c) **i)** 3 **ii)** –3 **iii)** 3
iv) $\frac{\pi}{2}$ **v)** 0 **vi)** (0, 0)
d) **i)** 0.16 **ii)** –1.16 **iii)** 0.66
iv) π **v)** 1 **vi)** (0, 0.5)
e) **i)** 2 **ii)** 0 **iii)** 1
iv) $\frac{2\pi}{3}$ **v)** 1 **vi)** $\left(\frac{2}{3}, 1\right)$

2. a) maximum = 3, minimum = –3, amplitude = 3, period = 2π, median value = 0, start point = (0, 0)
b) maximum = 1, minimum = –1, amplitude = 1, period = 2π, median value = 0, start point = (1.6, 0)
c) maximum = 3, minimum = –3, amplitude = 3, period = 4π, median value = 0, start point = (–2, 0)
d) maximum = 1, minimum = –1, amplitude = 1, period = 2π, median value = 0, start point = (–0.8, 0)

3. Answers may vary. Expand the graph vertically by a factor of 1, compress it horizontally by a factor of 0.5, and shift it down by 1.

4. a) Answers may vary. As the octaves decrease, frequencies decrease. The higher the octave the higher the pitch, the faster the tuning fork vibrates. The period is the only thing that changes from octave to octave (for the same note).
b) The three graphs have the same maximum and minimum values, amplitude, median value, and start point. The period of each octave is different. The higher the frequency, the smaller the period.

5. Answers may vary. The amplitude is decreasing over time. The period is decreasing over time. I could tell this was happening because the sound would get quieter and quieter.

6. a) 21.5 **b)** 364 **c)** 18°C, –25°C
d) –3.5°C **e)** July 3 **f)** January 4

7. a) Tuning fork: **i)** 1 **ii)** 0.0025;
Clarinet: **i)** 1.5 **ii)** 0.0025
b) The amplitude and the period would increase.

Tutorial 5.5

Exercises, page 244

Checking Your Skills

1. a) 0.4 m, 1.2 s, 0.5, 0.3 s **c)** 0.5 m, 0.5 m
d) 0.43 s

2. a) amplitude = 2, maximum value = 4, minimum value = 2, median value = 3, period = 12 h, start point = (6.5, 3)
b) About 3.35 m

3. a) $y = 13.26 \sin(0.56x - 2.61) - 1.31$
b) –13.56°C
c) The actual mean minimum monthly temperature for January, 1999 was –15.5°C. These values are different because the equation of best fit does not include the values below –14.5°C. The minimum value for the equation of best fit is –14.5°C.
d) $y = 13.2 \sin(0.57x - 3.05) + 0.87$

4. a) 40 cm, 0 cm, 0.05 s **b)** 72 000

5. a) About –2.49°C **b)** 4 months

6. a) $y = 34.6 \sin(0.262x - 1.64) - 14.85$
b) 7:57 am, 4:34 pm **c)** 516 min
d) No. The angles of elevation never reach 90°.

Extending Your Thinking

7. c) $a \doteq 0.875, b \doteq 0.5047, c \doteq -4.1888, d \doteq 1.9675$
d) $y = 0.875 \sin(0.5047x - 4.1888) + 1.9675$
f) $y = \sin(0.5x - 4.75) + 2$ **g)** About 1.33 m
h) A sinusoidal regression equation cannot be found using the calculator because there are not enough data points.
i) Answers may vary. Trial and error takes more time. The sinusoidal regression equation may not touch all the points.

Consolidating Your Skills

What Should I Be Able To Do?, page 251

1. a) About 10:15 am and 3:35 pm
b) $y = 16.226 \sin(0.2664x - 1.865) - 12.735$
c) 10:23 am, 3:36 pm; These values are close to the predictions from part a.
d) Answers may vary.

2. *a*

3. a) $y = 0.72 \sin(0.57x - 0.34) + 3.6$
b) From the equation, the depth of the water is never below 1.3 m. However, the data shows that the depth of the water was below 1.3 m after about 10:36 pm.

4. a) amplitude $\doteq$ 3 stellar magnitudes, period $\doteq$ 340 d, median value $\doteq$ 6.3 stellar magnitudes
b) About 4.75 stellar magnitudes

5. b) About 20.5°C
c) $y = 6.21 \sin(0.0187x - 2.292) + 12.78$

d) 18.7°C

e) Answers may vary. The regression equation is useful to help make predictions as long as the equation fits the data well. It is not very good if you want a very accurate result.

6. b) $y = 20.08 \sin(0.503x - 2.066) - 3.973$

c) $y = 21.175 \sin(0.478x - 1.807) - 5.342$

d) 1997 was not unusually different from 1967 to 1990.

Chapter 6 Patterns

Tutorial 6.1

Investigation 1, page 261

1. 1, 2, 4, 8, 16, 32, 64, 128, 256, 512

2. This is a geometric sequence because each term is multiplied by a common ratio to find the next term.

4. Answers may vary.

5. 12.5 h

7. Answers may vary.

8. Answers may vary. Bacteria die, are killed, or eaten.

Investigation 2, page 261

7. and 8. Answers may vary. One set of values may be as follows:

Bounce	Maximum Height	Difference in Heights	Ratio of Heights
1	0.741 671	—	—
2	0.564 698	–0.176 973	0.761 386
3	0.462 55	–0.102 148	0.819 110
4	0.380 173	–0.082 377	0.821 907
5	0.314 546	–0.065 627	0.827 376

9. a) Answers may vary. There does not seem to be a common difference. The difference in heights ranges from –0.18 to –0.07.

b) The common ratio seems to be 0.8.

c) The height of the ball is better described by a geometric sequence since there seems to be a common ratio. The height of the ball after bounce 6 is about 0.25.

Exercises, page 264

Checking Your Skills

1. a) 31 b) 885 735 c) –15.9

d) 61.509 e) 0.000 046

2. b) 15th week

d) Use the equation

Earnings = \$275 + (\$50 × Number of Cars Sold)

3. a) 400 b) 1600

4. b)

Bounce	Height (m)
1	1.5
2	1.125
3	0.843 75
4	0.632 812 5
5	0.474 609 375

c) 8 d) 26

5. 5 d

6. 23 d

7. a) 10% b) geometric c) \$67 275

8.

Year	Salary
1	\$40 000.00
2	\$42 000.00
3	\$44 100.00
4	\$46 305.00
5	\$48 620.25
6	\$51 051.26
7	\$53 603.83
8	\$56 284.02
9	\$59 098.22
10	\$62 053.13

9. 3

Extending Your Thinking

10. 114.464 m

Tutorial 6.2

Investigation 1, page 266

2. 6024

4. Answers may vary. The growth of the forest gradually declines over time. The forest size will stabilize after 27 years when it contains 6000 trees.

5. Answers may vary. Begin with a forest of 7000, which is harvested at a rate of 35% and 1750 new trees planted.

Investigation 2, page 267

1. $u(n) = u(n - 1) \times 0.95$

4. The values on the horizontal axis represent the number of minutes the insulin has been in the bloodstream. The values on the vertical axis represent the amount of insulin in the bloodstream.

5. 2.5 units

6. 13.5 min

7. 67.5 min

10. Answers may vary. The insulin lowers the blood sugar level. Food needs to be eaten otherwise the blood sugar level will get too low. The person could have a reaction (hypoglycemia). By eating soon after taking insulin, the blood sugar levels are maintained at a safe level.

Exercises, page 270

Checking Your Skills

1. a) $u(n) = u(n - 1) \times 0.88$, $u(n\text{Min}) = 101.3$, $n\text{Min} = 0$, $n\text{Max} = 50$

c) About 5.4 km

2. b) 5445 c) 4023

3. a) 49.9 mL b) 11 doses; 44 h c) 50 mL

4. a) 0.39% b) About 2.35 mg c) 602.35 mg

5. a) About 27.53 kg b) About 30 kg

6. a)

	Lake A	Lake B
5 years	327.07 tons	423.91 tons
20 years	222.93 tons	802.09 tons

b) The pollutant in Lake A will stabilize at about 455 tons after about 64 years. The pollutant in Lake B doesn't stabilize because it continually gets pollutants from Lake A but doesn't get rid of any.

Extending Your Thinking

8. a) 122 582

b) In the 11th year the fish stabilize at 122 581 trout.

c) 121 999

d) The population stabilizes after 14 years at 61 290 trout. There are fewer trout and it takes longer to stabilize the population.

e) a) 287 101

b) In the 25th year the fish stabilize at 287 805 trout.

c) 267 989

d) The population stabilizes after 22 years at 143 902. It takes less time to stabilize less trout.

9. a)

Time	Penicillin Remaining (mg)
8 am	150.00
9 am	90.00
10 am	54.00
11 am	32.40
12 pm	19.44
1 pm	11.66
2 pm	7.00

c) 150×0.6^x

d) Answers may vary. The data follows an exponential curve. The penicillin is decreasing exponentially.

e)

Time	Penicillin Remaining (mg)
8:00 am	150.00
8:15 am	132.02
8:30 am	116.12
8:45 am	102.26
9:00 am	90.00
9:15 am	79.21
9:30 am	69.71
9:45 am	61.36
10:00 am	54.00
10:15 am	47.53
10:30 am	41.83
10:45 am	36.81
11:00 am	32.40
11:15 am	28.52
11:30 am	25.10
11:45 am	22.09
12:00 pm	19.44
12:15 pm	17.12
12:30 pm	15.06
12:45 pm	13.25
1:00 pm	11.66
1:15 pm	10.27
1:30 pm	9.03
1:45 pm	7.95
2:00 pm	7.00

The half-life of penicillin is 1 h 21 min.

Tutorial 6.3

Investigation 1, page 276

6.

	Number of Additional Boxes	Length (cm)	Width (cm)	Height (cm)
Original	1	10.8	7.0	7.0
Iteration 1	2	5.4	3.5	3.5
Iteration 2	4	2.7	1.75	1.75
Iteration 3	8	1.35	0.875	0.875
Iteration 4	16	0.675	0.4375	0.4375

	Volume of One New Box (cm^3)	Volume of All New Boxes (cm^3)	Total Volume of All Boxes (cm^3)
Original	529.2	529.2	529.2
Iteration 1	66.15	132.3	661.5
Iteration 2	8.268 75	33.075	694.575
Iteration 3	1.033 593 75	8.268 75	702.843 75
Iteration 4	0.129 199 218 8	2.067 187 5	704.910 937 5

7.

	Number of Additional Boxes	Length (cm)	Width (cm)	Height (cm)
Iteration 5	32	0.3375	0.218 75	0.218 75

	Volume of One New Box (cm^3)	Volume of All New Boxes (cm^3)	Total Volume of All Boxes (cm^3)
Iteration 5	0.016 149 902 3	0.516 796 875	705.427 734 4

8. $t_n = 2^{n-1}$; 512

9. $t_n = 529.2 \times \left(\frac{1}{8}\right)^{n-1}$; 6.16×10^{-8} cm

Tutorial 6.4

Investigation 1, page 284

Part A

2.

Iteration	Number of Sides	Length of Each Side (cm)	Total Perimeter (cm)
Original	3	9	27
1	12	3	36
2	48	1	48
3	192	$\frac{1}{3}$	64
4	768	$\frac{1}{9}$	$85\frac{1}{3}$

3. 151.70 cm

4. The perimeter will continue to increase.

Part B

1. 81 square units

2.

Iteration	Area of One New Triangle	Number of New Triangles Added	Total Area Added	Total Area
Original				81
1	9	3	27	108
2	1	12	12	120
3	$\frac{1}{9}$	48	$5\frac{1}{3}$	$125\frac{1}{3}$
4	$\frac{1}{81}$	192	$2\frac{10}{27}$	$127\frac{19}{27}$

3. The area of a new triangle is reduced by a factor of 9.

5. The total area increases and approaches 130 square units as the iterations increase indefinitely.

Investigation 2, page 285

Part A

1.

Iteration	Number of Shaded Triangles	Length of Each Side	Total Perimeter
Original	1	16	48
1	3	8	72
2	9	4	108
3	27	2	162
4	81	1	243

2. This is a geometric sequence. $t_n = 3^n$

Part B

1. The area of one small triangle is $\frac{1}{4}$ the area of the original triangle. Answers may vary. I know this because there are four identical triangles formed in the first iteration and they make up the original triangle when pieced together.

2., 3., 4.

Iteration	Number of Smallest Triangles	Area of Smallest Triangle	Total Area
Original	1	64	64
1	3	16	48
2	9	4	36
3	27	1	27
4	81	$\frac{1}{4}$	$20\frac{1}{4}$

5. Total Area $= 3^n \times 4^{3-n}$

Exercises, page 289

Checking Your Skills

1. b)

Iteration	Length of Each Segment	Number of Segments	Total Length
Original	243	1	243
1	81	5	405
2	27	25	675
3	9	125	1125
4	3	625	1875

c) Number of Segments $= 5^n$; 15 625

d) The total length increases at a greater rate.

2. a)

Iteration	Side Length of New Square	Area of a New Square	Number of New Squares	Total Area of New Squares	Total Unshaded Area
1	9	81	1	81	648
2	3	9	8	72	576
3	1	1	64	64	512
4	$\frac{1}{3}$	$\frac{1}{9}$	512	$56\frac{8}{9}$	$455\frac{1}{9}$

b) As the number of iterations increases the total unshaded area decreases towards 0 and the total shaded area increases towards the area of the original square.

3. a)

Iteration	Number of New Circles	Radius of One New Circle	Increase in Circumference	Total Circumference of All Circles
1	2	1	12.57	25.13
2	4	0.5	12.57	37.30
3	8	0.25	12.57	50.27
4	16	1.25	12.57	62.83

b) As the number of iterations increases indefinitely, the total circumference will approach infinity.

c)

Iteration	Area of 1 New Circle	Number of New Circles Added	Increase in Area	Total Area
1	3.14	2	6.28	18.85
2	0.79	4	3.14	21.99
3	0.20	8	1.57	23.56
4	0.5	16	0.79	24.35

d) Yes. There will be a maximum area of 25.5 cm^2.

4. a) 12.57 cm^2; S.A. $= \pi \times 2^{16-2n}$

b) 33.51 cm^3; $V = \frac{1}{3}\pi \times 2^{23-3n}$

5. Answers may vary.

Extending Your Thinking

6. $3x(2^{2+n})$; $3x^2(2^{2n+1})$; $(2^{3n})x^3$

Consolidating Your Skills

What Should I Be Able To Do?, page 295

2.

Iteration	Length of Side of Newest Square (cm)	Increase in Area of Fractal (cm)	Total Area (cm^2)	Increase in Perimeter (cm)	Total Perimeter (cm)
Original	64	—	4096	—	256
1	32	2048	6144	192	384
2	16	512	6656	96	448
3	8	128	6784	48	480
4	4	32	6816	24	496
5	2	8	6832	12	504

b) 508 cm; 6834 cm^2

c) As the number of iterations increases indefinitely, the perimeter will get closer to 6835 cm and the area will get closer to 512 cm^2.

3. a) 0.34 m
b) 7

4. a)

Number of Hours	Time	Amount of Amoxicillin Remaining (mg)
0	7 am	250
2	9 am	125
4	11 am	62.5
6	1 pm	31.25
8	3 pm	15.625

b) 50.78 mg

5. a) 98 mg
b) 400 mg

6. a) 2922
b) 12 years

7. b) 448 mm

8. a) 56.3 cm^3
b) 2127.1 cm^2

Chapter 7 Vectors

Tutorial 7.1

Exercises, page 307

Checking Your Skills

1. a, d, g, h

2. a, b, d, e, f, g

3. a) 23 m/s east
b) 15.5 m west
c) 270 km/h south
d) 300 m/s north
e) 65 N east
f) 230 m/s^2 west

4. a) 40 m/s north
b) 21 m south
c) 33 km/h southeast
d) 49 m/s^2 northeast
e) 97.5 knots east
f) 220 m northwest

6. a) 18 m/s [west]
b) 14 km [218°]

Extending Your Thinking

9. 90 km/h

Tutorial 7.2

Investigation 1, page 312

3. 21 paces toward 258°

4. Answers may vary. They would walk 21 paces at a bearing of 258° from North.

Investigation 2, page 313

3. $(\overrightarrow{AC} + \overrightarrow{CD})$ = 785 km/h toward 100.5°,
$(\overrightarrow{AB} + \overrightarrow{BD})$ = 785 km/h toward 100.5°

4. $(\overrightarrow{AC} + \overrightarrow{CD}) = (\overrightarrow{AB} + \overrightarrow{BD}) = \overrightarrow{AD}$

5. Relative to the ground, the aircraft is travelling at 785 km/h toward 100.5°.

Investigation 3, page 314

1.

Mass (g)	Force (N)
300	3
400	4
500	5

3.

Mass (g)	Force on First Scale (N)	Force on Second Scale (N)	Measure of θ (degrees)
300	1.9	1.9	75
400	2.5	2.5	75
500	3.2	3.2	75

4.

Mass (g)	Force on First Scale (N)	Force on Second Scale (N)	Measure of θ (degrees)
300	2.1	2.1	90
400	2.6	2.6	90
500	3.4	3.4	90

5.

Mass (g)	Force on First Scale (N)	Force on Second Scale (N)	Measure of θ (degrees)
300	1.8	1.8	68
400	2.4	2.4	68
500	3.0	3.0	68

6. I would use the wire with more slack. Answers may vary. The angle is smaller and so each side of the wire, from the hanging point, would have less force.

8. Answers may vary. By decreasing the angle, the force in each string decreases. As the mass increases, the amount of force on each string also increases.

Exercises, page 318

Checking Your Skills

1. **a)** 11.5 km **b)** 8.5 km **c)** 208° SE
2. 2150 km [107°]
3. 21.6 m [148°]
4. 0.74 km [167°]
5. **b)** 325 N [196°]
6. **b)** 790 N [080°]
7. 780 N [312.15°]
8. 270 N [228.5°]
9. 8700 N [20.5°]

Extending Your Thinking

10. 80 km [207.8°]
11. **b)** 32 N [98°] **c)** 44.5° from North

Tutorial 7.3

Investigation 1, page 320

3. [090°]; 9 cm
5. [090°]; 15 cm
7. [270°]; 3 cm
9. [270°]; 3 cm

Investigation 2, page 321

4. Square; 6 cm; EFGH surrounds ABCD
5. Square; 6 cm; EFGH is outside of ABCD
6. Square; 6 cm; EFGH and ABCD share one side.
8. Answers may vary. Draw vectors that are $\frac{1}{3}$ in length from point P to each vertex.

Exercises, page 322

Checking Your Skills

4. **a)** $k = m, k > 0$ **b)** $k = -m, k > 0$ **c)** $k = -m, m > 0$ **d)** $k = m, k < 0$

Extending Your Thinking

5. **a)** $k = 2.1, m = 2.1$ **b)** $k = -3.6, m = -3.6$ **c)** $k = 1.55, m = 5.8$ **d)** $k = -5, m = -5.6$ **e)** $k = -3.4, m = 4.6$

Tutorial 7.4

Practise Your Prior Skills, page 324

1. 0.4, 22.6°, 13.0 cm
2. 0.4, 26.1°, 22.4 cm
3. 4.0 cm, 2.4 cm
4. 6.5 cm, 59°
5. 20.48°, 8.1 cm

Exercises, page 328

Checking Your Skills

1. 2.7 m/s toward 55.7° southwest
2. 254 km/h toward 10.2° northwest
3. 21.2 m/s toward 19.3° northeast
4. 133.3 km/h toward [242°]
5. 74.3 N toward [42.27°]
6. 911.1 m/s toward [56°]
7. **a)** 313.2 km/h toward [301.7°] **b)** 391.5 km
8. 5.1 km/h toward [343°]
9. 109.7 N toward [43.2°]

Extending Your Thinking

10. 0.9 m/s toward [56.31°]
11. 402.5 km/h toward [186.42°]

Tutorial 7.5

Exercises, page 328

Checking Your Skills

1. The net force is 5821 N acting at 59° from the vertical toward the west.

2. The net force is 176.14 N acting at 6.52° from the vertical toward Buster.

3. The total force is 2263.85 N acting 62° from the horizontal toward a bearing of [135°].

4. The resultant velocity is 9.6 m/s at 31° from the horizontal [14° southeast].

5. a) 105.32 m/s **b)** 22.5°

Extending Your Thinking

6. The resultant acceleration is 4.59 m/s^2 at an angle of 29.3° from the horizontal toward a bearing of [234.5°].

7. Yes.

8. The resultant velocity is 351 km/h at an angle of 7.37° from the horizontal toward a bearing of [031.36°].

Consolidating Your Skills

What Should I Be Able To Do?, page 341

3. 92 paces [017.5°] **4.** 5.3 m/s [341°]

5. 2.8 m/s [264.3°] **6.** 4.7 km/h [043.1°]

7. a) 8.1 N **b)** Player B's side

8. 379.4 km/h [082.8°]

9. The resultant force is 11.8 N acting at 25° from the horizontal toward a bearing of [095°].

Index

Numbers and Symbols

A

B

C

D

N

O

P

Q

R

S

T

U

V

Y

Z

PHOTO CREDITS AND ACKNOWLEDGMENTS

The publisher wishes to thank the following sources for photographs, illustrations, articles, and other materials used in this book. Care has been taken to determine and locate ownership of copyright material used in the text. We will gladly receive information enabling us to rectify any errors or omissions in credits.

PHOTOS

Cover, p. i A-Maywald/First Light; **p. v** (centre left), Tony Stone; **p. v** (bottom right), Simon Battensby/Stone; **p. vi** (top centre), Yokohama/First Light; **p. vii** (top left), John Maher/Stock Boston; **p. vii** (bottom right), Frank Siteman/Stock Boston; **p. vii** (bottom right), Artbase Inc.; **p. viii** (top right), Gloria H. Chomica/Masterfile; **p. viii** (centre left), Artbase Inc.; **p. ix** (top left), J. Swider/Armstrong Robert/Comstock; **p. ix** (bottom right), Peter Griffin/Masterfile; **p. x** (bottom left), Arthur Tilley/Masterfile; **p. x** (top right), Artbase Inc.; **p. xi** (centre), Mug Shots/Firstlight.ca; **p. xi** (centre right), Andrew Brookes/Firstlight.ca; **p. xi** (centre), Bob Daemmrich/Stock Boston; **p. xi** (centre), Bruce Coleman/Firstlight.ca; **p. xiii** (top right), Artbase Inc.; **p. xiii** (centre), Erik Anderson/Stock Boston; **p. xvii** (top right), Artbase Inc.; **p. xviii** (centre left), Pat Lam/Spectrum Stock; **p. xviii** (centre right), Rick Fisher/Masterfile; **p. 1** (centre left), 1997 Jose L. Pelacz/The Stock Market/Firstlight.ca; **p. 1** (top right), Rick Fischer/Firstlight.ca; **p. 2** (top left), Artbase Inc.; **p. 2–3** (centre), Lightscapes Inc./Firstlight.ca; **p. 4** (top right), Tony Stone Imaging; **p. 6** (top right), Stuart McClymont/Stone; **p. 10** (bottom centre), Simon Battensby/Stone; **p. 17** (top right), Artbase Inc.; **p. 19** (top right), 1997 Jose L. Pelacz/The Stock Market/Firstlight.ca; **p. 26** (top left), Artbase Inc.; **p. 26–27** (centre left), Rick Fischer/Masterfile; **p. 40** (top left), Artbase Inc.; **p. 40–41** (centre), Artbase Inc.; **p. 49** (top right), Artbase Inc.; **p. 49** (bottom left), Authur Tilley/Masterfile; **p. 50** (top left), Artbase Inc.; **p. 50–51** (centre), Artbase Inc.; **p. 58** (centre left), Douglas E. Walker/Masterfile; **p. 65** (centre right), Yokohama/Firstlight.ca; **p. 66** (centre right), Meteorological Service of Canada, Environment Canada; **p. 68** (centre right), Artbase Inc.; **p. 68–69** (centre left), Artbase Inc.; **p. 70** (background), Artbase Inc.; **p. 70–71** (centre left), Artbase Inc.; **p. 70–71** (centre), Artbase Inc.; **p. 72** (top left), Daryl Benson/Masterfile; **p. 77** (centre left), Thomas Kitchin/Firstlight.ca; **p. 78** (centre), David Lawrence/the Stock Market/Firstlight.ca; **p. 83** (bottom right), Jose l. Perlaez/The Stock Market/Firstlight.ca; **p. 84** (centre right), Arthur Tilley/Masterfile; **p. 86** (centre), Artbase Inc.; **p. 86–87** (background), Artbase Inc.; **p. 95** (top right), Artbase Inc.; **p. 96–97** (background), Spencer Grant/Stock Boston; **p. 109** (top right), Artbase Inc.; **p. 111** (bottom centre), Eric Anderson/Stock Boston; **p. 121** (bottom right), Albert Kozak/Sovfoto/ East foto/Picture Quest; **p. 124** (bottom right), Frederick G Cattroll/Spectrum Stock; **p. 126** (top right), Bob Daemmrich/Stock Boston; **p. 130** (centre), Artbase Inc.; **p. 131** (top right), Artbase Inc.; **p. 134** (centre left), Artbase Inc.; **p. 134** (centre), Article courtsey the Vancouver Sun; **p. 139** (bottom right), Brian Bahr/Allsport USA; **p. 140** (centre right), Roy Ooms/Masterfile; **p. 142–143** (background), Jon Gray/Spectrum Stock; **p. 149** (top right), Michael S. Yamashita/Corbis/Magma; **p. 149** (centre left), Frank Siteman/Stock Boston; **p. 149** (centre left), Artbase Inc.; **p. 150** (top left), Barbara J. Feigles/Stock Boston; **p. 151**(top left), Artbase Inc.; **p. 151** (centre), Artbase Inc.; **p. 151** (bottom left), Artbase Inc.; **p. 166** (top right), Michael S. Yamashita/Corbis/Magma; **p. 168** (centre), Artbase Inc.; **p. 173** (bottom right), Arthur Tilley/Masterfile; **p. 182** (top left), Barbara J. Feigles/Stock Boston; **p. 182–183** (centre left), Artbase Inc.; **p. 188** (centre), Frank Siteman/Stock Boston; **p. 188** (centre), Artbase Inc.; **p. 191** (top right), John Maher/Stock Boston; **p. 197** (centre), Artbase Inc.; **p. 198** (top left), Barbara J. Feigles/Stock Boston; **p. 198–199** (centre), Gloria H. Chomica/Masterfile; **p. 198–199** (bottom centre), Frank Siteman/Stock Boston; **p. 207** (top right), Firstlight.ca; **p. 207** (bottom left), Artbase Inc.; **p. 208–209** (bottom right), Bob Daemmrich/Stock Boston; **p. 208–209** (top left), Mug Shots/Firstlight.ca; **p. 208–209** (centre), Andrew Brookes/Firstlight.ca; **p. 221** (centre left), Bruce Coleman/ Firstlight.ca; **p. 230** (centre right), Gunter Marx Photography/Corbis/Magma; **p. 241** (centre right), Artbase Inc.; **p. 243** (top right), Gloria H. Chomica/Masterfile; **p. 245** (bottom left), John Eastcott/YVA Momatiuk/Stock Boston; **p. 248–249** (background), Steven Frame/Stock Boston; **p. 249** (top left), Peter Menzel/Stock Boston; **p. 249** (bottom right), David R. Austen/Stock Boston; **p. 254** (bottom centre), Mike Grandmaison; **p. 257** (top right), Robert Aschenbrenner/Stock Boston; **p. 257** (bottom centre), Artbase Inc. **p. 258** (top left), Artbase Inc.; **p. 260** (bottom right), Artbase Inc.; **p. 267** (centre right), Artbase Inc.; **p. 270** (centre right), Robert Aschenbrenner/ Stock Boston; **p. 271** (top right), LWA/Stephen Welstead/Firstlight.ca; **p. 272** (centre right), Mark Burnett/Stock Boston; **p. 273** (top right), Artbase Inc.; **p. 274** (top left), Artbase Inc.; **p. 274–275** (centre top), Arthur Beck/Firstlight.ca; **p. 283** (top right), Reader's Digest Atlas of Canada; **p. 283** (centre left), Reader's Digest Atlas of Canada; **p. 283** (centre right), Reader's Digest Atlas of Canada; **p. 292** (top left), Artbase Inc.; **p. 292–293** (background), Artbase Inc.; **p. 297** (centre), Ed Gifford/Masterfile; **p. 301** (top right), Artbase Inc.; **p. 301** (centre left), Bill Brooks/Masterfile; **p. 302** (top left), Mark Dawney/Lucid Images/Picture Quest; **p. 302–303** (background), CanPress Photo; **p. 302–303** (centre left), Artbase Inc.; **p. 303** (top right), Artbase Inc.; **p. 303** (bottom right), Artbase Inc.; **p. 306** (centre), George Hall/ CORBIS/MAGMA; **p. 310** (top left), Mark Downey/Picture Quest; **p. 310–311** (centre left), Ken Straiten/Firstlight.ca; **p. 313** (centre right), George Hall/CORBIS/MAGMA; **p. 318** (centre right), J. Swider/H. Armstrong Robert/Comstock; **p. 323** (top right), Artbase Inc.; **p. 336** (bottom right), Bill Brooks/Masterfile; **p. 338** (top left), Mark Downey/Picture Quest; **p. 338–339** (centre left), OFA/e Stock Photography/PictureQuest; **p. 342** (centre right), Peter Griffin/Masterfile

SET-UP PHOTOS

Dave Starrett; xv, 9, 15, 45, 46, 47, 48, 93, 94, 95, 96, 98, 100, 104, 129, 132, 142, 147, 148, 193, 205, 206, 208, 220, 248, 255, 256, 276, 277, 299, 300, 343, 344, 346, 348, 350, 358

ILLUSTRATIONS

Margo Davies Leclair/Visual Sense; 23, 33, 39
Dave McKay; 244, 325
Jack McMaster; 72
Jun Park; 246
Greg Stevenson; vi, viii, xii, 117, 133, 153, 258-259

TEXT

pp. 246, 251 data obtained with permission from: US Naval Observatory Astronomical Applications Department's web site at aa.usno.nava.mil/AA/data/